The British Dam Society

ice
Institution of Civil Engineers
publishing

Dams – Benefits and Disbenefits; Assets or Liabilities?

Proceedings of the 19th Biennial Conference of the British Dam Society held at Lancaster University from 7–10 September 2016

Edited by Andrew Pepper

Cover photograph of Ogden Reservoir, Haslingden, Lancashire by Suave Aerial Photographers, courtesy of United Utilities

Conference organised by the British Dam Society

www.britishdams.org

Organising committee: Pam Rigby (Chair), Dr Andy Hughes, Andrew Pepper, Andrew Thompson, Barbara Sharp

Published by ICE Publishing, One Great George Street, Westminster, London SW1P 3AA.

Full details of ICE Publishing sales representatives and distributors can be found at:

www.icebookshop.com/bookshop_contact.asp

Other titles by ICE Publishing:

Maintaining the Safety of our Dams and Reservoirs (18th British Dam Society Conference). British Dam Society. ISBN 978-0-7277-6034-0
Operational Safety of Dams and Reservoirs. DND Hartford, GB Baecher, PA Zielinski, RC Patev, R Ascila and K Rytters. ISBN 978-0-7277-6121-7
Floods and Reservoir Safety, Fourth edition. Institution of Civil Engineers. ISBN 978-0-7277-6006-7

A catalogue record for this book is available from the British Library

ISBN 978-0-7277-6133-0

ICE Publishing is a division of Thomas Telford Ltd, a wholly-owned subsidiary of the Institution of Civil Engineers (ICE).

Commissioning Editor: Laura Balchin
Production Editor: Tamsin Cousins
Market Development Executive: Elizabeth Hobson

Typeset by Manila Typesetting Company
Printed and bound by CPI Group (UK) Ltd, Croydon CR0 4YY

Preface

This book contains the proceedings of the 19th Biennial Conference of the British Dam Society, *Dams – Benefits and Disbenefits; Assets or Liabilities?,* held at Lancaster University in September 2016.

The 38 papers of these proceedings are grouped into eight chapters, although many papers cover more than one chapter topic area.

The papers have been written by professionals in the industry with practical experience in the design and construction of a range of dam types and their appurtenant structures, both in the UK and around the world.

Guidance for all involved in dams and reservoirs continues to grow. Recent examples of guidance are reviewed, along with a critique of how such guidance may be used. Further insights into methods of risk analysis are also provided.

Earth embankment dams form the highest percentage of dam types in the UK, and are also the oldest type. Understanding the geotechnical issues of such dams, with limited records and no 'as-constructed' drawings, is often a challenge, and a number of papers cover the investigations needed to ensure the stability of embankment dams.

A complete chapter is devoted to the discontinuance of reservoirs, an area of interest that is likely to grow as smaller reservoirs now come within reservoir safety legislation in three of the four countries of the UK. Examples of the technical and environmental issues involved in discontinuing a reservoir are provided.

Emergency drawdown of reservoirs is now receiving greater attention, and another chapter covers techniques and examples of how emergency drawdown can be practiced, as well as the environmental aspects that must be considered.

The risks associated with mechanical control items at dams are reviewed, and examples are given of the refurbishment of old systems. A serious incident caused by mechanical equipment operation is also covered.

Major works to upgrade existing reservoirs or to construct new ones require full consideration of the environmental issues. The papers covering such works explain how such aspects, in addition to engineering parameters, have to be included in the design process to maximise the benefits and minimise the disbenefits of any major works.

The conference included the presentation of the biennial Geoffrey Binnie Lecture by Dr Peter Mason, entitled *'Dam Engineering: the last 50 years and the next 50 years'*. This 2016 lecture is published in the Society's technical journal *Dams and Reservoirs.*

Chapel house,

Contents

4. Discontinuance of Reservoirs and the Environmental Benefits

5. Experiences with Drawdown Exercises

6. Issues with Mechanical Components of Dams

ice
Institution of Civil Engineers
publishing

Section 1:

Legislation and risk

Dams – Benefits and Disbenefits; Assets or Liabilities?
ISBN 978-0-7277-6133-0

doi: 10.1680/dbdal.61330.003

Amendments to Reservoirs Act 1975 in Wales and Natural Resources Wales potential reservoirs project

M O'BRIEN, Natural Resources Wales
S MORRIS, Natural Resources Wales
A R DAVIES, Natural Resources Wales

SYNOPSIS This paper provides an update on the implementation of the amendments to the Reservoirs Act 1975 (HMSO, 1975) in Wales and summarises the principal changes brought in with the amendments. In addition, the paper describes the project that Natural Resources Wales (NRW), as a reservoir undertaker, embarked on to identify and register its reservoirs.

INTRODUCTION AND OBJECTIVES

The Reservoirs Act 1975 (the 1975 Act) is the law which lays down the minimum requirements for the safe operation and management of large raised reservoirs in Wales. The Flood and Water Management Act 2010 (HMSO, 2010) (the 2010 Act), arose from the recommendations made by Sir Michael Pitt following extensive flooding in 2007. Schedule 4 of the 2010 Act allowed changes to the 1975 Act and provided the opportunity for the relevant Ministers to make regulations to support it.

In Wales the Minister for Natural Resources approved amendments to the 1975 Act and its regulations, which came into force on 1 April 2016. The steps taken by Welsh Government to amend the regulations are a reaffirmation that reservoirs hold a public safety risk which justifies its own primary legislation.

NRW is the enforcement authority for the Reservoirs Act 1975 in Wales, however it is also important to note that NRW acts as undertaker for many reservoirs serving a variety of purposes including flood alleviation, conservation and amenity. NRW identified a number of potential reservoirs in their ownership or management. Local knowledge was used to determine whether they are likely to possess a raised volume capacity in excess of the new 10,000m³ threshold and the likely risk classification for these structures. An initial study showed that NRW had 47 sites that required assessment, but this soon rose to 74 as the project started. A contract was awarded for an All Reservoirs Panel Engineer (ARPE) to visit each site to assess if the structure is a raised reservoir, to assess if the raised volume is likely to be above 10,000m³ and provide a provisional risk category for each structure. It was also an opportunity to identify any obvious works that would be required to bring the structures into compliance of the 1975 Act if deemed high-risk and over 10,000m³.

NRW maintains a separation of duties between these two roles and this paper provides a summary of the new regulations and describes the actions that NRW has taken to implement the changes as undertaker.

REGISTRATION OF RESERVOIRS WITH A CAPACITY OF 10,000 CUBIC METRES

The key determining factor for a large raised reservoir is the capacity to store water above adjacent natural ground level, and which may escape if the impounding structures are removed. Following public consultation, Welsh Government has made new regulations which specify this capacity threshold to be 10,000m³, above which raised reservoirs are to be regulated. This replaces the previous threshold of 25,000m³.

The new regulations place a duty on the undertakers to register their reservoirs. There is a period of six months from registration during which prescribed information for the register of large raised reservoirs must be provided. The prescribed information provides details of the reservoir's location, identity of the undertakers, basic measurements and, where available, the history of the reservoir with reference to engineers' certificates and reports.

Undertakers for previously registered reservoirs with a capacity of 25,000m³, in a valid inspection regime, are not required to submit a fresh registration but have been asked to confirm the currently recorded details as correct or to provide changes.

High-Risk Designation

Newly registered reservoirs do not need to appoint engineers for supervision or inspection unless they receive notice from NRW confirming their reservoir as high-risk. There are occasions in which engineers are required and these are described later.

NRW has a new duty to consider whether large raised reservoirs are *high-risk reservoirs*. A high-risk reservoir is defined in section 2C of the 1975 Act (as modified by the 2010 Act) as one where "*in the event of an uncontrolled release of water from a reservoir, human life could be endangered*". This definition is a consequence based assessment without regard to likelihood.

There are three stages to NRW's designation process:

1. Provisional Designation based upon currently held information;
2. A three month period for representations – this is the point at which undertakers are invited to engage with the designation process and provide evidence which better informs the final designation;
3. Final Designation – NRW reviews all the information and notifies the undertakers with their determination; this will be one of the following:
   i. High-Risk Reservoir. Undertakers for these reservoirs must observe and comply with all requirements of the law including supervision and inspection by qualified civil engineers;
   ii. Large Raised Reservoir. Construction, alteration and decommissioning activities will be controlled, but continual supervision and periodic inspection by qualified civil engineers is not required.

In some cases, where there is already sufficient evidence that human life would not be endangered, an undertaker may receive a letter at Stage 1 informing them that NRW do not consider their reservoir to be high-risk. A designation may be reviewed at any time, usually following a change in the information available. Where an undertaker is dissatisfied with a designation, there is provision to appeal via an independent review managed by the Welsh Ministers.

Following confirmation that any reservoir is not considered high-risk, it remains registered as a large raised reservoir, but the undertakers no longer need to appoint Supervising or Inspecting Engineers, nor maintain a Prescribed Form of Record.

<u>Responsibilities of qualified civil engineers</u>
There are changes in the 1975 Act and in the supporting regulations which affect some of the duties of qualified civil engineers.

All large raised reservoirs, regardless of their risk designation, must appoint qualified civil engineers, and notify NRW, for the following activities:

- Construction
- Alteration to increase or decrease capacity
- Abandonment
- Re-use of an abandoned reservoir
- Discontinuance

There is a new requirement on Inspecting Engineers to provide recommendations as to any measures to be taken in the interests of maintenance. Further to this, Supervising Engineers must include reference to these in their annual statement to the undertaker.

Where an Inspecting Engineer makes recommendations as to measures to be taken in the interests of safety, the engineer must also specify the period within which the measures should be taken.

Section 20 of the 1975 Act has been updated to specify the certificates and reports that need to be copied to NRW, these are:

- Any certificate of an engineer acting for any purpose of the 1975 Act;
- All reports by Inspecting Engineers, or engineers acting under section 8, 9 or 14; this now includes reports where no measures in the interests of safety have been made;
- The decisions of any referee appointed under the 1975 Act;
- The explanation of a Construction Engineer for the deferment of a final certificate;
- A Supervising Engineer's recommendation for an inspection;
- The advice of a Supervising Engineer which draws the undertakers' attention to a breach of sections 6(2) to 6(4), Section 9(2) or Section 11;
- Written statements of the Supervising Engineer made under Sections 12(2) and 12(2A);
- Directions prescribed by a Supervising Engineer made under section 12(6) for the undertaker to undertake a visual inspection.

There are transitional provisions in the amendments which determine that some of the requirements only come into force following the designation of a reservoir as a high-risk reservoir or the 1st April 2017, whichever is sooner.

Incident Reporting

The amendments introduce a new obligation on undertakers to report incidents that affect the safety of large raised reservoirs. Incidents that occur at reservoirs in Wales should be reported to NRW in the first instance. There is also a requirement for the undertaker to provide a full report on the incident within one year, providing the details of the incident and any lessons to be drawn from it.

NRW - RESERVOIR UNDERTAKER

The amalgamation of the three legacy bodies into NRW resulted in an increased portfolio of reservoirs under its management. Additionally, with NRW managing approximately 7% of Wales' total area, it was assumed that many reservoirs above 10,000m³ would be identified. NRW wanted to ensure that it had the full picture of its total reservoir stock before the changes in legislation came into force.

A full time Project Manager assignment was the most suitable way to investigate the potential reservoir stock, and in particular to identify NRW owned or managed sites that may fall under an amended Reservoirs Act. The project involved desk top studies, site visits and surveys. Following this an All Reservoir Panel Engineer was engaged to inspect each site and provide individual reports, including opinion as to the approximate volume and condition of the reservoir, and any recommendations.

Information Gathering

Prior to the creation of NRW, Environment Agency Wales had two in-house Supervising Engineers and a Reservoirs Act Coordinator who all transferred over to NRW. The Project Team possessed a background in Flood Risk Management ensuring the prompt identification of existing flood alleviation schemes and flood storage reservoirs, which were likely potential reservoirs. The majority of these sites had been previously managed within the spirit of the 1975 Act.

The next stage was to ascertain land ownership and identify any bodies of water, regardless of size, and compile a comprehensive list of all NRW reservoirs, irrespective of volume.

At this time, initial discussions identified a core of staff who had experience of the 1975 Act or who were currently managing or operating reservoirs. A workshop was held in June 2014 where the identified potential reservoirs were discussed and likely future actions identified. From this initial meeting, 45 sites were identified from all three legacy bodies as being potential reservoirs with a capacity greater than 10,000m³.

The promotion of the project through engagement with staff from the varied departments of the new organisation was invaluable, as many more potential sites were highlighted and identified for further investigation. Some records for many of the forest reservoirs were already in existence, as some of the sites were historically inspected annually as part of Forestry Commission Wales general asset management activities. Some existing information was also available for many of the water impounding structures located in the National Nature Reserves to facilitate water level management, some of which had surface areas exceeding a square kilometre.

Sites Visits

A desktop exercise was carried out to prioritise potential sites for the initial visits according to surface area, location and purpose. At this stage investigation was prioritised to those sites with a surface area of 5000m² and above, with the smaller reservoirs being investigated at a later date. In some instances local staff had requested that visits to certain sites were prioritised due to concerns regarding deterioration of the structures and the perceived consequence of failure.

Each of the identified potential reservoir sites was visited, in order of priority, to establish whether it was raised, and to ascertain true surface area, dam length, dam height and top water level. From these investigations it was concluded that many of the identified sites would benefit from a visit by an ARPE to better determine capacity and risk category.

It was also recognised that many of the identified sites would not conform to current reservoir safety standards and therefore would require varying degrees of remedial work.

The original project brief was to identify potential reservoirs with a volume greater than 10,000m³, but initial surveys revealed that many of the sites potentially had a capacity of greater than 25,000m³.

ARPE site visits

In parallel with the initial site investigations a tendering process was underway to engage an ARPE to inspect and assess all of the identified sites and produce an investigation report for each site.

The brief comprised the following:-

- Confirmation of whether the structure is a raised reservoir

- Assessment of the raised capacity
- Provisional risk category
- Identification of works required
- Summary of findings and recommendations.

Following the ARPE visits, it was confirmed that many of the reservoirs on the potentials list had a capacity greater than 25,000m³. These reservoirs were subsequently registered as large raised reservoirs, Section 10 reports produced and Supervising Engineers appointed.

Findings

The NRW reservoir stock is varied in terms of purpose, including:

- Flood storage
- Conservation, habitat creation & water level management
- General amenity
- Historical and heritage structures
- Water supply

The distribution of NRW reservoirs can be seen in Figure 1.

The flood risk and conservation sites were found to be fairly evenly spread across Wales. The vast majority of NRW reservoirs within forestry areas are located in North and Mid Wales. This is because the slate and metal mining industries were in decline or abandoned by the time land was purchased for the purposes of forestation in the early part of the 20th century. In contrast at the same time in South Wales, reservoirs were still in use and not acquired by the Forestry Commission and therefore not passed into NRW's management.

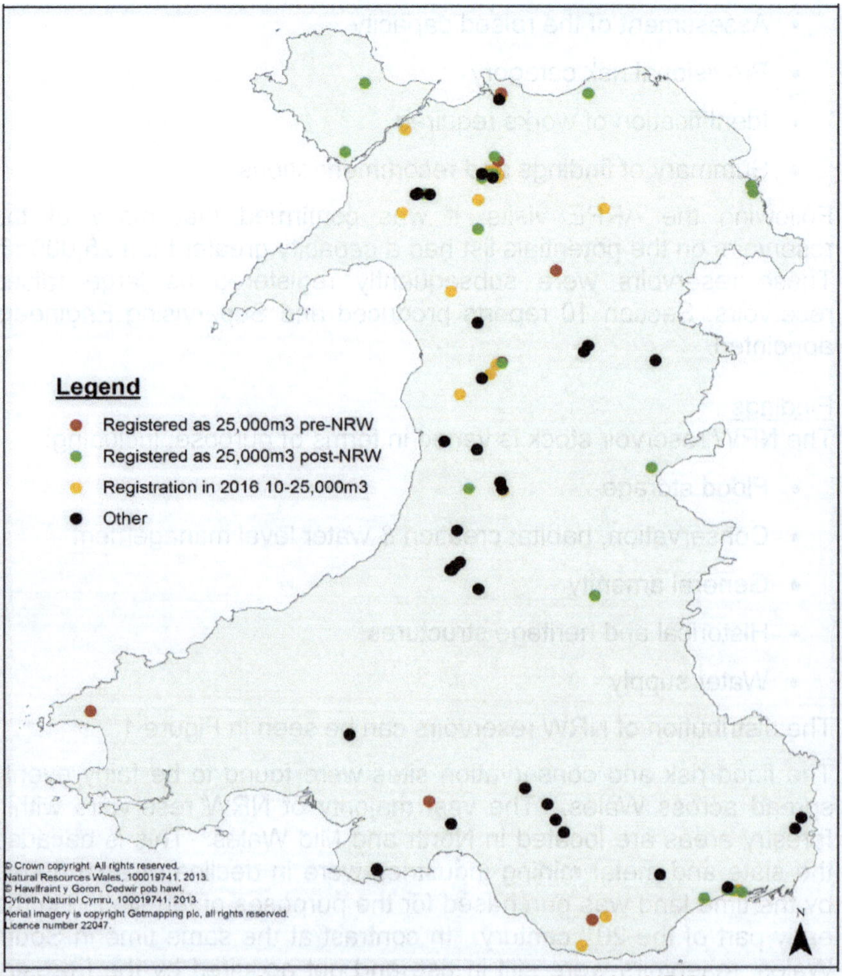

Figure 1. Distribution of reservoirs owned or managed by NRW

Examples

The following examples illustrate the range of reservoirs that NRW manage and some of the discoveries made during the project so far.

Llyn Fuches Las – Mining/Flood Storage

Llyn Fuches Las was discovered when reviewing aerial imagery of forestry in an area famous for its mining heritage. Discolouration of open land prompted a site visit, where upon a spillway and valves were discovered. Often empty, the reservoir has been seen to fill during periods of wet weather.

Table 1. Llyn Fuches Las data

Surface area	18,000m²
Dam	Height 4m; Length 100m
Volume	Greater than 27,000m³
Spillway	Broad crested masonry weir across masonry walls 1.2m by 1.2m
Pipework and valves	Masonry chamber with v-notch into 600mm pipe

Figure 2. View during impoundment Dec 2015

Figure 3. View of basin and SE standing on bottom outlet

New Pool – Post medieval Fish Pool

From investigations and discussions with local people and area staff it became apparent that the reservoir had been emptied before the land was purchased by the Forestry Commission – the bottom outlet valve was removed during the 19th century. The basin was used to plant trees but at some point in the late 1970s or possible early 1980s the bottom outlet became blocked resulting in an unplanned filling. Worried locals contacted the Forestry Commission who lowered a section the dam to reduce water levels, creating the present spillway.

Table 2. New Pool – Post medieval Fish Pool

Surface Area	8,943m²
Dam	Height 14m; Length 250m
Volume	Current 44,500m³ Estimated Original 75,000 m³
Spillway	3m wide breach
Pipework and valves	Not visible

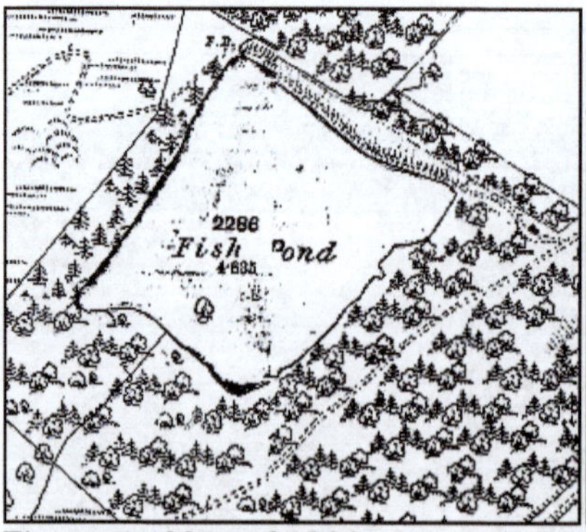

Figure 4. Map of 1886

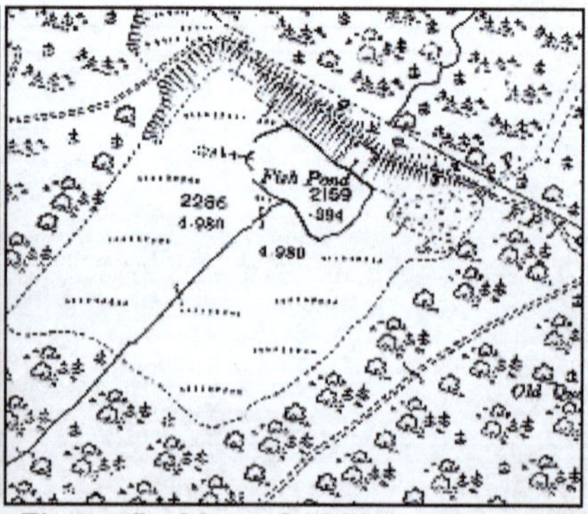

Figure 5. Map of 1906

Figure 6. View of crest as found

Figure 7. View of crest following vegetation clearance

CONCLUSIONS

NRW is both the enforcement authority for Wales and a reservoir undertaker, therefore it was important that, before the enactment of Schedule 4 of the Floods and Water Management Act, a thorough and comprehensive project was undertaken to establish the full extent of NRW's liabilities. It was essential to engage with staff at all levels, in order that all potential reservoir sites were brought to the attention of the project team for further investigation.

The identified sites, which were subsequently registered as large raised reservoirs under the 1975 Act, were found to be in varying states of repair. Some sites required minor maintenance and improvement works, whilst the Section 10 inspections have highlighted measures in the interests of safety that must be carried out.

The additional stock of regulated reservoirs is being addressed under a separate project and funding stream, which will progress and deliver the recommended measures highlighted in the Section 10 reports. Some of the work carried out to date includes vegetation clearance, topographical and bathymetric surveys, flood studies, establishing records and implementing a monitoring regime. Further substantial works to the dams are likely to be identified for future years.

REFERENCES

HMSO (1975). *The Reservoirs Act 1975.* Her Majesty's Stationery Office, London

HMSO (2010). *The Flood and Water Management Act 2010.* The Stationery Office, London

Dams – Benefits and Disbenefits; Assets or Liabilities?
ISBN 978-0-7277-6133-0

doi: 10.1680/dbdal.61330.015

Updating the English reservoir flood maps

A DEAKIN, Environment Agency, Sheffield, UK
A J BROWN, Stillwater Associates, Redhill, UK
J L CHEETHAM, JBA Consulting, Warrington, UK
R I LEWIS, Environment Agency, Exeter, UK
P J ROBINSON, JBA Consulting, Warrington, UK
J ZOLNACZ, Environment Agency, Bristol, UK

SYNOPSIS The 2009 specification that was used to produce the English reservoir flood maps is being updated to take account of changing circumstances and technical advances. The scope of the review includes digital terrain modelling, dam breach hydrograph and initial conditions at time of failure, both at the reservoir site itself and in the upstream and downstream catchments.

INTRODUCTION

Reservoir flood maps are used to inform people about areas at risk of flooding in the event of a dam or reservoir failure and sudden uncontrolled escape of water.

The reservoir flood maps for England are currently being reviewed to take account of changes that have happened since 2009 in the light of developing uses, advances in flood modelling and the legal requirement for six-yearly review. This paper describes the outcome of the review and the plans for updating the maps.

PURPOSE AND USE OF RESERVOIR FLOOD MAPS

The current maps

In 2007 Sir Michael Pitt recommended national flood mapping of potential reservoir failure (Pitt, 2008). In 2008 a draft specification was developed and pilot study carried out, leading in 2009 to the final Reservoir Inundation Mapping (RIM), now termed Reservoir Flood Mapping (RFM) specification. This was used to produce flood maps for 2,007 reservoirs, including 1,800 in England and 207 in Wales. In 2013 in accordance with the EU Floods Directive (OJEU, 2007) and Flood Risk Regulations (HMSO, 2009), simplified depth and velocity maps for all regulated reservoirs in England were mapped. A further 225 reservoirs in England were modelled and mapped in 2015, using the 2009 RFM specification.

The flood outline maps, showing the area that would be flooded in the event of a dam or reservoir failure, and the simplified depth and speed maps are all available to the public on the internet at https://www.gov.uk/prepare-for-a-flood/find-out-if-youre-at-risk, as part of wider flood risk information.

In addition, more detailed maps, showing flow depth, velocity and hazard, together with detailed information about the potential impact of dam failure, are available to those with a need to know, primarily the emergency services.

Emergency planning

Local resilience forums (LRFs) are multi-agency partnerships made up of representatives from the emergency services, local authorities, the Environment Agency and others. These agencies are defined as Category 1 Responders in the Civil Contingencies Act (HMSO, 2004). They are supported by Category 2 Responders, which include water companies, other public utility companies and Highways England. LRFs have a legal duty to share information for risk management purposes. The geographical zones they cover are based on police areas.

The detailed depth, velocity and hazard maps are shared with LRFs on 'ResilienceDirect', an online private government network. LRFs use the detailed reservoir flood maps to assess risks and plan for contingency, warning and evacuation, to either prevent or mitigate the impact of any incident on their local communities.

Risk designations

The Reservoirs Act (HMSO, 1975) was amended in 2013 by the implementation of schedule 4 of the Flood and Water Management Act (HMSO, 2010). One of the amendments was a new requirement for the Environment Agency to determine whether there would be danger to human life in the event of a sudden uncontrolled escape of water from the reservoir. If such a danger exists, the reservoir may be designated 'high-risk', with consequential requirements for the undertaker (operator or owner) to have the reservoir supervised and regularly inspected by government accredited panel engineers. The reservoir flood maps provide useful information to support the designation process.

Other purposes
Reservoir owners, panel engineers and the general public may refer to the maps to help them to assess the possible consequence of dam failure, as part of reviewing design standards for spillways etc, and also for quantitative risk assessment.

Spatial planners may refer to the maps to assess the flood risk arising from proposed development and to help them to decide whether any reservoirs in the vicinity would need to have their spillway capacity increased to improve reservoir safety.

NEED FOR REVIEW

Legal requirements
The Flood Risk Regulations (HMSO, 2009) transposed the EU Floods Directive (OJEU, 2007) into law in England and Wales. The Directive aims to provide a consistent approach to flood risk management across all of Europe. The Regulations required the Environment Agency to prepare and publish, by December 2013, flood hazard maps relating to significant risk of flooding from reservoirs. The flood hazard maps must be reviewed, and if appropriate updated, at intervals of no more than six years, with the first review due by December 2019.

Changing needs and uses
In a culture of increasingly open data it has proved impracticable to limit the use of the maps, which were originally intended for emergency planning only. As a result, they have come under increasing scrutiny. The legal context and new research and technical advances in digital terrain models (DTMs), hydrology, dam break analysis and computational hydraulic modelling, have led the Environment Agency to review the 2009 specification for reservoir flood mapping, to assess where improvements are warranted.

Improving our understanding of flood risk
Continuous improvement requires the Environment Agency to invest in and improve public understanding of flood risk. The new specification and revised maps will deliver better evidence of the consequences in the unlikely event of dam failure.

Review process
The scoping review has been carried out by the authors of this paper and has included consultation with a group of selected Panel Engineers and other specialists.

RANGE OF CHARACTERISTICS OF UK DAMS

The wide ranges of dam height, reservoir volume and catchment area are shown in Figures 1 and 2. A challenge is ensuring that the specification is appropriate to the full range of these characteristics (noting the log-log scales, such that physical properties vary over a range of up to four orders of magnitude). Thus the approach to a 2m high dam is likely to be very different depending on whether it is a 25,000m³ or 2.5Mm³ reservoir, and also will vary depending on its catchment area.

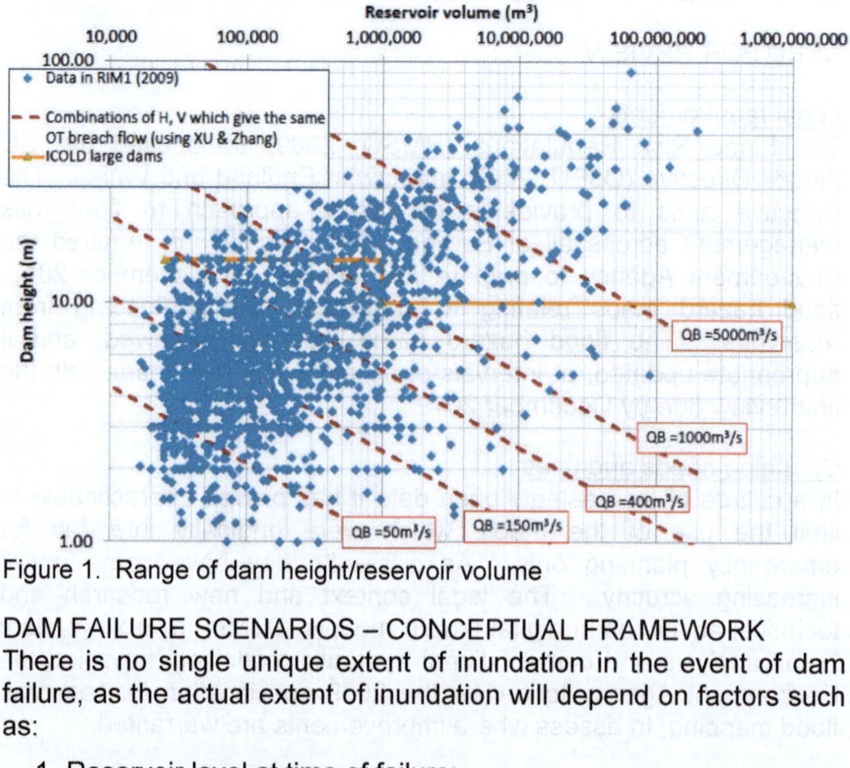

Figure 1. Range of dam height/reservoir volume

DAM FAILURE SCENARIOS – CONCEPTUAL FRAMEWORK

There is no single unique extent of inundation in the event of dam failure, as the actual extent of inundation will depend on factors such as:

1. Reservoir level at time of failure;
2. Inflows into reservoir, which depend on the severity (return period) of any storm being considered;
3. Peak breach discharge (which depends on erodibility of the dam and time to failure);
4. Flows in the water course downstream, which will depend on the magnitude and distribution of rainfall on adjacent and downstream catchments;
5. The extent to which bridges and other obstructions to flow trap debris and lead to elevated upstream water levels;

6. The extent of scour and debris created by the dam break flood wave, including the impact of this on flow paths;
7. The water level in downstream reservoirs at the time of failure, which depends on factors such as timing of storms on the reservoir and adjacent catchments.

A national flood mapping programme has therefore to define typical failure scenario(s) that are reasonably representative of the majority of reservoirs.

The other key issue affecting definition of the failure scenario(s) is the degree of conservatism in the estimate, which should be a reflection of the uncertainty. A discussion of tools to manage uncertainty in flood risk management is given in Section 3.6 of Volume 1 of RARS (EA, 2013), FD2302 (EA, 2003) and Reducing Risk, Protecting People (HSE, 2001, Paragraph 86 to 93 and Appendix 1). Combining uncertainties is complex and can include a simulation approach (e.g. a Monte Carlo analysis of input parameters such as peak breach flow, starting reservoir level etc, where the output would give a probability distribution of the key output parameters), or sensitivity testing to explore the importance of an individual input variable on the final output. If these were to be carried out on every individual dam they would be disproportionately costly for a national flood mapping scheme.

Cabinet Office guidance (2012) requires LRFs to assess the '*reasonable worst case scenario*', which is defined as '*a challenging manifestation of the scenario after highly implausible scenarios are excluded*'. This is considered to be a reasonable approach in terms of ensuring public safety and consistent with the definition of a high-risk reservoir in section 2C of the Reservoirs Act: '*in the event of an uncontrolled release of water from the reservoir, human life could be endangered*'.

It is also reasonable from the perspective of a reservoir owner in terms of risk management, because of the uncertainties in the consequences assessment and need for the owner to demonstrate that '*the procedures adopted for handling uncertainty are in line with the precautionary principle' (HSE, 2001)'*.

The 2009 specification included only a rainy day failure scenario, but the review has concluded that this should be extended to include both sunny and rainy day failure scenarios, as shown in Table 1. A quandary for the rainy day scenario is the extent of downstream flooding at the time of reservoir failure. On one hand, Dales and Reed (1989) suggest that if a probable maximum precipitation (PMP) storm occurred at a dam catchment, then the rainfall on the adjacent and downstream catchments would be of a similar magnitude, and this is supported by change in areal reduction factors with increasing catchment area in FEH (CEH, 1999). On the other hand, the T1000 may be a failure flood for low hazard (flood Category C/D) dams; additionally the extent of fluvial flooding with this annual chance is available nationally and thus allows ready comparison of the extent of flooding from reservoir failure with fluvial flood risk. The T1000 was therefore the rainy day flood severity selected for downstream fluvial flooding concurrent with dam failure.

Table 1. High level definition of failure scenarios

Condition	2009 Specification	Proposed revised scenarios	
		Sunny day	Rainy
Reservoir level at time of failure	0.5m above dam crest (to model downstream flooding)	At spillway crest	At dam crest (or higher on large catchments), subject to check that catchment large enough to produce this volume of runoff from a PMP storm
Inflow hydrograph	None	None	Allowance made as above
Downstream flooding	None (modelled by 0.5m above crest)	None	Downstream flooding to replicate T1000 flood extent (Flood Zone 2)
Dam breach hydrograph	1.5 times Froehlich peak flow	1.0 times Xu & Zhang peak flow for piping mode	As sunny day, but equation for overtopping mode
Consequences of failure	Dam failure	Dam failure	Consider two scenarios: with and without dam failure, to allow quantification of incremental effect of dam failure

Practical difficulties arise in the modelling simplifications and assumptions needed to deliver these scenarios and are discussed in the following sections. At a high level the national flood mapping can be seen as a Tier 2 simplified quantitative assessment, as defined in the Guide to risk assessment for reservoir safety management (RARS) (EA, 2013).

Although more detailed assessments are possible, they would

a) need increased data on the composition of the dam, spillways etc;

b) still be a significant simplification unless all possible combinations of factors such as rainfall return period, spatial distribution and downstream bridge blockage were modelled, and

c) would be disproportionally costly for a national flood mapping programme.

IMPLEMENTATION ISSUES: QUALITY OF DATA ON DAM AND RESERVOIR

Production of the existing reservoir flood maps was sometimes limited by data quality issues, with for example in some cases the calculated freeboard being greater than the dam height. Planned improvements to the quality of data held in the public register of large raised reservoirs should help address this issue. Regulation 5 of Statutory Instrument No.1677 (HMSO, 2013) requires reservoir owners to provide the Environment Agency with up to date information about their reservoirs within 28 days of it becoming apparent that the registered information is incorrect or incomplete. Inspecting and Supervising Engineers and reservoir owners are also urged to check the correctness of data in the Prescribed Form of Record.

IMPLEMENTATION ISSUES: MODELLING DAMBREAK

<u>Reservoir level at time of failure</u>

The 2009 specification assumed that water level was 0.5m above dam crest, understood to be an attempt to allow for flooding downstream prior to reservoir failure. This, however, has caused confusion and is often not a reasonable means to allow for downstream flooding. The updated specification has therefore split out initial downstream flood conditions from reservoir level.

The draft proposals are that in a sunny day the reservoir is at spillway crest, whilst in a rainy day it will depend on the catchment area. For the rainy day scenario it is always assumed that the spillway would be 100% blocked from the start of the storm on the basis that

- it is impractical in a national flood mapping programme to route floods through individual reservoirs, as many have multiple spillways, and spillway capacity is often governed by downstream control, such as a pipe, bridge piers or invert, rather than the weir itself;

- there is often no information on these constraints on spillway capacity in the Prescribed Form of Record;
- even large spillways can be blocked by trees or other debris, so this is consistent with the strategy of a 'reasonable worst case'.

Some time was spent debating whether the rainy day water level should be above the dam crest, the arguments for including this being that many dam safety incidents relate to overtopping of the crest, and many, but not all, failure modes for a rainy day failure require water above the crest to cause surface scour.

On reservoirs with very small catchments where runoff from the catchment in a rainy day is insufficient to fill the flood freeboard, the reservoir level will be determined by the available runoff, for example for a non-impounding reservoir reverting to the depth of PMP rainfall above top water level.

On large catchments, a check will be made that overtopping is physically credible, i.e. the PMF inflow is greater than the peak overtopping flow needed to cause reservoir failure, and the volume of runoff is sufficient to supply both the volume to overtopping level and the volume lost as the head builds up over dam crest. If the PMF runoff volume is sufficient then an overtopping depth of 0.3m would be used. On a 2H:1V downstream face this crest overtopping depth corresponds to a velocity of 4.5m/s on the downstream face, which is around twice the 'no-damage' value on average grass, and seems reasonable as an indicative failure condition on grass crests. This neglects the fact that greater overtopping depths may be required to cause dam failure by overtopping where the crest is protected by a tarmac road, there is a wave wall or it is a concrete dam, as being an appropriate simplification in a national flood mapping programme where there is a lack of reliable reservoir specific data on the vulnerability of the dam to overtopping.

Consideration was given to exemptions to an assumed 0.3m overtopping for Flood Category A dams, but rejected on the basis that for Category A dams the maps will not be used much for risk designation but mainly for emergency planning, and such an exemption would add more complexity (and uncertainties if there was some form of reduced exemption for category B).

Where the 0.3m depth of overtopping is not physically credible due to small catchment area, then a lower value of overtopping is proposed, based on linear interpolation of the actual catchment area between the minimum credible catchment area for 0.3m overtopping and the catchment area necessary to fill the reservoir to dam crest. An indication of the likely effect on UK reservoirs is shown on Figure 2.

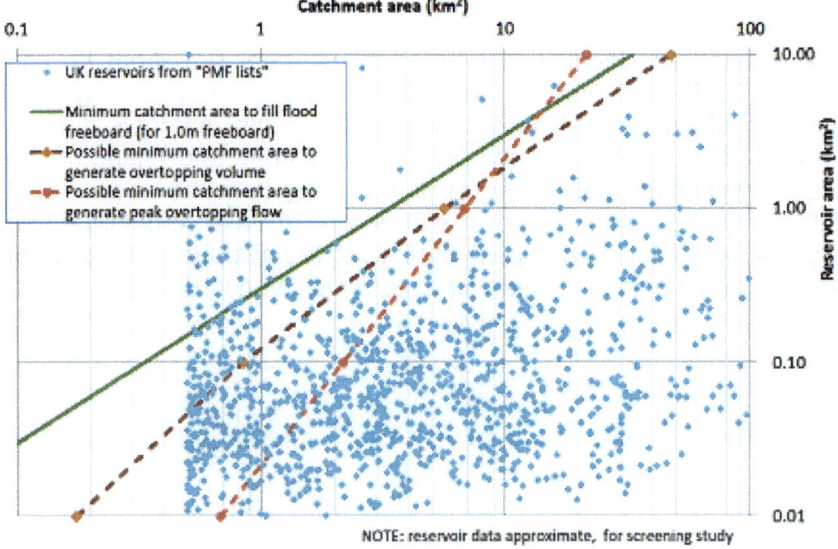

Figure 2. Indicative distribution of catchment and reservoir areas, and conditions when 0.3m overtopping may apply

<u>Breach hydrograph</u>

Definition of a breach hydrograph would ideally be based on the physics of breach development, taking into account the physical characteristics of the dam embankment or foundation. Although computer models are being developed to model both internal erosion and overtopping failures, they use parameters which are not readily linked to geotechnical parameters obtained from site investigation, and the output hydrographs often have poor correlation with observed hydrographs in historical floods. For a national flood mapping programme breach hydrographs have therefore to be based on available breach hydrographs from historic dam failures. The USBR has recently carried out a review of Embankment Dam Breach Equations (USBR, 2014) and concluded that

'Para 1 The evaluation showed that the Xu and Zhang (2009) breach height, breach width, and peak outflow equations produced reasonable predictions of observed breach parameters for medium and high-erodibility dams.

Para 2 The failure times predicted by the Xu and Zhang (2009) equations (both 'best' and 'best simple') were consistently and significantly longer than observed breach formation times..........

For predicting the breach formation time to be used as input to a parametric dam failure model, other existing equations should be utilized, such as Froehlich (2008), Von Thun and Gillette (1990), or others.'

The updated specification will therefore use peak flow from Xu and Zhang (2009) and time to peak flow based on Froehlich (2008). Because of the lack of geotechnical data on the materials forming most UK dams the High erodibility equations in Xu and Zhang will be used; anticipating that in the medium to long term the science and data available may progress sufficiently to take erodibility into account. A comparison of peak outflow for the average distribution of each of dam height and reservoir capacity is given in Figure 3.

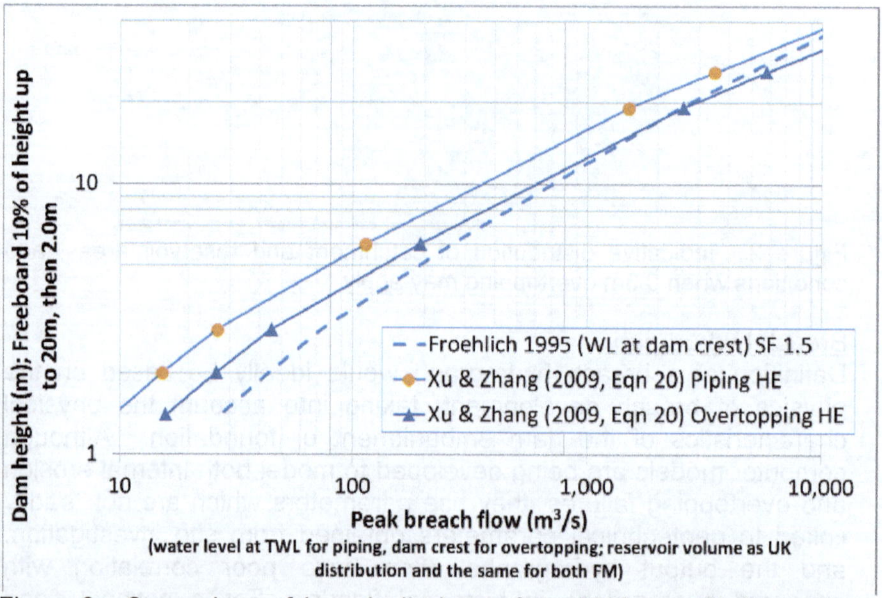

Figure 3. Comparison of breach discharge from 2009 specification with updated methodology

IMPLEMENTATION ISSUES: MODELLING AND MAPPING ROUTING OF BREACH FLOOD DOWN VALLEY

Improvements in ground data and modelling software since 2009

To determine the flood extent and the distribution of depth, velocity and hazard, the breach discharge hydrograph is used as an input to a 2-D flood routing model. Since the last national programme of reservoir flood mapping was carried out in 2009, there have been improvements in the computational power available, the quality, resolution and coverage of the topographical data (the Digital Terrain Map or DTM) and to the models themselves. As a result it is now feasible to simulate the flood routing at a resolution of 2m using a shallow water equation solver, whereas in the past this has been done at a resolution of between 5m and 20m or with an inferior solver. An illustration of the difference this can make to the predicted flood extent can be seen in Figure 4.

Using a 2m DTM and 2m 2-D cell size, the flood extent (blue outline) is largely stopped by a railway line, whereas at 10m resolution, the flow passes over the railway and into the valley beyond, apparently leading to a larger number of properties at risk. This difference in model prediction is explained by examining the profile (along the light blue line) through the two DTMs. In the 2m DTM, the railway line presents a barrier to flow around 2m high but in the 10m DTM, the vertical detail of the railway is smoothed out and presents no barrier to flow.

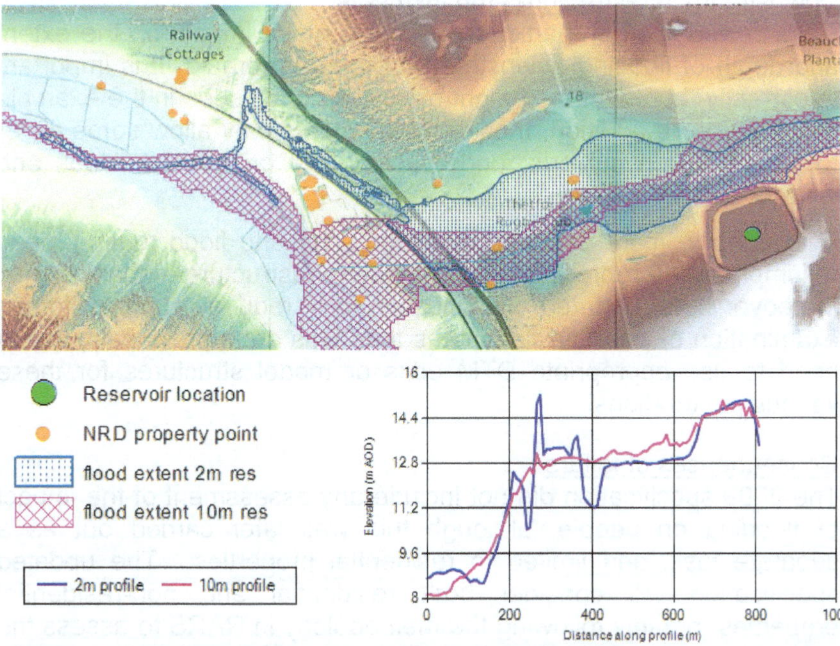

Figure 4. Comparison of flood extent and DTM profiles at 2m and 10m resolution. Contains OS data © Crown copyright and database right (2016)

Modelling the downstream T1000 fluvial flood extent

As summarised in Table 1, the 2009 specification did not call for the explicit representation of fluvial flooding in the valley downstream of the dam break, rather this was allowed for very approximately by an addition of 0.5m to the assumed reservoir water level at the time of breach.

The updated specification recognises that the previous treatment did not correctly deal with the large range of reservoir catchment to reservoir surface area ratios, or with the need to determine the incremental impacts of the dam break over the fluvial flooding event. Instead, in future there will be a requirement to represent the fluvial flooding more explicitly. A number of options have been considered and require further investigation with regard to the practicality of implementation over a wide range of possible topographies. At the time of writing, the likely candidates will either make use of the national flood zone 2 fluvial flood outline, a national set of fluvial depth grids (both 1000 yr event), a fluvial hydrograph at the dam and incremental hydrographs at suitable distances downstream or some combination of these treatments.

Downstream infrastructure embankments

Railway and road embankments have a major impact on the extent and depth of flooding downstream of the dam break. It is important to correctly represent the height of the barrier as seen in the example of Figure 4. In addition, these embankments may allow some water to pass through culverts, underpasses and bridges for roads and watercourses.

The use of a more detailed resolution DTM and flood routing mesh will improve the representation of these infrastructure embankments. Moreover, the updated specification will require a more formal examination of possible flow paths than was done previously, with a need to use appropriate DTM edits or model structures for these various penetrations.

Consequences of failure

The 2009 specification did not include any assessment of the impact of flooding on people, although this was later carried out as a separate task, and limited to residential properties. The updated specification will consider both residential and non-residential properties, broadly following the methodology in RARS to assess the time-averaged population at risk, average societal life loss and property damage.

DISCUSSION AND CONCLUSIONS

The 2009 reservoir flood maps have greatly improved public understanding of the potential consequences of dam and reservoir failures and are helping emergency planners, reservoir owners, panel engineers and the regulator to make informed decisions to minimise liabilities. Advances in technology and an increasing demand for more accurate maps have led to the current review. Future fitness for purpose reviews will be carried out on a six-year cycle.

It is always open to reservoir owners to use the flood maps specification to prepare their own flood maps if they wish to do so, which might be the case, for example, to understand the consequences if the dam is modified or new developments are planned in the valley downstream.

ACKNOWLEDGMENTS
The authors wish to thank Jonathan Hinks, Andy Hughes, Mark Morris and Alan Warren, who have all provided helpful advice during the review of the reservoir flood mapping specification.

REFERENCES

USBR (2014). *Evaluation of Erodibility-Based Embankment Dam Breach Equations.* Hydraulic Laboratory Report HL-2014-02. United States Bureau of Reclamation, Washington DC

Cabinet Office (2012). *National Risk Register of Civil Emergencies.* https://www.gov.uk/government/uploads/system/uploads/attachment_data/file/211858/CO_NationalRiskRegister_2012_acc.pdf

CEH (1999). *Flood Estimation Handbook (FEH).* Centre for Ecology and Hydrology, Wallingford.

Dales MY and Reed DW (1989). *Regional flood and storm hazard assessment.* IH Report 102. Institute of Hydrology, Wallingford, UK.

EA (2003). *Risk, Performance and Uncertainty in Flood and Coastal Defence – A Review.* Report FD2302. Environment Agency, Bristol

EA (2013). *Guide to risk assessment for reservoir safety management.* SC09000. Environment Agency, Bristol

Froehlich D (2008). "Embankment Dam Breach Parameters and Their Uncertainties." *J. Hydraul. Eng.*, 10.1061/(ASCE)0733-9429(2008)134:12(1708), 1708-1721. American Society of Civil Engineers.

HMSO (1975). *Reservoirs Act 1975.* Her Majesty's Stationery Office, London.

HMSO (2004). *Civil Contingencies Act 2004.* The Stationery Office, London.

HMSO (2009). *The Flood Risk Regulations 2009.* SI 2009 No.3042. The Stationery Office, London.

HMSO (2010). *Flood and Water Management Act 2010.* The Stationery Office, London.

HMSO (2013). *The Reservoirs Act 1975 (Capacity, Registration, Prescribed Forms, etc) (England) Regulations 2013.* SI 2013 No. 1677, The Stationery Office, London.

HSE (2001). *Reducing risks, protecting people.* Health & Safety Executive, Norwich.

NERC (1975). *Flood Studies Report, vols 1-5.* NERC, Wallingford, UK.

OJEU (2007). *Directive on the assessment and management of flood risks.* 2007/60/EC.

Pitt M (2008). *The Pitt Review – learning the lessons from the 2007 floods.* Cabinet Office, London.

Xu Y and Zhang LM (2009). *Breaching parameters for earth and rockfill dams. Journal of Geotechnical and Geoenvironmental Engineering.* 135(12):1957-1970.

Dams – Benefits and Disbenefits; Assets or Liabilities?
ISBN 978-0-7277-6133-0

doi: 10.1680/dbdal.61330.029

Managing the safety of very high consequence dams – is the UK doing enough?

A J BROWN, Stillwater Associates, Redhill
M HEWITT, Mott MacDonald, Glasgow

SYNOPSIS This paper explores key issues in relation to the management of the safety of very high consequence dams, where the very high consequences make it almost impossible to reduce the risk out of the Unacceptable zone without rebuilding the dam.

The issues arising include guidance to evaluate when risk has been reduced as low as reasonably practicable, the weighting that should be attached to deterministic and risk based approaches, the role for non-structural measures such as enhanced monitoring and surveillance, and the effort and detail that should be involved in periodic dam safety reviews

The paper concludes by identifying some potential improvements to current UK practice.

INTRODUCTION

Quantitative risk assessment (QRA) is now being used more routinely as part of dam safety management in the UK and internationally, including as part of portfolio risk assessment. This relates closely to legislative change in the UK, which is moving towards a risk based approach.

When carrying out reservoir safety reviews an Inspecting Engineer may be informed by risk assessments as well as a traditional deterministic-standards based approach. The paper discusses the practical implications of using standards and risk based approaches to inform such judgements, and comments on the advantages and disadvantages of each approach, as well as the challenges in making dam safety judgements for very high consequence dams.

VERY HIGH CONSEQUENCE DAMS

In terms of QRA, high risk dams are those in the Unacceptable zone of an FN chart, as shown on Figure 1. Figures 15.2 and 15.3 of the "Guide to risk assessment for reservoir safety" (RARS) (Environment Agency, 2013) suggest that around 15% of UK dams may have very high consequences of failure which include an average social life loss (ASLL) of in excess of 1000 deaths, whist the overall probability of failure of most UK dams is between 1 in 1,000 and 1 in 100,000. Potentially, this places a significant number of very high consequence dams within the Unacceptable zone, where the risk cannot be justified except in extraordinary circumstances (HSE, 2001). The challenge to the dam engineering profession is identifying practicable means to manage the risk of these very high consequence dams.

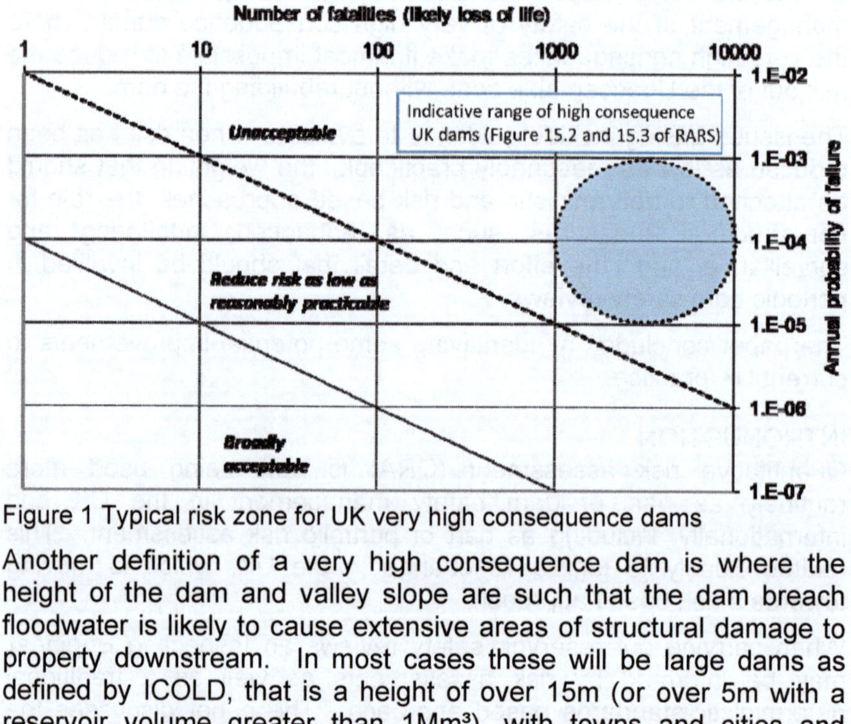

Figure 1 Typical risk zone for UK very high consequence dams

Another definition of a very high consequence dam is where the height of the dam and valley slope are such that the dam breach floodwater is likely to cause extensive areas of structural damage to property downstream. In most cases these will be large dams as defined by ICOLD, that is a height of over 15m (or over 5m with a reservoir volume greater than 1Mm³), with towns and cities and extensive infrastructure situated in the downstream valley.

INTERNATIONAL PRECEDENT

Process for periodic review of safety of high consequence dams

One of the areas that the UK dams industry can look to in managing the risk of very high consequence dams is international precedent, with current practice summarised in Table 1.

Table 1. Comparison of current practice in periodic dam safety review of high consequence dams

Country	Frequency	Description of process (source)
Australia	Comprehensive Inspection every 5 years. Safety review as required.	Dams engineer and specialist(s), inspection; evaluation of monitoring data; applying current and prevailing knowledge; possibly inspection of outlet/submerged works by draining/divers. (ANCOLD, 2003)
Canada	Dam safety review every 7 years for extreme consequences; 10 years for High. Not required for Significant/Low consequences	Collection of all available dam records; field inspections; detailed investigations and possibly laboratory testing. It then proceeds with a check of structural stability and operational safety of the dam, beginning with a reappraisal of basic features and design assumptions. (British Columbia, 2011)
USA	Every 5 years	Team of highly trained specialists. Includes a review to determine if the structures meet current accepted design criteria and practices, and are performing as designed; detailed inspection, includes underwater structures affecting integrity. Risk informed decision making adopted by FERC/USACE/USBR. (FEMA, 2004)

Tolerable Risk guidelines

The level for tolerable risk, as determined from a QRA, is not consistent internationally in the low probability high consequence zone, with some published guidelines set out in Figure 2.

ANCOLD's published societal risk (2003) includes a horizontal truncation at an Annual Probability of Failure (APF) of 1E-05 on the F-N chart. Their guidelines provide the following rationale:

"The horizontal truncations . . . are without precedent, but represent ANCOLD's present judgment of the lowest risks that can be realistically assured in light of:

- *Present knowledge and dams technology; and*

- *Methods available to estimate the risks.*

In the case of existing dams, many were built long ago using very poor technology. Whilst some aspects of safety can be improved, it is simply impracticable to bring such dams full up to the safety levels of a well designed and constructed modern dam. The choice is to either accept the horizontal truncation or to abandon the dam. Since dams are of significant benefit to society, it is considered that the horizontal truncation is justified." (ANCOLD, 2003).

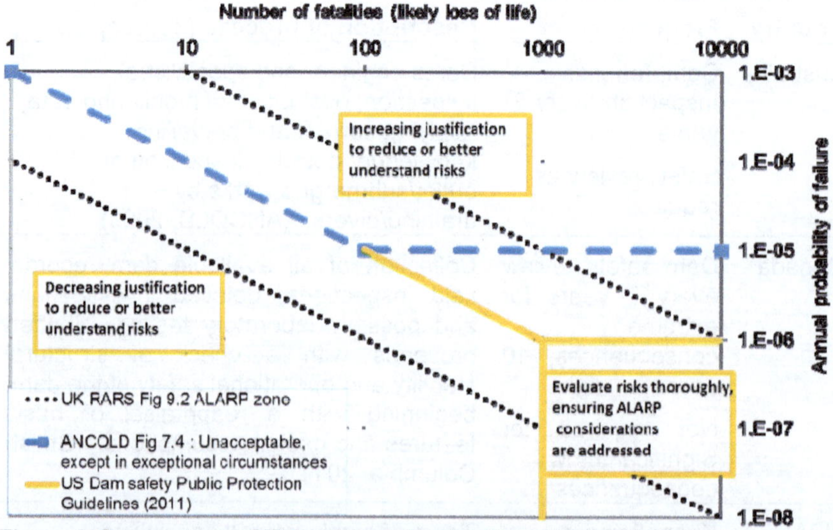

Figure 2 Comparison of Tolerable risk guidelines for existing dams

The issues that influenced ANCOLD's approach in 2003 in respect of the age and knowledge about a dam's composition are also true of many UK dams. It is understood that ANCOLD is currently reviewing the inclusion of the truncation in view of improvements in risk estimation methods, and to emphasise the need to carefully consider risks within that zone, rather than treating it as a limit of tolerability and ignoring them, which some had taken as the intention of the truncation.

Reclamations Public Protection Guidelines (USBR, 2012) is now to include an area bounded by 1E-06 and 1,000 lives within which risks are evaluated thoroughly and ALARP considerations are addressed. Similar zones are shown by USACE and others, as referenced in FEMA (2015). It is suggested that in this region the decision strategy changes to considerations of the more qualitative aspects of the structure and the hazards it poses. USBR's rationale for dealing with this area of risk is as follows:

"There is a lower bound of the likelihood of events beyond which results become unreliable. There is also a threshold beyond which the magnitude

of the consequences necessitates extraordinary measures to control risks. However, setting a horizontal threshold below which risk reduction measures need not be evaluated was not considered appropriate. Likewise, setting a vertical threshold to the right of which risks are unacceptable irrespective of the likelihood of the event could necessitate decommissioning projects whose societal benefits are extremely valuable. Therefore, it is appropriate to treat low probability and high consequence situations with care and ensure everything reasonable has been done to reduce risks. Decisions should be made in those cases considering all relevant information rather than using uncertain risk calculations to avoid a potentially difficult decision". (USBR 2012).

In the case of new dams and major rehabilitations, ANCOLD's and USBR's guidelines aim for risk reduction actions to be an order of magnitude below the tolerability guidelines, to ensure that uncertainty, increase in downstream consequences over time, deterioration and ageing and robustness are considered in the decision process.

<u>Exceptions</u>

Most tolerable risk guidelines suggest that except in exceptional circumstances, risks in the Unacceptable zone must be reduced irrespective of cost. "Exceptional circumstances" are not defined but it is clear that the societal benefits of, say, a large dam upstream of a town that provides flood storage or the only source of potable water may justify higher risk of failure in order to enjoy the benefits of reduced flood risk or health benefits of clean water. Similar considerations apply to construction of airports near major cities where the transportation benefits outweigh the increased risk of ground fatalities due to aircraft impact.

The difficulty in defining what constitutes exceptional circumstances is one of the reasons why HSE has moved away from the rigid ALARP lines shown on Figure 1, instead suggesting that in terms of societal risk "an accident leading to the deaths of 50 or more people in a single event is intolerable if the frequency is more than 1 in 5000 (HSE, 2001, para 136). Other than this point, risks are to be managed through an ALARP process to be tolerable, where the benefits enjoyed by society due to the presence of the dam outweigh the risks.

<u>Dam safety case</u>

This has developed as a means of managing risks of very high consequence dams, and is defined in publications such as Chapter 4 of the USBR's Best Practices in Dam and Levee Safety Risk Analysis (USBR, 2012) and Interim Dam Safety Public Protection Guidelines

(USBR, 2011). It is described as risk informed, such that in addition to quantitative estimates additional information is included to support the case for proposed actions (or non-action). This would normally include the cost effectiveness of candidate risk reduction measures, relevant recognised good practice, and societal concerns as revealed by consultation with the community and other stakeholders.

PRACTICABLE MEASURES TO REDUCE RISK

To reduce the risk of the high consequence dams shown in Figure 1 into the ALARP zone the overall likelihood of failure has to reduce to less than 1E-06, whilst to move into the broadly acceptable zone it would have to be reduced to less than 1E-08. Whether current understanding of, and tools to apply, the science of dam engineering are adequate for this are discussed in the next section.

In terms of practicable measures the dam owner can take to reduce the risk of dam failure, these are summarised in Table 2.

Table 2. Options to reduce risk of dam failure and release of reservoir

Option	**Example/ Comment**
Selected structural fixes	Enlarge spillway, increase freeboard, add filter berm on downstream face, increase drawdown capacity. To reduce risk it needs to correctly identify and address potential failure mode
Rebuild dam	Extreme case of above. Period of higher risk to downstream community while dam being demolished and rebuilt. Temporary environmental impacts
Non-structural	
Increase surveillance	Increases likelihood of early detection of any developing structural problem.
Install real time monitoring	In principle increases chance of early detection of structural problems but many practical difficulties in ensuring reliable equipment/setting trigger levels for alarms
On-site plan	Increases chances of successful intervention to prevent catastrophic failure
Off-site plan	Would need considerable effort, including regular exercising if to be sure of effective evacuation of downstream population in advance of dam failure wave reaching community at risk

DETERMINISTIC STANDARDS TO ASSESS DAM SAFETY: BENEFITS AND LIMITATIONS

A deterministic-standards based approach has been the traditional approach to dam engineering and dam safety assessments. It has developed over the last 200 years or so, through the evolution of dam design based on empirical evidence, mathematical and engineering considerations, recognised good practice and experienced judgement.

In this approach, safety is assessed against established rules for events and loads, structural capacity, factors of safety and defensive design measures. It has largely focused on the threat from natural hazards, such as floods and seismic events, and is not well suited to assessing some important dam safety issues.

It is now broadly accepted that even the most restrictive design standards result in structures that have some likelihood of failure, even though that likelihood may be very small. In the UK, for the highest consequence dams, the PMF forms the safety check flood condition. Whilst the 4th Edition of the Guide (ICE, 2015) introduced a risk based approach, it stated that the PMF should be retained as the most onerous inflow flood for UK dams. However, the Probable Maximum Precipitation (PMP) which is estimated as set out in the Flood Studies Report (NERC, 1975) has been exceeded in some observed storms (Collier *et al*, 2010).

It is therefore a misconception, and potentially an incorrect conclusion, that if a dam meets traditional engineering standards, then the residual risks are negligible or can be tolerated. Such a judgement can only be made from an informed opinion of what the residual risks are.

The likelihood of occurrence of deterministic criterion, such as PMF (probable maximum flood) or MCE (maximum credible earthquake) although expected to be very low, is unknown, as is the reserve capacity or likelihood of failure. This can result in uneven risk across failure modes and loadings, and inconsistencies when assessing the safety of a dam using a deterministic-standards based approach. For example, in the case of the most onerous criterion for UK dams, the MCE is quoted as having an estimated Annual Exceedance probability (AEP) of 3.3E-05; an order of magnitude greater than PMF, which is assigned an estimated AEP of 2.5E-06. Thus designing a spillway upgrade for an existing dam to PMF standards, and not providing a similarly low likelihood of failure from a seismic event, or any other credible threat, is illogical.

QUANTITATIVE RISK ASSESSMENT: BENEFITS AND LIMITATIONS

Risk assessment provides a means to describe and quantify how a dam may fail, including explicit consideration of modes of failure, and quantification of each of the steps in that process. It enables potential failure modes to be examined and dealt with in a systematic and rational manner, which is especially valuable where no traditional standards have been established. Examples include operational failures (e.g. failure to open spillway gates during floods).

One of the potential limitations of risk based methods is where the approach is based on historic rates of failure, such as the Tier 2 methodology for internal threats in RARS, as by definition the lowest probability that can be achieved with these methods is dictated by the data on historic failure rates, which limit the lower bound value, and may change with time. However, this is a screening level tool, and more detailed appraisal methods should give lower values when appropriate.

An important detail is whether fN, or FN (cumulative) curves are used, the former being plots of individual failure modes, each with its own consequences assessment. The former provide greater understanding of the risk, and are less conservative in terms of tolerability criterion, but require dambreak and consequences assessment for each failure mode and so greater inputs into the QRA.

Some remain doubtful of the advantages of a risk-informed assessment over traditional deterministic based analysis. It is accepted that there are challenges in the use of QRA including understanding and characterising dam system performance, and consequence estimates. However, the traditional based approach shares these, and has the additional significant limitation of an inability to define standards for a number of dam failure modes, leading to a gap in understanding and potentially an inappropriate assessment of safety (CDA, 2013).

DISCUSSION

The challenge for the UK dams industry is whether our existing tools and process ensure adequacy of control measures on our very high consequences dams. The various issues that this raises are discussed below.

Tolerability of residual risk

QRA has highlighted that the residual risk from very high consequence UK dams may lie in the Unacceptable region, even though they will have been found to be adequate when checked against maximum criterion using a standards based approach.

Others have addressed this issue by simply amending tolerable risk limits. However they found that such an approach may actually be detrimental to managing the safety of dams in this risk region by inferring that the dam is safe enough, and by curtailing the careful consideration and examination of risks that is required when making decisions in this region about safety. It would also not recognise the potentially misleading issue that in this risk region uncertainty, both in knowledge of an existing dam and modelling of its performance, is a key factor to be taken into account when reviewing likelihood of failure estimates.

Others are addressing this issue by taking the approach that for very low probability/ very high consequence events it is necessary to ensure that everything reasonable has been done to reduce risks, by applying ALARP and good practice to build a "Dam Safety Case". This is a logical set of arguments used to advocate a position that either additional safety-related action is justified, or that no additional safety-related action is justified. In certain cases the scale of the consequences may dictate that in order to retain the dam society demands that extraordinary measures are required, for example redundant defensive measures.

Work over the last few decades has provided an improved understanding of internal erosion, and concepts such as "the perfect filter" (Vaughan & Bridle, 2004) suggest that for new dams it should be possible to design for and achieve very low risk, provided there is high quality in production and placing of filter materials. However, the majority of UK embankment dams pre-date the rational design of filters, and in many cases basic geotechnical information, with which to make a preliminary assessment of the erodibility of the core or the filtering capability of the downstream shoulder, is not known. When assessing the vulnerability of our existing dams to internal erosion there will continue to be considerable uncertainty over the likelihood of failure, which can only be reduced by investing in geotechnical investigations and a risk based assessment of vulnerability to take account of uncertainty.

<u>Periodic safety review</u>

Under current UK reservoir safety legislation, a single expert engineer, appointed from a panel established under the legislation, carries out an inspection of the dam at periods of up to 10 years. Because of the age of most UK dams there are often no drawings or construction records, which inevitably leads to a reactive process, relying on current external condition and past performance, and most likely a simplistic assessment of intrinsic internal condition.

A key difference is that in some other countries owners are often required to produce a "Dam Safety Case" which presents a logical, transparent case, taking account of all available information, setting out why the dam is adequately safe, with the independent regulatory review being an audit of this safety case.

In the UK information presented to the Inspecting Engineer is normally limited to factual records with no assessment by the owner of the safety of their dam. The typical input by the Inspecting Engineer over the course of the inspection, even for very high consequence dams, is much less than that expended in producing a "Dam Safety Case".

Another key difference is that some other countries have now embraced a risk-informed decision-making process within the core of their approach to managing dam safety. Those that have embraced a QRA have found that it has greatly improved understanding of the safety of a dam, through the systematic analysis of the process of failure mechanisms. RARS (Environment Agency, 2013) notes that risk assessments can be an important part of carrying out such an inspection, and recommends that the reports should include at least a Tier 1 (qualitative) risk assessment. However based on the results of a questionnaire issued to Inspecting panel engineers in November 2015, this is has yet to become the norm; most considered that adequate consideration was given to assessing potential failure modes even though this was not explicitly recorded.

In practical terms the threat from floods to very high consequence dams from a PMF event should by now have been identified and reduced to the point at which they are no longer a significant failure mode. However, deficiencies are still being identified by subsequent Inspecting Engineers, despite the threat from floods being arguably the most mature in terms of the deterministic-standards based methods (hydrology, hydraulics, structural design) available to assess the dam and spillway response. Whilst some of the recent upgrade works may have arisen from improved knowledge, it seems that more critical failure modes had not been previously appreciated, but which might have been expected to have been identified by a QRA assessment.

Portfolio Risk Assessment (PRA)

Many major UK dam owners have applied risk assessments in some form varying from portfolio risk assessments to the more detailed

QRA during the decision-making process for upgrades. Some make use of portfolio risk assessments for continuous and progressive safety improvements of dams across their stock using QRA to demonstrate this, and provide outputs from QRA and PRA to the panel engineer during the course of an inspection.

One of the weaknesses of the UK system is that Inspections are carried out on a single dam, and take no account of the overall risk profile of all the dams owned by one organisation. This can lead to conflict between priorities as assessed on the whole portfolio, and recommendations made by an individual panel engineer on an individual dam, where "matters in the interest of safety" on a low or medium consequence dam may end up taking priority over risk reduction on a very high consequence dam. This was illustrated in the results of a questionnaire issued to Inspecting panel engineers in November 2015, where most indicated that they would not take an owner's PRA and proposed overall risk reduction into account when determining whether an issue should be addressed as a "measure in the interests of safety" but it might influence their view on the timescale for completion.

SYNTHESIS AND CONCLUSIONS

Use of quantitative risk assessment (QRA) has highlighted the particular issues posed by very high consequence dams in the UK, typically defined by an average societal life loss (ASLL) in excess of 1,000. Consideration of current international practice suggests that the UK would benefit from identifying these as a special case, and developing good practice at a national level to manage the risk.

The first step would be to agree modified tolerable risk guidelines for very high consequence dams, following the American approach, of a "special case" zone, where the approach to making decisions is based on a "dam safety case" rather than consideration solely on the basis of a standards or a QRA approach. In the light of UK dams typically being older than those in the US the current upper ALARP line could be adopted with the "special cases zone" defined to the right of an ASLL of 1000 and annual probability of 1E-05.

The "dam safety case" would be prepared by the owner, who it is to be expected would be driving the process of understanding and evaluating the risks posed by the dam, as ultimately he bears responsibility for whether the dam is adequately safe. It should be informed by the outputs from both deterministic and QRA assessments, consideration of good practice in relation to structural and non-structural measures, and ALARP principles, noting that the

UK definition of a safety case would need to take account of the UK's legal and cultural context.

The "dam safety case" would fit with the current statutory inspection process in UK, which is fundamentally an external audit, and would address the observation that currently the periodic review of very high consequence dams is more detailed in other countries.

It is suggested that preparation of such a dam safety case would require at least a basic level of information on the internal composition and geotechnical properties of a very high consequence dam to ascertain its vulnerability to internal erosion, which may have to be obtained by investigations.

Such a safety case would also logically be prepared in the context of a portfolio risk assessment, where a strategic and consistent approach could be taken, and economies gained in the approach to non-structural measures and prioritisation of reduction of risk across a number of dams.

One of the implicit features of a "dam safety case" type approach is that that very high consequence dams provide significant benefits to society, sufficient to meet the exceptions justification, and that where this is no longer valid they would logically be decommissioned.

REFERENCES

ANCOLD (2003). *Guidelines on Dam Safety Management.* Australian National Committee on Large Dams, Sydney, Australia

British Columbia (2011) *British Columbia Dam Safety Regulation, B.C. Reg 44/2000 & amendments.* Province of British Columbia, Victoria, Canada.

CDA (2013). *Dam Safety Guidelines.* Canadian Dam Association, Toronto, Canada

Collier CG, Morris DG & Jones DA (2010). Assessment of the return period of near PMP point and catchment rainfall for England and Wales in *Meteorological Applications, Volume 18 Issue 2, pages 155-162, June*

Environment Agency (2013) *Guide to risk assessment for reservoir safety (a) Volume 1: Guide - Report SC090001/R1 (b) Volume 2: Methodology & Supporting Information - Report SC090001/R2, (RARS).* Environment Agency, Bristol

FEMA (2004). *Federal Guidelines for Dam Safety.* Federal Emergency Management Agency, Washington DC, USA

FEMA (2015). *Federal Guidelines for Dam Safety Risk Management FEMA P-1025.* Available at https://www.fema.gov/federal-guidelines-dam-safety Federal Emergency Management Agency, Washington DC, USA

HSE (2001). *Reducing risks, protecting people HSE's decision-making process.* Health & Safety Executive, London

ICE (2015), *Floods and Reservoir Safety, 4th Edition.* Institution of Civil Engineers, London

NERC (1975) *Flood Studies Report.* Natural Environment Research Council, London.

USBR (2011). *Interim Dam Safety Public Protection Guidelines. A Risk Framework to Support Dam Safety Decision-Making.* United States Bureau of Reclamation, Washington DC, USA

USBR (2012) *Best Practices in Dam and Levee Safety Risk Analysis* United States Bureau of Reclamation, Washington DC, USA

Vaughan PR and Bridle RC (2004). An update on perfect filters, in *Long-Term Benefits and Performance of Dams, Proceedings of the 13th Biennial Conference of the British Dam Society, pp 516-531.* Thomas Telford, London

Dams – Benefits and Disbenefits; Assets or Liabilities?
ISBN 978-0-7277-6133-0

doi: 10.1680/dbdal.61330.043

ice
Institution of Civil Engineers
publishing

Building on RARS: development of key themes

A PETERS, Arup
T DOYLE, Arup
I CARTER, MWH
R COOMBS, MWH
A J BROWN, Stillwater Associates

SYNOPSIS In 2015 the authors undertook a portfolio risk assessment (PRA) across the full range of dams owned by a large water company, including earth embankment dams, concrete and service reservoirs. The PRA was based on a Tier 2 assessment as set out in the Guide to Risk Assessment for Reservoir Safety Management (Environment Agency, 2013).

This technical paper details the work undertaken to refine and extend the RARS Tier 2 assessment, ensuring a more comprehensive understanding of the safety of an asset and improved tool to manage the safety of a portfolio of dams.

INTRODUCTION

The Guide to Risk Assessment for Reservoir Safety Management (RARS) document provides a framework for assessing the risk posed by a single dam and the consequences of the failure of that dam. The guidance can be applied to a single dam, or across a portfolio of dams. By undertaking a portfolio risk assessment (PRA), the undertaker can determine the overall risk posed by their dams and can readily identify those of most concern where further investigation or capital expenditure should be prioritised. A PRA approach is therefore a useful tool in the active management of a portfolio of assets.

This paper is limited to a tier 2 assessment which provides a quantitative assessment of the probability and consequence of failure, providing the user with a value which can be plotted on a Fn chart, giving information on the balance of likelihood of failure and consequence.

BACKGROUND

A portfolio risk assessment was undertaken across 135 structures consisting of earth, concrete and service reservoirs, with four AR Panel engineers involved, each overseeing the assessment of around a quarter of the dams. Through a pilot stage, guidance was developed and refined to aim to improve consistency in application and was extended further than consideration of the core failure modes included in RARS. The project included independent assessment of a number of dams by teams from Arup and MWH, as verification of consistency of the methodology. This was a valuable process in terms of promoting consistency.

EMBANKMENT DAMS

The RARS document contains fairly comprehensive guidance on the application of the methodology to embankment dams, although lacks systematic worked examples, which can lead to inconsistency in application. Further development of the themes and methods of application have been made by the authors, with some of the significant changes summarised below.

Internal threats

Ensuring consistency in scoring across different dams, and by different assessors was identified as a challenge early on. This was managed by extending Table 8.3 in RARS, with an example of guidance shown in Table 1.

Other important refinements are contributory factors such as the frequency with which monitoring results are checked by an engineer, how fast the reservoir fills each winter, etc. These were included in the Interim Guide to Quantitative Risk Assessment, 2004 (Defra, 2004), but omitted from RARS. The authors have also therefore considered the impact of these factors in the evaluation of current condition.

Table 1. Extension of table 8.3 of RARS, on consistency in scoring uncertainty

<table>
<tr><th>Degree of uncertainty</th><th>Seepage/ deformation</th><th>Over conservatism due to double/ triple counting</th></tr>
<tr><td>Zero</td><td>Score as zero when being measured (or good visual inspection possible) and this shows
• No significant change in seepage/ rate of deformation over time;
• values normal (less than threshold for “large” defined in RARS);
• no fines indicated by turbidity monitoring or settling out box for fines on a V notch</td><td rowspan="4">There are circumstances when several indicators of poor performance may occur simultaneously, and give an unreasonably high current condition score, compared to a dam where manifestation of a problem is limited to one symptom. This can be corrected by the “user adjustment” row at the bottom of the sheet. The need for judgement and obtaining advice of a AR Panel engineer is emphasised.</td></tr>
<tr><td>Unlikely (but no definitive evidence that absent)</td><td>Not measured, but reasonable visual inspection possible and suggests
• no significant change over time or
• values ≤ RARS threshold value</td></tr>
<tr><td>Unknown</td><td>Not measured, visual inspection constrained in some way</td></tr>
<tr><td>Likely (but not certain)</td><td>Not measured , visual or other indicators suggest likely</td></tr>
</table>

Structural integrity of spillways

Masonry spillways have been shown to be vulnerable to structural failure (Environment Agency, 2010), but RARS does not include a methodology to assess the likely annual chance of failure in operation.

Therefore, in addition to consideration of the core failure modes included in RARS, the project team also considered the failure of masonry spillway channels. The team developed guidance and an assessment tool to calculate the probability of failure, enabling the threat from structural failure of the spillway lining to be compared with other threats. Although initially the method was developed for masonry spillways, it was later extended to reinforced concrete and hybrid masonry concrete structures.

The methodology is based on separating intrinsic properties and current condition, following a similar approach to other internal (deterioration) threats considered in RARS. The guidance is provided with examples of when a low or high score may be appropriate. An outline of the scoring process for masonry spillways is shown in Tables 2 and 3.

For concrete spillways the approach to assessing current condition was similar, but the intrinsic condition was based on shear restraint at joints; quality of concrete and joint sealant/ waterstop; control of groundwater (e.g. underdrainage, weepholes) and velocity.

Table 2. Outline of scoring system for Intrinsic condition of masonry spillways

Feature	Guidance	Example score
Lining thickness (for masonry with concrete backing, combined depth)	This looks at potential for block plucking, under groundwater/ static head; expressed as head below top of wall relative to the weight of the lining (density of water to density of block)	Zero if ratio of lining thickness to wall height <10/22; 4 if <0.5 x 20/ 22
Erodible foundation / backing to blocks	Is the foundation material vulnerable to interval erosion/ surface scour by water?	0 for rock, 4 for soil
Geometric irregularities	Surface irregularities can introduce negative pressures (and positive (stagnation) pressures which transfer behind facing) which can lead to block plucking	1 for step < 100mm, 3 for steps< 500mm, 4 for upstand at downstream end
Joint width / geometry and mortar grade	The durability of the mortar is important to the integrity of the wall, and determines whether water pressures can be transferred to the back of the blocks	Zero for tight even joints, 2 for tight but trapezoidal block so joint width increases with distance from facing
Velocity	Velocity is a measure of the potential for stagnation pressure to develop. The range of velocities is based on current practice but also takes into account the range on the client's spillways.	Zero if < 8m/s, 4 if > 24 m/s

Note Foundation and geometry given a weighting of 1, other factors a weighting of 2.

Features that were not included in the assessment included: the nature of the backfill to the walls; details where the invert was dished and when the spillway was located on the mitre. These were either included elsewhere in the risk assessment process, or were deemed of less significance than other indicators.

Table 3. Outline of scoring system for Current condition of masonry spillways

Feature	Guidance	Example score
Water ingress/ egress	Evidence of water passing through the spillway wall introduces potential for material to be washed through and suggests pressure can access back of blocks	0 for dry, 4 for significant quantities of water
Loss of pointing	Critical areas are where there is any potential for water under pressure to enter invert/ wall lining. Groundwater ingress/ egress can be used as surrogate for current condition of mortar	4 for major loss, or omission of pointing i.e. > 50% of area
Verticality of walls	Walls leaning into the spillway channel suggest ground movement and risk of structural movement. Risk of walls collapsing into channel during flood event allowing erosion	2 for resultant in middle third, 4 for resultant within section
Cracking	Cracking can imply structural distress and risk of walls collapsing into the spillway chute	2 for 2mm cracks, 4 for severe distortion/ missing sections

Note: all features given a weighting of 2, other than verticality which has a weighting of 1.

The anchor points for best condition and worst condition dams were selected as an annual chance of failure of 1E-3 and 1E-5 for masonry spillways, with the best case anchor point reducing to 1E-6 for mass concrete and massive masonry (300mm (single layer) thick invert or wall), and 1E-7 for reinforced concrete. These anchors were agreed through detailed discussions with the QCE's involved in the development and were ratified through piloting of the assessment on a number of spillways with known issues.

Low level outlets

A low level outlet facility does not in itself create a risk of failure however it does affect the ability to avert failure when a structural problem develops. Hence in order to provide allowance for this factor in the risk assessment, and thus provide a business case for increasing the capacity, a method was developed to assign an "index" annual probability of failure related to the presence and capacity of low level outlets.

The method for estimating annual probability of failure, in line with the client's target drawdown capacity (1m/day), is set out in Table 4. The speed of failure was assessed following Box 8.2 of RARS. Where no information was available on the embankment material then the online resource BGS geo index was used to infer likely soil properties at given sites, based on local solid geology in proximity to the dam.

Table 4. Method for estimating probability of failure due to insufficient drawdown capacity

<table>
<tr><th></th><th colspan="2">Embankment speed of failure</th></tr>
<tr><th></th><th>Slow</th><th>Medium or Fast</th></tr>
<tr><td rowspan="3">Index Annual Probability of failure</td><td>1 in 10,000 where no bottom outlet.
1 in 10,000,000 where the low level outlet facility is greater than or equal to 1m/day.</td><td>1 in 1,000 where no bottom outlet.
1 in 10,000,000 where the low level outlet facility is greater than or equal to 1m/day.</td></tr>
<tr><td colspan="2">Interpolation between high and low anchor based on actual capacity as a percentage of target capacity.</td></tr>
<tr><td>For example, a bottom outlet with 0.5m/day capacity would have a 1 in 320,000 annual probability of failure.</td><td>For example, a bottom outlet with 0.5m/day capacity would have a 1 in 100,000 annual probability of failure</td></tr>
</table>

This method only considers fixed capacity and neglects temporary pump capacity.

It was considered that 1 in 10,000,000 was a suitably low anchor based on the comparative range of PoF values assigned to individual threats across the portfolio. The upper anchor values were derived from what is considered unacceptable in terms of Average Societal Life Loss (ASLL):

- Slow speed of embankment failure – ASLL ≥ 10.
- Medium or fast speed of embankment failure – ASLL ≥ 100.

SERVICE RESERVOIRS

The RARS methodology for Tier 2 studies suggests implementation of event tree analyses to assess the annual probability of failure for service reservoirs. Limited guidance is provided within the RARS documentation, and additional guidance was therefore developed.

The first step is deciding which failure modes are credible and significant and likely to give the highest annual chance of failure. An additional screening assessment table was produced for service reservoirs which provides the user with a means to eliminate the need to analyse irrelevant failure modes. Consideration of the type of structure (e.g. vaulted brick arch, reinforced concrete etc) is given at all steps of the following process. In summary the screening assessment includes the following basic failure modes:

- Failure of the body of the service reservoir walls
- Failure or loss of support for wall foundations
- Undermining of structural rigidity and eventual failure due to loss of perimeter embankment.
- Landslides
- Deterioration in foundation materials
- Differential Settlement (as a result of mining activities)

The next step is to describe the failure process. This is complex for service reservoirs, as for the contents of the reservoir to be released both the perimeter wall and external fill which is normally present have to be removed over a width sufficient to allow release of a wall panel and a large flow of water. The standard eight step process in RARS was extended to 10 steps, with a wider range of mechanisms by which progression could occur, as summarised in Table 5. In order that the user applies the methodology in a consistent manner, a flowchart of appropriate node choice for steps 4 to 6 (progression) was developed for a number of different initiation mechanisms, with an extract in Figure 1.

Each mechanism has an associated table giving guidance to allow consistent scoring of the conditional probability of failure of that step. These are scored using metrics which are directly comparable with the observations one might expect to find within statutory reporting. These specifically focus on the adverse conditions listed in CIRIA R138: Underground Service Reservoirs, Waterproofing and Repair manual (Johnson *et al*, 1995).

Table 5. Mechanisms of progression of failure of service reservoir (updated version of RARS Table 8.12)

Ref	Phase	Comment
1	Reservoir level	
2	Initiation	8.11 modified
3	Continuation	
4 to 6	**Progression – mechanisms vary as below**	
A	Scour of supporting fill	New Guidance table, based on unit discharge and fill type
B	Instability of supporting fill	No change
C_SR	Crack propagation through structural medium	Replaces crack on lift joint
D	Failure in foundation propagates	Updated (same table as gravity dams)
E	Internal erosion in foundation	Withdrawn – would be a Tier 3 analysis
F	Stability (shape) factor	New figures to replace figs 8.12 to 8.16
G	Structural movement to rupture watertight element	Updated (applies to mass concrete only)
G_SR	Structural Failure of homogenous structural medium	New guidance to replace that for gravity dams
G_Comp	Structural failure at heterogeneous structural interface	Likelihood depends on materials in base/ wall
SR3	% loss of support to perimeter wall, following slope instability in perimeter bank	Depends on % of wall height exposed
SR4	Wall panel/perimeter wall fails outwards	Likelihood depends on spacing of vertical joints and wall type
SR5	Wall panel/perimeter wall fails into reservoir	Depends on type of material in wall
SR6	Progression of Landslide	Depends on steepness of slope, depth of soil, rock joint spacing and whether there is historic slope instability in the area

Ref	Phase	Comment
7	Detection/ Intervention	No change
8	Breach	Extended/ modified

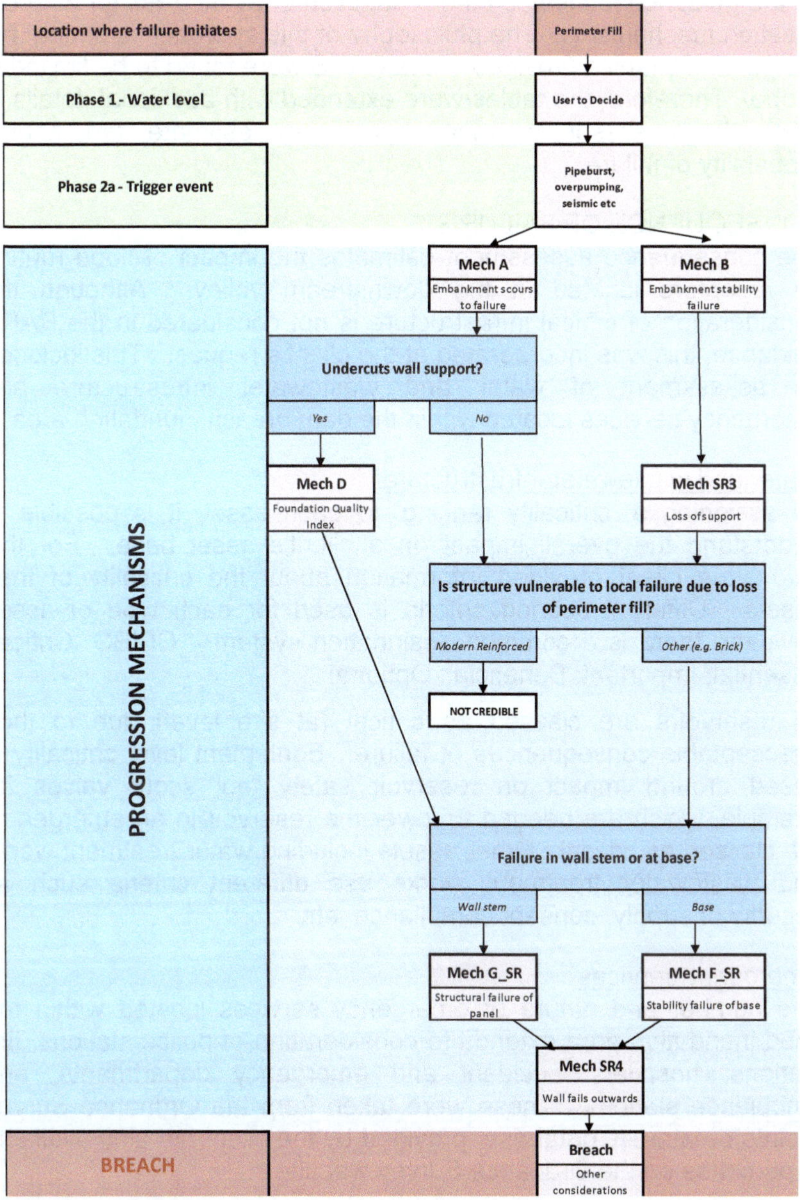

Figure 1. Example of flow chart describing common failure process for service reservoirs

CONCRETE AND MASONRY GRAVITY DAMS

As with the Service Reservoir approach given above, the RARS Tier 2 methodology employed for concrete and masonry gravity section dams was reviewed prior to implementation. The method promoted by the guidance focuses on the production of event trees for different initiation mechanisms. The philosophy of this approach is sound, but the guidance tables provided in the RARS were found to be limited in scope. Therefore, the tables were extended with additional details to allow the assessor to more specifically estimate the annual probability of failure.

CONSEQUENCE OF FAILURE

The consequence assessment estimates the impact of flood routing on receptors located in the downstream valley. Although the consideration of critical infrastructure is not considered in the RARS guidance, this was incorporated at the client's request. This included an assessment of water and wastewater infrastructure and emergency services located within the dam breach inundation area.

Water and wastewater infrastructure.

By assigning a 'criticality ranking' to each asset, it is possible to understand the overall impact on a client's asset base. For this study, the client provided information about the criticality of their assets. Different scoring criteria is used for each type of asset; however, there is a common designation system – CEIBO: Critical; Essential; Important; Beneficial; Optional.

All reservoirs are classed as critical (at site level) due to their unacceptable consequences of failure. Equipment level criticality is based around impact on reservoir safety too; scour valves for example, which are needed to lower the reservoir in an emergency, are classed as critical. Other assets including water treatment works and wastewater treatment works use different criteria such as security of supply, consent compliance, etc.

Emergency services

The number and nature of emergency services located within the flood inundation area extends to consideration of police stations, fire stations, hospitals, accident and emergency departments, and ambulance stations. These were taken from the Ordnance Survey 'Points of Interest' database, provided by the client but also available to purchase on the Ordnance Survey website.

IMPACT OF REFINEMENTS TO RARS ON LEVEL OF RISK

The impact of the refinements to RARS stretch across a range of measures, including improved consistency in scoring, greater confidence that the most significant failure modes have been incorporated and annual likelihood of failure estimated, and thus greater confidence in the output for use in decision on dam safety. The collaboration between Arup, MWH and Stillwater Associates has led to improvements to the RARS process and provided increased assurance on the quality of the output.

One of the measures of these refinements is to compare the distribution of likelihood of failure for different threats across the portfolio, as shown in Figure 2. This shows that the greatest threat to the safety of embankment dams is masonry spillways, followed by internal threats in embankments; buried structures; then chute overtopping. This appears a reasonable conclusion. The generally low threat from crest overtopping is noted and reflects the significant investment in spillways over the last few decades.

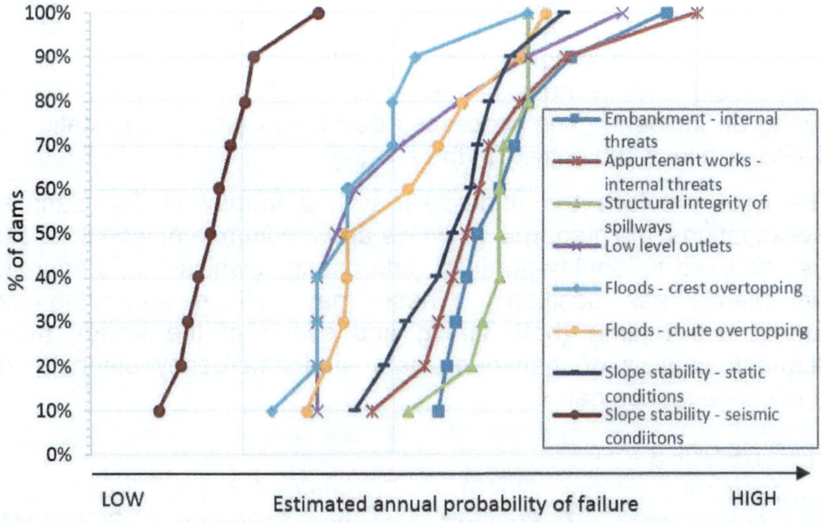

Figure 2. Comparison of different threats to embankments across the portfolio

A separate measure of the reasonableness of the output is how the tools provided in RARS rank concrete gravity dams and service reservoirs compared to embankment dams, with the distributions shown in Figure 3. Again the output is considered reasonable, with service reservoirs typically an order of magnitude safer than embankment dams, and concrete dams even more so.

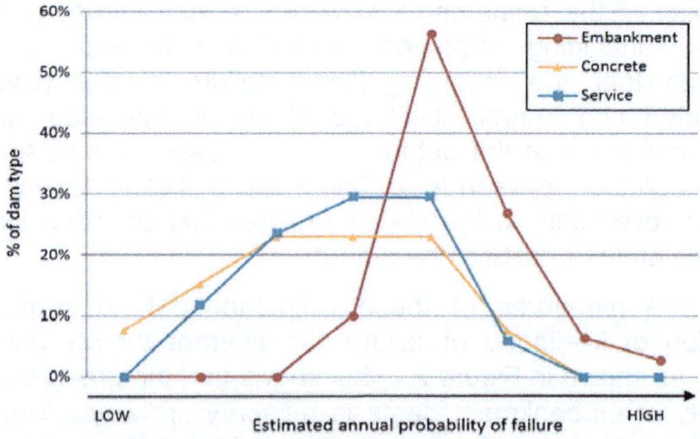

Figure 3. Comparison of likelihood of failure of embankment dams with concrete gravity dams and service reservoirs

CONCLUSIONS

The advances in the application of the RARS guidance as described above provided the undertaker with an overview of its full portfolio, which is extended to consider the condition of spillway channels, as well as the effectiveness of low level outlets on the performance during an incident. The processes developed enable application of RARS with less subjectivity in the scoring.

The PRA provides the undertaker with a framework for planning investigations, maintenance activities and medium term works. It can also be used to identify portfolio wide changes which would improve the client's risk position. These may include amending the surveillance regime (both timing and scope of the visits); more frequent analysis of monitoring data, and emergency planning for high consequence dams.

REFERENCES

Environment Agency (2010). *Guidance for the Design and Maintenance of Stepped Masonry Spillways. SC080015.* Environment Agency, Bristol

Environment Agency (2013). *Guide to risk assessment for reservoir safety management. SC09000.* Environment Agency, Bristol

Defra (2004). *Interim Guide to Quantitative Risk Assessment for UK reservoirs.* DEFRA, London

Johnson, RA, Leek, DS, King, ES and Dunne, HB, (1995). *Report 138: Underground Service Reservoirs – Waterproofing and Repair Manual.* CIRIA, London.

Section 2:

Guidance for reservoir engineers

Dams – Benefits and Disbenefits; Assets or Liabilities?
ISBN 978-0-7277-6133-0

doi: 10.1680/dbdal.61330.057

ice
Institution of Civil Engineers
publishing

Guides and Guidance: A "Luddite" view of guidance

C W SCOTT, Black & Veatch, UK

SYNOPSIS

Luddite (noun) - *a member of any of various bands of workers in England (1811–16) organized to destroy manufacturing machinery, under the belief that its use diminished employment.* (OUP, 2010)

Are guidance documents an unalloyed boon to reservoir engineers? This paper will consider whether the reservoir community's current addiction to guidance is helpful or not. Does guidance improve public safety? Does working within an ever widening network of guidance make reservoir engineering an attractive choice for talented young engineers?

Provoke (verb) - *make angry; arouse action; produce a reaction or effect* (OUP, 2010)

INTRODUCTION

The Guide to Risk Assessment for Reservoir Safety Management (EA, 2013) contains the following sentences referring to the report produced following a statutory inspection:

> "The report should state explicitly the significant failure modes identified through a potential failure mode identification process. ... Although not a legal requirement, it is recommended that it should also include the equivalent of a Tier 1 qualitative risk assessment."

This statement is supported neither by the Statutory Instruments that define the statutory content of reports nor by the Inspection Report contents published in the First Edition of the Guide to the Reservoirs Act (ICE, 2000) which were current at the time of publication, (although the suggested contents list in the Second Edition of that Guide (ICE, 2014) has been amended to include this "requirement").

No evidence is provided to support the reason for or purpose of the instruction in the Risk Assessment Guide (notwithstanding the preamble regarding its evidence based approach). The statement appears to be an opinion masquerading as a fact, or, perhaps more accurately, an opinion presenting itself to an instruction! Is that the purpose of Guidance Documents?

Attention is drawn to this example merely as the most egregious recent example of how guidance documents are going well beyond the original intent of guides and guidance and potentially seeking to constrain and define how Inspecting Engineers exercise their judgement in fulfilling their statutory role.

SOME DEFINITIONS

A quick reminder on definitions is perhaps a helpful starting point to this discussion.

Guide (noun) - *a book, document or display providing information on a subject* (OUP, 2010)

Guidance (noun) - *advice or information aimed at resolving a problem or difficulty* (OUP, 2010)

PURPOSE OF GUIDANCE DOCUMENTS

The first significant "modern"[1] reservoir related guidance document produced was the first edition of Floods and Reservoir Safety: an engineering guide, published in 1978 (ICE, 1978). This document was truly a document to provide "advice or information aimed at resolving a problem or difficulty". The problem or difficulty in question was how Reservoir Engineers were to respond consistently to the Flood Studies Report of 1975 (NERC, 1975).

The preface to the first edition of Floods and Reservoir Safety makes clear that the guidance, suggestions and recommendation contain in the guide were the product of lengthy gestation involving discussions at a variety of fora (including the BNCOLD Symposium at Newcastle - the forerunner to this conference). (BNCOLD, 1975)

Notwithstanding the wide engagement with practitioners, the guide states that "the recommendations made here are in no way mandatory" going on to state "the working party suggests that where an engineer feels it is right to depart from its recommendations the fact should be recorded in the inspection report". Accepting that 1978 was perhaps a less bureaucratic and prescriptive age, this

[1] Documents such as the 1933 report *Floods in relation to reservoir practice* have not been considered part of the current guidance culture.

seems an appropriate position for a guide to take: it is, after all, offering guidance rather than instruction!

The introduction to the guide contains a sentence that still encapsulates what should be the goal and purpose of guidance documents:

> "This guide is intended to assist those individuals who bear the personal responsibility that comes with being appointed to the statutory panels of engineers qualified to design and inspect impounding reservoirs."

If one adds undertakers and others with responsibilities for statutory reservoirs and widens the scope to include all reservoirs that fall within the ambit of the relevant reservoir safety legislation applicable across the United Kingdom one has a good working definition of the purpose of guides:

> "Guides are intended to assist panel engineers, undertakers and others with roles related to statutory reservoirs to fulfil the responsibilities of their particular roles."

Given that these roles are largely defined by legislation the overriding purpose of reservoir legislation also needs to be borne in mind: public safety.

GUIDANCE EXPANDS

The following table seeks to capture all the guides and guidance documents relevant to UK reservoirs since 1978.

Much of this guidance has been the product of the reservoir research programme funded successively by the Department of Environment (DoE), Department of Environment, Transport and Regions (DETR) and Department for Environment, Food and Rural Affairs (Defra). This research programme now continues as part of the joint Environment Agency / Defra Flood & Coastal Erosion Risk Management (FCERM) research and development programme.

In a paper presented to the BDS Conference in 1992, regarding the reservoir research programme, Wright, Coats and Charles observed:

> "The research is designed to provide Panel Engineers with an appropriate and consistent background for carrying out their duties under the Act." (Wright, Coats & Charles, 1992)

That is, research (and the resulting Guides and Guidance) is intended to provide background and context rather than to mandate or instruct.

Table 1. Reservoir Guides and Guidance

Year	Title
1978	Floods and Reservoir Safety: an engineering guide
1987	Design of reinforced grass waterways.
1989	Guide to analysis of open-channel spillway flows.
	Floods and Reservoir Safety. 2nd Edition
1990	An engineering guide to the safety of embankment dams in the United Kingdom
1991	An engineering guide to seismic risk to dams in the United Kingdom.
1992	Design of flood storage reservoirs.
1994	Register of British dams.
1995	Performance of blockwork and slabbing protection for dam faces.
1996	Small embankment reservoirs.
	Engineering guide to the safety of concrete and masonry dam structures in the UK.
	Investigating embankment dams: a guide to the identification and repair of defects.
	Reservoir dams: wave conditions, wave overtopping and slab protection.
	Bibliography of British dams.
	Floods and Reservoir Safety. 3rd Edition
1997	Valves, pipework and associated equipment in dams: guide to condition assessment.
1998	An application note to 'An engineering guide to seismic risk to dams in the United Kingdom'.
1999	An engineering guide to the safety of embankment dams in the United Kingdom. 2nd Edition.
2000	Guide to the Reservoirs Act 1975.
	Risk management for UK reservoirs.
2001	Climate change impacts on British reservoirs
	Sedimentation in storage reservoirs
2002	Floods and reservoir safety integration
2003	Early detection of internal erosion
2004	Interim guide to quantitative risk assessment for UK reservoirs.
	Revised guidance to Panel Engineers on FEH
2006	Engineering guide to emergency planning: Volumes 1 to 3
	Supplement No 1 to Interim Guide to Quantitative risk assessment for UK Reservoirs
2010	Masonry spillway guidance
2012	Regulation and risk assessment of reservoir releases
2013	Guide to risk assessment for reservoir safety management
2014	A Guide to the Reservoirs Act 1975. 2nd Edition.
2015	Floods and reservoir safety. 4th Edition.

To some extent, all of these publications address the purpose defined earlier. Challenges can arise in the detail and, as identified in the Introduction, in overreaching the scope and purpose of a Guide. Some have required "clarification": for example the Application Note of 1998 clarifies the Seismic Guide of 1991.

Discussion of the draft Guide to Drawdown Capacity for Reservoir Safety and Emergency Planning at the Inspecting Engineers' Forum in November 2015 agreed with the research contractor that the most appropriate guidance approach was to allow for judgement by panel engineers within a consistent approach.

DOES GUIDANCE IMPROVE PUBLIC SAFETY?

The answer to this question is "not necessarily". It is a struggle to conclude that including "the significant failure modes identified through a potential failure mode identification process" within the report on a Statutory Inspection does anything to improve public safety. Should the Inspecting Engineer consider the possible failure modes of the structure in determining its safety? Of course! Does the absence of a formal statement of the modes considered within the report mean that did not occur? No. it doesn't. It could be argued that including this requirement is a check box exercise with no direct implication regarding the quality or thoroughness of the inspection.

Conversely, the clarity and consistency of approach provided by successive Editions of Floods and Reservoir Safety has provided a strong framework for discourse on flood safety and overflow capacity that has contributed to improved public safety. It should be noted that the dam community have benefitted from extensive well-funded research on UK flood hydrology (notwithstanding some issues arising from the publication of the Flood Estimation Handbook in 1999 (IoH, 1999).

Other Guides have met the test defined by Wright, Coats and Charles and provided "an appropriate and consistent background" to practitioners. Absent the unnecessary strictures regarding matters outside its scope, the 2014 QRA Guide provides a consistent and logical framework to undertake risk assessments when they are deemed necessary.

DOES GUIDANCE MAKE RESERVOIR ENGINEERING AN ATTRACTIVE CHOICE FOR TALENTED YOUNG ENGINEERS?

There is a danger that an ever-widening network of Guidance could make reservoir safety engineering an unattractive choice for talented

Engineers when compared to other areas with scope for the application of imagination and creativity.

The ultimate question is what skills and capability are deemed necessary in the reservoir engineers of the future:

1. an ability to assimilate and apply extensive prescriptive guidance in a manner that will pass external audit, or
2. the ability to develop imaginative solutions that fulfil the needs of the reservoir owner while maintaining the highest standard of public safety.

The guidance and creativity are not necessarily incompatible provided the guidance has an appropriately "light touch".

ARE GUIDANCE DOCUMENTS AN UNALLOYED BOON TO RESERVOIR ENGINEERS?

The answer to this question is tied up with the nature of the guides and guidance produced.

- A document providing "information on a subject" or "advice or information aimed at resolving a problem or difficulty" is a boon.
- A document that provides prescriptive solutions and requirements is not a boon.

Reservoir safety legislation in the UK places the onus on making an assessment of the safety of reservoirs on individuals appointed to the statutory panels of engineers. Those individuals bear the personal responsibility for those assessments. Legislators have, thus far, chosen not to tell those individuals how to fulfil that responsibility

The challenge sits with Funders, Steering Committees and the Authors of Guides and Guidance to resist the temptation to tell people what to do and rather to provide "appropriate and consistent background for carrying out their duties".

REFERENCES

BNCOLD (1975), *Inspection, Operation and Improvement of Existing Dams*, BNCOLD/University of Newcastle, Newcastle

EA (2013), *Guide to Risk Assessment for Reservoir Safety Management,* Environment Agency, Bristol

ICE (1978), *Floods and Reservoir safety: an engineering guide*, Institution of Civil Engineers, London.

ICE (2000), *Guide to the Reservoirs Act,* Thomas Telford, London

ICE (2014) *Guide to the Reservoirs Act, Second Edition*, ICE Publishing, London

IoH (1999), *The Flood Estimation Handbook (5 volumes),* Institute of Hydrology, Wallingford

NERC (1975), *The Flood Studies Report (5 volumes),* Natural Environment Research Council, Institute of Hydrology, Wallingford

OUP (2010) *Oxford Dictionary of English, Third Edition*, Oxford University Press, Oxford

Wright, CE, Coats, DJ and Charles, JA, (1992) The reservoir safety research programme of the Department of the Environment, *Water Resources and Reservoir Engineering - Proceedings of the 7th BDS Conference*, Thomas Telford, London.

Dams – Benefits and Disbenefits; Assets or Liabilities?
ISBN 978-0-7277-6133-0

doi: 10.1680/dbdal.61330.065

Operating procedures for ensuring Reservoir Safety - How do you do it – too much or too little?

Dr A K HUGHES, Atkins, Epsom, United Kingdom

SYNOPSIS This paper will describe the process carried out with one large water company to decide the level of resourcing to try to ensure reservoir safety.

This study not only looked at the impact of different types of structure; but the likely modes of failure, access problems, geographical constraints, the frequency of visits, what is carried out on each visit, how information is obtained and recorded, how quality is controlled, how information is analysed.

The paper will also open up the debate on valve operations, scour discharges, the frequency of reading instruments, the designation of confined space, and lone working procedures.

INTRODUCTION

As a dam owner you are often faced with conflicting demands on resources but clearly an owner has responsibilities and liabilities, whether they are liabilities driven by the law – criminal law associated with the Reservoirs Act (HMSO, 1975), and common law associated with the law of tort (Rylands v Fletcher) (UKHL, 1868) or responsibilities to keep a dam safe and to ensure it does not fail – because it is an asset supplying water, irrigation, hydropower, flood protection, amenity, navigation, environmental benefit etc etc. If it fails you not only lose the asset but you might lose your business. So how much does an owner have to do? How much 'work' is enough?

BACKGROUND

An owner has to try to 'ensure' reservoir safety, to manage reservoir safety with all that it entails. For those reservoirs subject to the Act that means supervision, inspection, record keeping, maintenance and regular monitoring and surveillance.

After privatisation of the water companies it was noticeable that there was a move to a marked reduction in manpower levels and outsourcing of many activities including maintenance. It was noticeable that levels of maintenance reduced to a level which was deemed unacceptable in the late 1990s, and which resulted in the inclusion within the legislation of maintenance items which are 'enforceable',

Another 'problem' has arisen due to a lack of succession planning and knowledge transfer in most of the water companies, again because of a lack of money, a move towards different forms of procurement (frameworks etc) and a generation of ageing engineers. The profession has not been particularly successful in encouraging the young into engineering, partially because of the general 'status' of the engineer in society.

Knowledge transfer before people retire, succession planning and training of the young must be one of the highest priorities for the future. Unless we do this the prospects for our profession are dire. Owners and procurement departments must accept that any project must have an element of training within it which must be paid for by the project. I have suggested to the World Bank that they should explicitly define an element of training in every large project which must be priced and paid for.

Organisations who train the young have to take a responsibility for the future. Organisations which are a group of individuals who do not provide that succession planning cannot help in this role.

JUDGEMENT BY THE ENGINEER

So with many aspects of the Act the Engineer is asked to exercise judgement. In my opinion this is a valuable part of our legislative framework and one to be preserved. I certainly agree with Chris Scott (Scott, 2016) that we must not get to a stage where we're becoming bound up with prescriptive rules defined by a minority. Many of the documents produced to support our profession are 'Guidance', and that is precisely what they are – they are not mandatory and in the prefaces of many of them it clearly states that.

It is up to the engineer whether he takes and applies that guidance or not – and I believe that is how it should be. However, my advice, having given evidence in a number of court cases, is that if you depart from that guidance then you must have valid reasons, and must record those reasons in writing.

JUDGEMENT BY THE REGULATOR

There are many areas within our legislation where the legal definition is not well defined or not clear. One of these phrases is 'as soon as practicable', but other phrases include 'reasonable' etc.

Some years ago the Regulator sought clarity as to what 'as soon as practicable' means as far as the implementation of matters in the interests of safety.

The legal interpretation which came back from a barrister was for Category A it meant 3 years; for Category B, it meant 4 years and fro Categories C and D it meant 5 years. With the recent changes to the Reservoirs Act 1975 from the Flood and Water Management Act 2010 (TSO, 2010) Inspecting Engineers are required to state dates against individual recommendations. Care and judgement must be examined because there is little point setting a date which cannot be achieved – a project could involve planning applications, public consultation, appeals, as well as the visual design, tender documentation, tender appraisal, contractor selection phases – all of which take time.

JUDGEMENT BY THE OWNER

There is no doubt the owner is responsible for safety. Once again, there are conflicting demands on his resources whether those be people and/or money but it is a fact that if something goes wrong the owner's actions will be compared with the actions of similar organisations and what is deemed to be 'best practice' found in organisations in this country and abroad.

Some of these areas include surveillance and monitoring, and valve operations/scour discharges, all of which involve employees going to do the works. The questions posed by one water company was 'Are we doing enough?' How often should we go – because we are actually not carrying out the visits that the Inspecting Engineer requested? How many staff do we need?, and what do we want people to do when they are on site?

How do we think about the frequency of monitoring and supervision at a reservoir?

FREQUENCY OF SUPERVISION AND MONITORING AT RESERVOIRS

Monitoring and Surveillance is an important and integral part of achieving reservoir safety and the continuing safety of dams relies to a large extent on field observations. Surveillance usually describes visual observations, whilst monitoring describes reading and interpretation of instruments.

The continuing safety of an embankment must rely on a programme of field observations carried out within an appropriate safety management framework.

The appropriate level of monitoring and surveillance, including the frequency of visits by responsible personnel, depends on a number of functions including type, size and condition of the dam, and the potential hazard that may be present. Other considerations may be associated with loss of reputation for the owner should a failure occur. The higher the hazard usually the more comprehensive the surveillance, but equally a reservoir with a high operational value could warrant a level of surveillance greater than that required purely for safety reasons.

Surveillance involves a range of activities carried out by different personnel. These include periodic visits and report by Inspecting Engineers, Supervising Engineers and Reservoir Keepers.

Surveillance and Monitoring

It is known that if one is able to detect a problem with a dam, then there is more time to deal with the problem; to do works to avert failure; and to provide a warning in time to enable the loss of life to be avoided or at least reduced. In this way the liabilities and risk to the company can be reduced. This can reduce insurance premiums.

Surveillance and monitoring is usually tailored to different types of dams, with different frequencies related to likely modes of failure and the speed in which failures would progress. The objective of the process is to detect a change which would indicate that a failure process had started – it is a means of 'buying time' to be able to take action to reduce the risk, perhaps by reducing water levels or other means of mitigation. It also enables the company to react in time to warn others and try to protect persons and property downstream.

Types of Dam

Obviously, some dams are more susceptible to failure than others – some dams have many more failure modes than others. For example the modes of failure for earth dams include overtopping, instability, and internal erosion, and the failure processes can be quite rapid, whereas a concrete dam is susceptible to sliding or overturning, and perhaps in some rare situations foundation seepage. Rockfill structures would be somewhere in between in terms of number of failure modes and speed of failure. Service reservoirs have very few failure modes associated with them. However, each site and type of dam must be considered on its own merits.

Approach

The approach by one company was to identify the levels of surveillance requirements in order to develop a set of generic standards based on the tasks to be carried out and the frequency of the visits needed to carry out those tasks.

These have been defined as follows:

Table 1. Levels of surveillance requirements

Ref	Standard	Frequency
'A'	Take water level and do walk around consisting of walk over of crest (looking at upstream face), mitres, downstream face (where applicable) and toe.	Every visit
'B'	'A' standard plus weekly readings of instruments and look/listen down shafts and up tunnels.	Every week
'C'	'A' standard plus monthly readings, weekly readings and walk through tunnels and shafts.	Every month

Other factors which might cause the generic standard to be changed would include:

Consequence of failure
Where the consequence of failure is particularly high or indeed particularly low – as long as the company can sustain any attack on its reputation, i.e. should failure of a low-consequence dam occur.

Single source supply
If the reservoir is the only supply source to an area it would be important to prevent failure otherwise the cost of supplying water to the customers could be extremely costly.

People and property
Related to consequence of failure, the legislative framework seeks to protect persons and property against an escape of water. Clearly the number of persons at risk can vary from one to many thousands.

Recreation
If the reservoir is used for recreation the loss of the reservoir will attract negative publicity but on the other hand 'club' members can be used as an additional level of surveillance.

Failure modes and the influence on frequency

Earthfill dams can fail relatively quickly, in as little as perhaps 4 or 5 hours, or as long as 4 or 5 days depending on the mode of failure and the energy supplied to the failure process. Historically, earthfill dams in the UK were visited daily but financial constraints over the years reduced this to only two times a week in some companies. After a major incident in 2002, most companies reintroduced a regime where earthfill dams were visited at least three times a week and some every 48 hours. Where the consequence of failure is extremely high some embankments are still walked on a daily basis (Thames Water).

Concrete dams are really only subject to failure modes from extreme events such as overturning and sliding. Failure of the foundation caused by erosion would be monitored by measuring the quantity and turbidity of leaking waters. This failure process is likely to be an extended process. Karst foundation or soluble materials in the foundation could affect concrete dams but these would have to be considered on a site-by-site basis.

Concrete dams perhaps therefore need only be visited once a week if there are known leakages and once a month if there are no concerns about performance.

Rockfill dams would fall somewhere between earthfill and concrete – perhaps being less resilient than concrete dams but more resilient than an earthfill dam; thus the frequency of monitoring would be less than an earthfill dam but more than a concrete dam.

There will be other forms of construction which need consideration, for example an earthfill dam with concrete core – if intact a concrete core would be less erodible than a puddle clay core – if cracked with significant leakage then the stability of the downstream shoulder could be compromised. Thus the frequency of monitoring and surveillance could be varied according to the situation.

Other tasks and staff availability

Valve operation is usually undertaken every six months as a planned activity involving increased effort above the normal surveillance levels, thus involving a greater staff attendance.

In general, it would be beneficial if the staff who carry out the surveillance were able to do minor works e.g. pulling out saplings, putting back displaced pitching stones etc.

It is recognised that some staff have other tasks to do. Operations staff and perhaps reservoir safety is seen as a secondary task – an unidentified company will have to consider the results of such an organisational plan.

The resources needed will have to be evaluated with regard to a number of issues including:-

- Geography, size of patch, spacing between reservoirs
- Access
- Transport
  - Availability of appropriate vehicles (may vary according to weather)
  - Breakdown
  - May require men to walk in bad weather
- The need for double manning
- Sites where there are particularly concerns about performance
- Where the IE has directed more frequent visits
- The occurrence of extreme events – floods/seismic
- Confined spaces
  - Many of the designated confined spaces may in fact not be confined spaces and their designation needs to be revisited. This is necessary because a confined space designation usually requires extra men and resources and often a confined spaces team. Many spaces do have the potential for slips, trips and falls but no chance of the ingress of gas.
  - Improved lighting and access could be an investment in health and safety which could lead to a change in confined space designation.
  - Once the number of confined spaces has been defined it may be possible to 'plan a route' for a confined space team to enable the efficient use of resources in order that all the sites can be visited together.
- Bad weather can slow down the 'round' of reservoirs.

<u>Resource Planning</u>
In resource planning for the future I believe any company should:-

- Consider the organisational structure and where those that do the reservoir surveillance/monitoring report to, e.g. Operations / Reservoir Safety.

- A resource plan needs to recognise travel times and access to the site (which might vary at different times of year and different weather conditions).
- Allowance must be made for leave and bank holidays.
- Recognise succession planning is essential for the future sustainability of the business.
- Consideration must be made for forthcoming retirements.
- Shadowing of staff before they leave is necessary
- Recruitment needs to be targeted within the geographic areas being considered
- Training and retraining is required
- Overlap from one area to another enables one staff member to cover another's if they are absent for whatever reason
- Good signage should allow a number of the public alert the company if they find a problem
- Grass cutters/contractors should be told to report visual features – new wet patches etc.

<u>Resource Plan</u>
A database on site should be assembled which seeks to review the existing levels of monitoring and surveillance at each site and the levels proposed as defined by the standards set.

The company then needs to plan routes and times and thus the number of staff and vehicles required to provide the level of cover required.

Failure to provide sufficient cover will put the company at risk, particularly as when a failure occurs anywhere in the UK, comparisons will be made with the 'standards' applied by other similar companies.

<u>Benefits of Surveillance</u>
The analysis of the risk of failure of dams shows that if surveillance is increased then that element of the probability of failure changes. For example, moving from daily monitoring to weekly monitoring would move from a failure mode, say by piping, being much less likely to develop to one where it is more likely to develop, whereas for general factors moving from daily to monthly again moves the likelihood of failure from much less likely to more likely.

CONCLUSION

In conclusion I believe that companies need to look at a number of impacts on their businesses and make sure that they are matching 'best practice'. Failure to do so could make them vulnerable if challenged.

Failure to recognise the challenges of the future could also make companies ineffective!

REFERENCES

HMSO (1975). *Reservoirs Act 1975*. Her Majesty's Stationery Office, London.

TSO (2010). *Flood and Water Management Act 2010*. The Stationery Office, London.

Scott (2016). 'Guides and Guidance: A "luddite" view of guidance' in *Dams – Benefits and Disbenefits; Assets or Liabilities – Proceedings of the 19th Biennial Conference of the British Dam Society*. ICE Publishing, London

UKHL (1868). *Rylands v Fletcher, Judgment of the Court of Exchequer Chamber affirmed.* House of Lord's Journals, 17th July, 1868, London

Dams – Benefits and Disbenefits; Assets or Liabilities?
ISBN 978-0-7277-6133-0

doi: 10.1680/dbdal.61330.075

Reservoir Panel Membership: Is the end nigh?

P KELHAM, Arup, UK
I SCHOLEFIELD, United Utilities, UK
C W SCOTT, Black & Veatch, UK
A L WARREN, Mott MacDonald, UK

SYNOPSIS Concerns have been raised over a number of years about the falling number of engineers on the various panels operated under the auspices of the Reservoirs Act 1975. Numbers have undoubtedly fallen over the 30 years since the Act came into force. But is there a 'crisis'? Why have numbers diminished?

This paper examines the current situation, seeking to quantify the challenge, identifies the factors affecting the reduction in numbers and considers what steps could be taken to address those factors. The authors will draw on work completed under the auspices of the Institution of Civil Engineers (ICE) Reservoirs Committee and other work.

The purpose of the paper is to engender some discussion and debate within the reservoir community.

INTRODUCTION

The issue of the declining numbers of people on the panels established under the Reservoirs Act 1975 has been the source of discussion and concern for many years and yet the crisis of having insufficient qualified resources to inspect reservoirs has not occurred.

The issue has been the subject of study by the Reservoirs Committee twice in the past 5 years: 2011 and 2015. Within the past 18 months it has been discussed at both the Supervising Engineers' and the Inspecting Engineers' Fora. This paper seeks to set some context for the discussion, identify some of the relevant factors and discuss the potential steps that can be made.

A BIT OF HISTORY

Panels of reservoir engineers came into being with the enactment of the Reservoirs (Safety Provisions) Act 1930 (HMSO, 1930). The initial panel structure was of two panels:

- Panel A: comprising engineers qualified to design and to supervise the construction or modification of reservoirs and to undertake periodic statutory inspections of all reservoirs
- Panel B: comprising engineers qualified only to undertake the periodic statutory inspections.

In 1946, following experience with the operation of that panel system, the panel structure was amended. Three new panels were added (II, III & IV): Panel A was renamed Panel I and appointments to Panel B were suspended. The Panels were constituted as set out in the table below.

Table 1. Panels constituted under Reservoirs (Safety Provisions) Act 1930

Panel	Composition
I	Engineers qualified to design and to supervise the construction or modification, and to undertake periodic statutory inspections of all reservoirs.
II	Engineers qualified to design and to supervise the construction or modification, and to undertake periodic statutory inspections of all non-impounding reservoirs.
III	Engineers qualified to design and to supervise the construction or modification, and to undertake periodic statutory inspections of, non-impounding reservoirs of less than 50 million gallons capacity (227,000m³ approx.).
IV	Engineers qualified to design and to supervise the construction or modification, and to undertake periodic statutory inspections, of non-impounding reservoirs made of brickwork, masonry, concrete or reinforced concrete.

The consultations associated with the implementation of the Reservoirs Act 1975 (HMSO, 1975) and the requirements of that Act resulted in further amendments to the panel structure as follows:

- All Reservoirs Panel (AR): qualified to design and supervise the construction and alteration of, to inspect and report upon, and to act as supervising engineers for, all statutory reservoirs.
- Non-Impounding Reservoirs Panel (NIR): qualified to design and supervise the construction and alteration of, to inspect and report upon, and to act as supervising engineers for all statutory non-impounding reservoirs.

- Service Reservoirs Panel (SR): qualified to design and supervise the construction and alteration of, to inspect and report upon statutory service reservoirs, and to act as supervising engineers for all statutory service reservoirs.
- Supervising Engineer Panel (SupE): qualified to supervise all statutory reservoirs when no construction engineer is employed.

In the period following the enactment of the Reservoirs Act 1975, when it appeared that it may not be implemented, there was a concern that a shortage of Panel I engineers may arise and the decision was made to restart appointments to Panel B. (Agnew, 1984)

In 1984, Michael Kennard noted that there were 253 engineers appointed to the various panels constituted under the Reservoirs (Safety Provisions) Act 1930. This is a slightly deceptive number as appointments under the 1930 Act were for life, so a proportion of those on the panel were either retired or inactive. (Kennard, 1984)

Table 2. Panel membership in 1984

Panel	Total No of Engineers	Percentage Retired or Inactive	'Active' Panel Members
I	96	33%	64
II	41	38%	25
III	41	22%	32
IV	68	16%	57
B	9	55%	4

CURRENT PANEL NUMBERS

The ICE Reservoirs Committee has reviewed the issues around succession planning and numbers twice in the past five years. As part of that exercise the number and age profile of engineers on the panels was assessed. The numbers are summarised in the table below.

Table 3. Current Panel Numbers

Panel	2011	2015
AR	42	35
NIR	5	1
SR	6	4
SupE	160	148

HOW MANY PANEL ENGINEERS ARE NEEDED?

Currently there are a total of 2,892 registered reservoirs which fall within the ambit of the unamended Reservoirs Act 1975 (i.e. capable of holding 25,000m³ of water) in England, Scotland and Wales.

'Inspecting Engineers' are those on panels AR, NIR and SR. Assuming that the required inspections are evenly spread and are uniformly at 10-year intervals, 289 inspections are required each year. With 40 Inspecting Engineers that equates to 7.2 inspections per Inspecting Engineer per year. If only All Reservoirs Panel Engineers are considered that increases to 8.3 inspections per year.

Anecdotally, it is understood that the majority of inspections are undertaken by approximately 15 Inspecting Engineers, implying an average of 19 inspections per year by those "active" Inspecting Engineers.

The Flood and Water Management Act 2010 (HMSO, 2010) amendments to the Reservoirs Act 1975 in England and Wales and the other legislation enacted by devolved governments mean that the number of reservoirs requiring inspections is currently in flux.

Using the following assumptions, a tentative number of reservoirs requiring inspections in the future has been calculated:

- The reservoir capacity threshold is reduced to 10,000m³ in Wales and Scotland
- The capacity remains at 25,000m³ in England (i.e. Phase 2 of the Flood and Water Management Act 2010 amendments is not implemented)
- 90% of reservoirs with capacity greater than 25,000m³ are designated 'High Risk'
- 50% of reservoirs in Scotland and Wales with capacities between 10,000m³ and 25,000m³ are designated 'High Risk'
- 150 reservoirs in Northern Ireland require inspection.

Table 4. Number of reservoirs requiring statutory inspections

	Capacity	Total Number	'High Risk' Number
England	>25,000m^3	2,001	1,801
	>10,000m^3 and <25,000m^3	1,200[1]	0
Northern Ireland	>10,000m^3	150	150
Scotland	>25,000m^3	686	617
	>10,000m^3 and <25,000m^3	1,536	768
Wales	>25,000m^3	205	185
	>10,000m^3 and <25,000m^3	351	176
Total		**6,129**	**3,697**

Again, assuming that the required statutory inspections are all at 10 year intervals and are spread uniformly that equates to 370 inspections per annum. (This is recognised to be a flawed assumption considering that batches of new reservoirs will require inspection when designated as 'High Risk' reservoirs with capacities between 10,000m^3 and 25,000m^3 .) Assuming that the size of AR panel required should be set based on the panel's 'active' members, the minimum panel size to serve this level of activity is considered to be approximately 20. Implementation of Phase 2 of the Flood and Water Management Act 2010 would not likely to be a significant factor as, given the assumptions made, the number of active Inspecting Engineers required would increase to 23.

FUTURE SCENARIOS – INSPECTING ENGINEERS

Attendees at the Inspecting Engineers' Forum in November 2015 were polled to ask when they anticipated that they would cease to be working members of the Panel to which they belonged. 31 out of the total of 39 Inspecting Engineers attended the Forum and responses were additionally gained from most of the absent engineers through correspondence. A scenario analysis was done using this data and looking at three scenarios of the average rate of new appointments to the Inspecting Panels:

1. Average rate of one new Inspecting Engineer every two years
2. Average rate of one new Inspecting Engineer every year

[1] Estimate for reservoirs of capacity greater than 25,000m^3 and less than 10,000m^3 vary from 1200 to 4700. The number used is that provided to the Reservoirs Committee by the Environment Agency.

3. Average rate of two new Inspecting Engineers every year.

The chart below summarises the running of those three scenarios combined with the retirement profile derived from the survey undertaken at the Forum. An average term as an Inspecting Engineer of 20 years has also been assumed for the new appointments (i.e. appointment on average in late 40s).

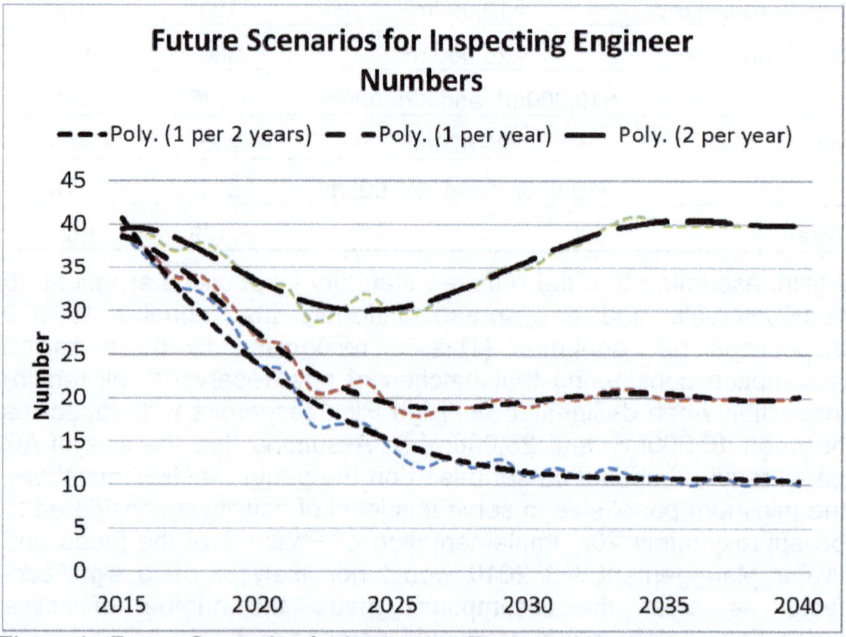

Figure 1. Future Scenarios for Inspecting Engineer Numbers

The resulting long-term numbers are:

Table 5. Long Term Inspecting Engineer Numbers

Scenario	Long-term Number of Inspecting Engineers
1	10
2	20
3	40

It should be noted that the long-term numbers are a direct product of the assumption on the term of newly appointed Inspecting Engineers. If the average term is reduced to 15 years (i.e. average age at appointment in early 50s), the numbers decrease to 8, 16 and 31 respectively across the three scenarios.

The appointment rate to the All Reservoirs Panel over the past 10 years has been most close to Scenario 1 (seven new appointments since 2006 with an average age in the late 40s). Thus, there is a *prima facie* case that the current rate of replenishment is not adequate to meet future needs.

COMMERCIAL CONSIDERATIONS – INSPECTING ENGINEERS
Simple economics would suggest that there is currently an oversupply of Inspection Engineers. Evidence for that view would be the level of competition for appointments, the pricing being offered and the extent to which customers are able to dictate commercial terms.

There is little or no commercial incentive for an organisation with investment choices to select reservoir engineering as the area in which to invest. Much of the work is, in essence, "lone expert" work. However, the work, by and large, does not command "lone expert" prices. Commercial terms are unattractive (e.g. a recent water company procurement required a £10 million limit of liability on assignments that were likely to have a value of much less than 0.05% of that number).

The structure of the market means that competition includes a breadth of organisations from major national and international engineering consultants, to sole traders and small specialist organisations. It is not realistic to expect one part of that market to bear the cost of the future development of the necessary base talent.

OTHER CONSIDERATIONS – INSPECTING ENGINEERS

Career Choices
The requirements of the routes to panel membership mean that appointment to one of the Inspecting Panels occurs later in a career than advance to such a senior role in other branches of engineering. This is exacerbated by paucity of design and construction experience opportunities within the UK. That said there remains a pipeline of engineers keen to be involved in reservoir engineering and aspiring to panel membership.

Routes to Empanelment
Concerns have been expressed about the routes to and requirements for panel membership. Discussions occurred on this subject at both the 2015 Supervising Engineers' Forum and Inspecting Engineers' Forum. Such was the breadth of opinion expressed that it would be impossible for the process to satisfy all. On one side is the call for greater clarity and objectivity in the assessment of appointments and reappointments; on the other is outrage that a well-respected engineer has had some challenges in getting reappointed. The two positions are not compatible: either

assessment is objective on the basis of submitted material and performance at interview, or it is subjective on the basis of knowledge of the individual. In reality the former is the only morally satisfactory approach.

The ICE Reservoirs Committee is reviewing and updating the requirements for panel membership to be better aligned with the approach adopted for applications for membership of the ICE. The revised approach will provide better clarity of the competencies required. This work is being undertaken by others and it is not the purpose of this paper to address this matter.

POSSIBLE SOLUTION - ALTERNATIVE PANEL STRUCTURE

The necessity of finding a solution depends on the extent which the potential reduction in Inspecting Engineer numbers present as a problem. Market economics would suggest that as numbers reduce the attractiveness of the discipline would increase and this would result in an increase in numbers. As noted above, the evidence suggests that 40 Inspecting Engineers represents oversupply!

Taking a more interventionist view an option would be to adjust the panel structure, following precedent of the late 1970s when there was concern about numbers. The decision could be taken to introduce an intermediate panel of Engineers qualified to undertake inspections. The most logical way to do that would be some form of resurrected Panel B (i.e. a panel comprising engineers qualified to undertake the periodic statutory inspections).

This would form a stepping stone between the Supervising Engineers' Panel and All Reservoir's Panel.

The question arises as to whether the members of such an 'Inspecting Panel' should be qualified to undertake QCE work arising from statutory inspections. Following the logic of the Panel A and Panel B structure would suggest not. Considering the fact that works in the interests of safety can be extensive and involve modifications to spillways, low level outlets, the impermeable barrier, etc. it seems logical that the Engineer overseeing such work should have similar skills, abilities and experience to one supervising modifications that alter the level at which water can be stored. This logic would suggest that the QCE role should remain with members of the AR Panel.

Given the current rate of appointment, the All Reservoirs Panel may reach a steady state somewhere between 10 and 15. What would be a sufficient number of engineers to fulfil the needs of Construction Engineer and QCE appointments? The analysis above does not address this question but if one assumed that 50% of inspections gave rise to works requiring a QCE, then a little less than 200 QCE appointments would be required, or between 14 and 20 per AR Panel member: a number which does not appear excessive.

A further alternative would be to create a panel similar in scope to the original Panel III, whereby engineers can be appointed to a second tier inspecting panel who would be qualified to inspect statutory reservoirs of any type up to a certain reservoir volume threshold. The same question arises whether such engineers should be allowed to act as a QCE on such reservoirs. Similar reasoning would appear to apply suggesting a similar answer.

The use of consequence category in place of volume as the basis for determining which reservoirs could be inspected by "Panel III" engineers is attractive. However, this could lead to complications as one of the factors to be considered by the Inspecting Engineer is the dam category. For example, what happens if the Panel III engineer deems that the category of the reservoir under inspection should be increased such that she/he is no longer qualified to undertake the inspection? Thus, on balance, volume is a criterion with less scope for complication although this dilemma perhaps raises questions about the underlying logic of the "Panel III approach".

Neither of the above options is desirable if one considers that construction experience is important to the role of carrying out periodic inspections.

Regardless of the option selected the authors consider that the independence of the inspecting engineer from the reservoir undertaker must be preserved.

SUPERVISING ENGINEERS

The situation with Supervising Engineers is different. With 148 engineers on the Supervising Panel and roughly 3,700 reservoirs requiring supervision means an average of 25 reservoirs per Supervising Engineer. The calculation is approximate as the number of reservoirs will be higher as the Medium Risk reservoirs in Scotland will require supervision and Inspecting Engineers can also act as Supervising Engineers. For the purposes of the "pub maths" used in this paper, those two factors have been deemed to be self-cancelling.

It is interesting to note that when the implementation of the 1975 Act was being planned it was thought that about 175 Supervising Engineers would be required for the Water Authorities (as they were then) in England and Wales (Jollans, 1984).

New appointments to the Supervising Engineer's Panel run at approximately 10 to 15 per year.

Supervising Engineers broadly fit into two types:

- those considering the role as a potential stepping stone to membership of other panels, and
- those viewing it as an end in itself.

The former group are likely to be holding a relatively small number of appointments while working on other work in studies, design and construction supervision. They are most often employed by engineering consultants and, on average, will be towards the younger end of the age distribution.

The second group will include those working for undertakers, independents and small specialist consultancies. They will typically have a larger number of appointments and potentially come from a much wider educational and experiential background.

Major undertakers have indicated that they are finding it difficult to identify and retain professionals to fulfil the Supervising Engineer role and are developing staff in-house. This is a function of their organisational requirements as well as the qualifications and experience necessary to obtain panel membership.

It has been suggested that Supervising Engineers should be Incorporated Engineers. On the basis of information on the website of the Engineering Council, this would require one of the following minimum academic qualifications:

- accredited Bachelors or Honours degree in engineering or technology
- accredited HNC or HND in engineering or technology (for programmes started before Sept 1999)
- HNC or HND started after Sept 1999 or a Foundation Degree in engineering or technology, plus appropriate further learning to degree level
- NVQ4 or SVQ4 that has been approved for the purpose by a licensed engineering institution, plus appropriate further learning to degree level.

Without a degree individuals would have to follow the Technical Report route to gain Incorporated Engineer status. This would be under the supervision of a mentor who would have to vouch that they are working at a Bachelor's degree level of competence. Is this too high a bar and would it deter many suitable candidates from developing to become Supervising Engineers? Initial consultation with reservoir owners suggests that perhaps this is too high in general and that perhaps Engineering Technician may be a more appropriate grade if a minimum professional membership level is to be applied. Others have questioned as to what is wrong with the current system of gaining appropriate knowledge, skills and experience, application to the panel, interview by experienced members of the Reservoirs Committee and the recommendation for appointment to the Panel. That is perhaps a topic for a separate paper after further consultation. However, the minimum qualification requirements will have a significant bearing on the ability to recruit and develop capable and competent candidates.

The wider powers and responsibilities implied by the changes in the legislative framework also have an impact on the necessary qualifications, skills and experience. It is important that the minimum qualifications required for Supervising Engineers keeps in step with current and emerging legal powers which might include overseeing the preparation of statutory flood plans (emergency plans).

A further challenge will be the ability of the existing Supervising Engineers to service the regional distribution of future 'High Risk' reservoirs (3,697) from their home base as shown in the table below. The main challenge being a potential lack of locally based Supervising Engineers in Scotland and Northern Ireland.

Table 6. Reservoir and Supervising Engineer distribution

Country	Proportion of 'High Risk' Reservoirs	Proportion of Supervising Engineers
England	49%	73%
Scotland	37%	17%
Wales	10%	9%
Northern Ireland	4%	<1%

CONCLUSION

This paper has presented some data and analysis investigating the question of whether there is an imminent succession crisis with both the Inspecting and Supervising Panels.

With regard to the Inspecting Panels, the conclusion is an emphatic 'maybe'. There appears to be a scenario where there would potentially be a shortage of Inspecting Engineers in about 10 years. The analysis ignores any changes that occur as a consequence of reduced numbers and so can only be considered tentative at best. Options for possible adjustment to the inspecting panel structure that could in some way ameliorate the problem have been discussed.

The situation with Supervising Engineers is less clear. However, some water companies have experienced severe difficulties in attracting full time Supervising Engineers and are now developing staff in-house. Insisting on Incorporated Engineer as a minimum qualification may have a significant impact on their current trainee's progression to Supervising Engineer status. There is a trend of reduction with 12 fewer panel members in 2015 than in 2011. Quality candidates continue to come forward and appointments are continuing at a rate of between 10 and 15 per year.

So is the end nigh? Probably not. But that is no reason not to consider how things could be changed for the better, recognising that any changes will take many years to implement and due consideration should be given to the potential impacts of making the current situation worse.

REFERENCES

Agnew SC (1984). Reservoir Safety Legislation, *Symposium on Water Management: A review of Current Issues.* Institution of Water Engineers and Scientists, London

HMSO (1930). *Reservoirs (Safety Provisions) Act 1930.* His Majesty's Stationery Office, London

HMSO (1975). *Reservoirs Act 1975.* Her Majesty's Stationery Office, London

HMSO (2010). *Flood and Water Management Act 2010.* Her Majesty's Stationery Office, London

Jollans, WM (1984) Implementation of the Reservoirs Act 1975: Practicalities of Implementation – A Water Authority's View. *Symposium on Water Management: A review of Current Issues.* Institution of Water Engineers and Scientists, London

Kennard MF (1984). Implementation of the Reservoirs Act 1975: Practicalities of Implementation – A Consultant's View. *Symposium on Water Management: A review of Current Issues.* Institution of Water Engineers and Scientists, London

Dams – Benefits and Disbenefits; Assets or Liabilities?
ISBN 978-0-7277-6133-0

doi: 10.1680/dbdal.61330.087

Reservoir Flood Estimation: Time for a Re-think

D FAULKNER, JBA Consulting
J BENN, JBA Consulting

SYNOPSIS Design floods for reservoirs in the UK and Ireland are currently estimated using a method that in many respects has not changed since the Flood Studies Report of 1975. Although estimates of design rainfall for reservoirs in the UK have recently been updated, other aspects of the design method and the estimation of probable maximum precipitation (PMP) are dated. Methods for river flood estimation have moved on since the 1970s and there are new and longer-term sources of hydro-meteorological data. Research has shown instances of both PMP and probable maximum flood estimates being exceeded.

This paper gives an overview of aspects of the design flood estimation procedure that are in need of an update. Discrepancies are identified between the different methods used to calculate percentage runoff and time to peak for the 10,000-year flood and the probable maximum flood. The pros and cons of adopting the newer Revitalised Flood Hydrograph rainfall-runoff method for reservoir safety work are discussed and suggestions offered for development of an up-to-date method for reservoir flood estimation that builds on existing methods, with the aim of improving understanding of the liabilities associated with dams and reducing the risk of dam failures.

INTRODUCTION

There are currently about 2500 regulated large raised reservoirs in the United Kingdom. There are also new reservoirs being constructed (mainly for flood storage), alterations being made to existing dams, and re-assessments as part of the decennial inspections. All require flood estimates.

With the definition of a large raised reservoir changing in Wales and Scotland, and a new regulatory regime being implemented in Northern Ireland, the number of statutory reservoirs is likely to increase over the coming years. Many of the reservoirs being newly regulated will require flood estimates as part of their risk assessment. A substantial proportion of these reservoirs will be less than 25,000m³ in volume and are likely to have small contributory catchments.

METHODS OF UK FLOOD ESTIMATION

Floods and Reservoir Safety Fourth edition (ICE, 2015) gives a guide to methods currently recommended for reservoir flood estimation. Details of the methods are presented in several publications, the most comprehensive being Volume 4 of the Flood Estimation Handbook (FEH) (IH, 1999) which contains a re-statement of the Flood Studies Report (FSR) rainfall-runoff model (NERC, 1975).

The current situation is rather complicated, with different generations of methods being recommended for different purposes. Also the timing of the publication of the ICE (2015) guide was rather unfortunate, being just a few months before the release of the latest rainfall depth-duration-frequency (DDF) statistics, FEH13 (Stewart *et al*, 2013).

There are two rainfall-runoff models recommended by the guide:

- The FSR/FEH rainfall-runoff model, used for estimating the Probable Maximum Flood (PMF) and the 10,000-year and 1000-year floods.
- The Revitalised Flood Hydrograph (ReFH) model (Kjeldsen *et al*, 2005), recommended as an alternative to the FSR model for the 150-year flood. The guide hints that ReFH will be extended in future to enable its application for more extreme floods.

Three generations of design rainfall statistics are mentioned:

- The FSR (NERC, 1975), the only source of probable maximum precipitation (PMP) estimates and also used for estimating the 10,000-year flood and (along with FEH rainfalls) the 1000-year flood, until the release of FEH13.
- The FEH (IH, 1999) used for the 150-year flood and (along with FSR) the 1000-year, until the release of FEH13.
- FEH13, which after many years in gestation was released to practitioners via the FEH web service in November 2015. ICE (2015) suggests that FEH13 is used for estimating all except the PMP, the implication presumably being that FSR/FEH rainfalls are completely superseded.

REVIEW OF EXISTING METHODS

Six specific areas where current methods are in need of review and improvement have been identified:

1 PMP estimates are dated and have been exceeded.

Since the present PMP estimates were published in 1975 there has been much research, in the UK and overseas, into PMP. The establishment of the weather radar system in the UK has provided large amounts of data on the structure of rainstorms, which was not available for the FSR research. Austin *et al* (1995) derived new estimates of PMP for north-west England using radar data for convective storms. Their results were higher than the PMP estimates for durations longer than 12 hours. Clark (1995) has criticised the method used in the FSR for estimating PMP, claiming that it tends to underestimate.

One pressing reason for a re-examination of PMP is that there are instances of the FSR estimates having been exceeded. Stewart *et al* (2013) mention five such events (Table 1).

Table 1. Storms that have exceeded the FSR PMP, from Stewart *et al* (2013)

Date	Location	Rainfall (mm)	Duration (hours)	% of FSR PMP
28 Jul 1917	Bruton, Somerset	243	8	102%
18-19 Aug 1924	Cannington, Somerset	225	5	103%
19 Jul 1955	Martinstown, Dorset	280	15	101%
7 Oct 1960	Horncastle, Lincolnshire	184	3	101%
19 May 1989	Halifax, West Yorkshire	193	2	119%

In four of the five cases, the rainfall total was only marginally in excess of the estimated PMP. However, it is possible that larger depths of rain fell at locations away from raingauges. For example, Clark (2005) estimated that around 350mm fell at the centre of the Martinstown storm in 1955, from detailed analysis of informal measurements and the structure of the storm. This is 17% above the FSR estimate of PMP at that location.

In contrast to the persistence of the FSR procedures for estimating PMP, methods for estimating less extreme rainfall have been through two succeeding generations since the FSR. As well as being based on new science and rainfall records that are more numerous and longer, these new methods also benefit from digital terrain data that allow the results to be interpolated with a great deal more spatial detail, following the local topography and avoiding subjective and error-prone interpolation of rainfall contour lines on maps.

As discussed later, operational estimates of PMP in Australia, Canada and many American states are based on much more recent research and data than those currently used in the UK.

In places where the FSR method underestimates PMP, there is a distinct possibility that PMF is underestimated as a result. This may mean that some Category A dams are not as safe as they are currently thought to be.

2 The FSR rainfall-runoff model has been superseded

In 2005 a replacement was released for the FSR rainfall-runoff method for design flood estimation (Kjeldsen *et al*, 2005). The Revitalised Flood Hydrograph (ReFH) method aimed to address several problems that had been identified with the FSR model, including:

- General overestimation of design flows;
- The need to re-examine the composition of the design flood event to accommodate the transition from FSR to FEH design rainfall statistics;
- The relatively small size of the calibration floods.

The ReFH method was rapidly adopted for estimation of design floods for river flood studies in England, Wales and Northern Ireland, and in 2015 an update was released, ReFH2 (WHS, 2015).

In their current state the ReFH or ReFH2 models may not necessarily be suitable for estimating the 10,000-year flood or PMF. There are some limitations of the method, for example as discussed by Faulkner and Barber (2009). It would be necessary to consider how the method scaled up. A version of ReFH2 has been tested in conjunction with the FEH13 rainfalls for return periods up to 1000 years. This contrasts favourably with the longest return period of 10 years for which the performance of the FSR design event was assessed (NERC, 1975).

3 Flood response times may be much quicker than thought in extreme events on some catchments

One of the parameters of the FSR and ReFH rainfall-runoff models is the time to peak of the unit hydrograph, Tp. It affects both the peak flow and, via its effect on the storm duration, the volume of the flood hydrograph. A shorter Tp results in a higher peak flow.

Tp is generally assumed to be a constant for a given catchment. An exception is made when estimating the PMF, for which Tp is reduced by a third. This factor was derived from the ratio of minimum to mean observed Tp for each gauge in the FSR flood event archive. No such adjustment is made when estimating the 10,000-year flood, for which response time is assumed to be identical to minor floods. The FSR found no statistical evidence that Tp was dependent on storm characteristics over the dataset as a whole, although on some individual catchments there was a strong tendency for Tp to decrease with rainfall intensity. The ReFH research (Kjeldsen *et al*, 2005) found that 15 of the 20 largest events showed a faster than average response.

On some catchments there is clear evidence of a trend towards shorter Tp for more intense rainfall. Figure 1, from the data used by Wass *et al* (2008) shows how the lag time (related to Tp) drops dramatically as the maximum 15-minute intensity increases for three catchments on the North York Moors. During the June 2005 flood on the River Rye (which led to spillway damage at Boltby Reservoir) Tp was one third of its average value. A postulated physical explanation was that the extreme rainfall intensity led to overland flow, concentrated into erosion gullies that extended the channel network, making the delivery of rainfall to the river more efficient. Another example of this effect, for a small lowland catchment in Suffolk, is included in Figure 1. A fifth example can be found at Boscastle in Cornwall, where HR Wallingford (2005) found it necessary to reduce the Tp parameter by at least 50% when simulating the August 2004 flood.

Wass *et al* (2008) point out the implications for reservoir flood estimation. If the assumption of a fixed Tp is wrong for floods smaller than PMF, then the design floods could also be wrong. There is also a concern that the one third reduction in Tp for the PMF may be insufficient for some types of catchment.

Research is needed to identify the types of catchment that show this behaviour. Most of the published examples are on small catchments, often steep, of the sort that drain into many upland reservoirs. Reed and Field (1992) expressed concern that response times may not be well estimated on small steep areas with shallow soils that are typical of reservoir catchments. On larger lowland catchments the opposite effect is sometimes reported: larger floods can show a slower response. Possible explanations include an extension into the headwaters of areas that contribute rapid runoff and/or increased attenuation via floodplain flow.

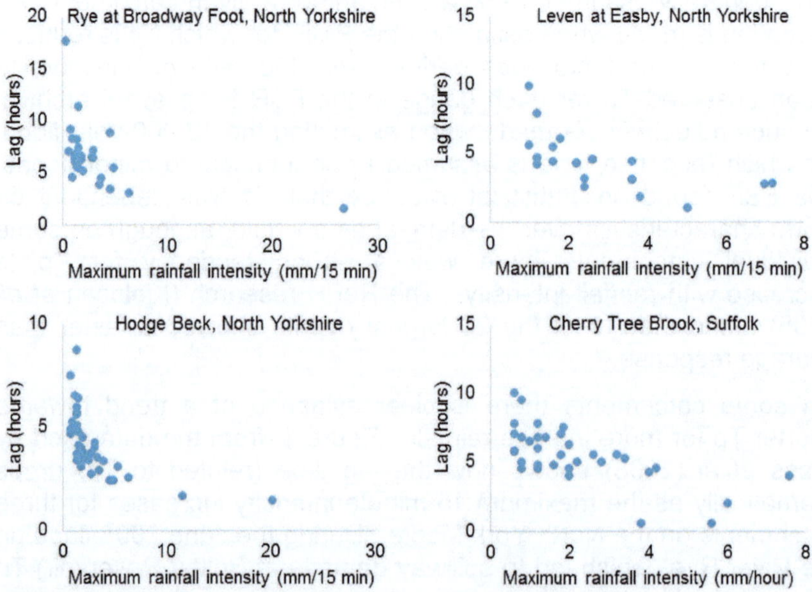

Figure 1. Lag time as a function of maximum rainfall intensity for four catchments. Each symbol represents a different flood event.

The phenomenon of time to peak reducing during extreme rainfall was postulated by Acreman (1989) as one possible explanation for the fact that six historical floods in the UK were thought to have exceeded the FSR estimate of the PMF, five on catchments smaller than 10 km² (Table 2). In three cases the estimated peak flows were at least twice the estimated PMF. It should be noted that most of the flow rates were estimated using approximate hydraulic methods rather than measured at gauging stations.

Recent research has developed a modification of the ReFH model with a dynamic link between rainfall depth and the shape of the unit hydrograph, allowing for a faster response for larger events. Kjeldsen *et al.* (2016) describe the modified model, developed using runoff data from South Korea, and show that the consequences include an 80% increase in the estimated PMP. This finding illustrates the potentially alarming consequences for reservoir flood estimation.

Table 2. Floods that have exceeded the FSR PMF, from Acreman (1989)

Date	Location	Catchment size (km^2)	Estimated peak flow (m^3/s)	% of FSR PMF
17 Aug 1917	Red-a-ven, Devon	4.0	110.4	141%
12 Aug 1948	Stobshiel, East Lothian	4.1	40.8	111%
8 Aug 1967	Claughton, Lancashire	2.3	66.6	200%
16 Aug 1970	Dorback, Morayshire	365	1939	139%
13 Jun 1980	Caldwell Burn, Dumfriesshire	5.7	189	381%
12 Jul 1982	Chulmleigh, Devon	1.7	68	270%

4 There is inadequate guidance on spatial variation in snowmelt.
When estimating a PMF for the winter season, snowmelt is added to the design storm. It is necessary to estimate the depth of snow and the rate at which it melts.

The map of 24-hour snowmelt rate in Floods and Reservoir Safety (ICE, 2015) is not very helpful. It shows one contour, for a melt rate of 42mm per day (1.75 mm/hour) and notes that a higher, but unspecified, value might apply at certain mapped upland locations. This hardly seems a satisfactory basis for inspecting engineers to judge the safety of an upland reservoir during a winter PMF. The guide recommends that in the upland areas practitioners should make their own estimate of the melt rate, for example using information from Hough and Hollis (1997).

There were two components to the FSR snowmelt study, one based on meteorological data, carried out at the Met Office, and the other on snowmelt runoff, based at Newcastle University. The results from these studies were widely divergent and the disparate views have not yet been resolved. The work at Newcastle found melt rates in excess of 5mm/hour and these findings were supported by observations by observations of snowmelt runoff in the northern Pennines (Archer, 1981) and theoretical studies (Mawdsley *et al*, 1991).

In the authors' opinion it is time that observations and scientific findings made 20-40 years ago are translated into useful and accessible guidance for practitioners estimating the PMF in upland areas.

5 Inconsistency over frozen ground allowance

ICE (2015) repeats a suggestion from the FSR to allow for the possibility of frozen ground by increasing the standard percentage runoff (SPR) parameter when estimating the PMF for the winter season. This is in line with the "worst possible scenario" philosophy of the PMF concept, although the need for this adjustment is acknowledged to be a matter for judgement given the conservatism of other components of the design event (NERC, 1975).

No frozen ground adjustment is recommended when estimating floods smaller than the PMF. Yet some flood studies in the UK have included a frozen ground allowance for the 10,000-year flood. The difficulty with making such an adjustment is that it breaks the relationship between the return period of the input rainfall and the return period of the output flood flow. There can no longer be any confidence that the method is yielding the 10,000-year flood when the input is a 10,000-year rainfall. This does not cause a difficulty in estimating the PMF, when the aim is to consider the worst possible scenario, regardless of probability.

One particular drawback of applying the 10,000-year rainfall in conjunction with frozen ground is that it relies on the implicit assumption that the 10,000-year rain will always occur during a period when soils are frozen across the whole of the catchment. This assumption is not borne out by the findings of Stewart *et al* (2013) which show that summer rainfall depths are generally more extreme than winter rainfalls, at least for storm durations up to around two days.

There has been some research into the joint probability of extreme rainfall and snowmelt for reservoir safety studies (Reed and Anderson, 1992) although it appeared not to lead to any general solutions for practitioners.

One reason why some practitioners may be tempted to include frozen ground in estimation of the 10,000-year flood is that on more permeable catchments there can be a large discontinuity between the percentage runoff assumed for the PMF and that calculated for lesser floods. For a catchment with permeable soils in a low rainfall area, the estimated percentage runoff could in theory be as low as around 12% for a 10,000-year storm with duration 6 hours. In contrast, the percentage runoff used for estimating the winter PMP would be in the region of 65% thanks to the frozen ground allowance.

This may lead to understandable concern that there is a risk of under-estimating the 10,000-year flood.

Mis-application of the frozen ground allowance is not a drawback of current procedures, but of the way they are sometimes implemented. It may lead to over-conservative estimates of the 10,000-year flood. Strengthening the guidance may help to avoid this. The authors recommend that future research includes a check for discontinuities between estimates of the PMF and the 10,000-year flood.

6 No accounting for climate change

ICE (2015) does not give definitive guidance on whether and how to allow for the potential effects of climate change in reservoir flood estimation. It refers to change factors for extreme rainfall and river flood flows published by the Environment Agency in 2011 (which have been revised in February 2016), while noting that “extreme” in the context of fluvial flooding refers to much more frequent events than those often considered for reservoir safety.

There has been limited UK research on the impacts of climate change on PMP or other extreme rainfalls (Babtie, 2002; Atkins, 2013). Collier (2009) noted that theoretical considerations suggest that air can hold more moisture in a warmer climate, but there is evidence that this increase does not continue at high temperatures due to a concomitant increase in the speed of movement of atmospheric systems. The paper reached an interim conclusion that, as the climate warms, current UK estimates of PMP (referring to research carried out in the 1990s and 2000s) remain valid. However, it was recommended that further detailed analysis was urgently needed to confirm this conclusion.

In contrast, using simulations from seven climate models, Kunkel *et al* (2013) concluded that changes in air movement associated with a warming atmosphere were too small to offset the increase in moisture, and hence climate change will increase PMP globally.

Given that the effects of climate change are now being detected in the signatures of floods such as the winter 2013-14 event in the south of England (Schaller *et al*, 2016), the need to investigate the impacts of climate change on PMP (and other extreme rainfalls) is even more urgent than when it was recommended in 2009. In the authors’ opinion it is inconsistent that flood risk assessments for sites such as housing estates have long since routinely included an allowance for climate change, while design floods for reservoir safety have not.

REVIEW OF SELECTED INTERNATIONAL PRACTICE

In some other developed countries techniques used to estimate floods for reservoir safety are based on more recent analyses of PMP and methods of flood modelling. Examples include:

The USA, where guidelines by FEMA's National Dam Safety Program (FEMA, 2013) point out that there have been significant technological and analytical advances since the late 1970s when most previous federal and state guidelines on hydrology for dam safety were written. PMPs were estimated in a series of Hydrometeorological Reports by the National Weather Service (NWS) between 1963 and 1999, along with many more recent state- or site-specific studies.

The FEMA guidelines recommend that the effects of climate change on PMP are considered, referring to research showing increases in peak moisture content of 10% every few decades that would correspond to 10% increases in PMP (Easterling & Kunkel, 2011). However it is noted that at present it would be difficult to quantify the increase in the PMP due to climate change.

Canada, where provincial guidelines in Alberta (Alberta Transportation, 2004) recommend using new PMP estimates developed in 2004-5, along with snowmelt. The guidelines point out that the lag time for extreme floods may be smaller than observed for lesser floods. They stop short of recommending increasing estimates of PMP or PMF based on historical data to account for climate change, instead stating that hydrologists should be prepared to consider the possibility of climate change affecting future extremes.

Australia, where the recent revision of the Australian Rainfall and Runoff guide (Nathan and Weinmann, 2015) recommends the use of PMP estimates developed by the Bureau of Meteorology in 2003 and 2006. For dams of lower category, the design flood is the 1000-year return period which is estimated from design rainfall statistics published by the Bureau of Meteorology in 2015. Comprehensive guidance is provided on the hydrological modelling of losses and runoff routing, based on recent science.

The guide stresses that a key factor to be considered in modelling runoff processes for very rare or extreme floods is that the parameters found from calibration to observed floods cannot be assumed to apply to more extreme events, recommending that parameter selection is be guided more strongly by physical hydraulic consideration of the response of the catchment to extreme rainfall.

No allowance for the impact of climate change on the PMP or very rare rainfalls is currently recommended, in the light of work by the Bureau of Meteorology (Jakob *et al*, 2009) which noted that while it is likely that rainfall extremes will increase thanks to increased moisture availability, it is not currently possible to confirm that PMP estimates will definitely increase under a changing climate.

CONCLUDING REMARKS

Reservoir safety is now largely risk-based. Flood estimation is a fundamental input into risk assessment, whether connected with dam breach, spillway capacity assessment or ALARP-based decision making on levels of investment.

Existing methods of flood estimation for reservoir safety are overly generalised and probably providing overly high flows in some instances and too low in others. While in other areas of flood risk management this can simply be put down to 'estimation uncertainty', this is not the case for dams. The 2015/16 winter floods have shown how political and public scrutiny of extreme events is challenging previously accepted thinking.

The financial implications of changes to flood estimation are of course relevant: the change in flood estimation methods with the FSR resulted in large capital investment in upgrading spillways and "there is no financial rebate where the new procedure has indicated a smaller required spillway capacity" (Reed and Field, 1992). However, investment in new reservoirs continues, the number of 'high risk' reservoirs is likely to increase, and the potential lives at risk increases along with the general population.

At this conference 16 years ago, MacDonald and Scott (2000) wrote that "Reservoir engineers and other interested parties should not be lulled into a false sense of security that all is well with flood estimation and that no further improvements are required". They recommended that the PMP values generally adopted for UK should be reviewed in the light of current best practice. This review is long overdue.

With the development of the FEH13 rainfall statistics, Defra and the Environment Agency have made a good start in funding the development of reservoir flood estimation methods fit for the 21st century. In the authors' opinion the job needs to be completed now by upgrading the rainfall runoff model, the estimation of PMP and related guidance, which should cover topics including snowmelt and how to incorporate historical and palaeoflood data (e.g. sedimentary evidence of past extreme floods) into reservoir flood estimation.

The need to develop a rainfall-runoff method appropriate for modelling extreme floods was recognised in the Flood and Coastal Erosion Risk Management Research and Development Programme's previous Reservoir Safety R&D Strategy (Defra/EA, 2009) but has not yet been acted on. At the time the proposed project was ranked 8 out of 49.

A new version of the strategy is due for publication in 2016, and includes two relevant proposed projects: rainfall-runoff models for estimating extreme floods (Rank 1 on a list of projects still to be developed) and PMF estimation (Rank 3).

REFERENCES

Acreman MC (1989). Extreme historical UK floods and maximum flood estimation. *J. IWEM* **3**, 404-12.

Alberta Transportation (2004). *Guidelines on Extreme Flood Analysis.* Alberta Transportation, Canada.

Archer DR (1981). Severe snowmelt runoff in north-east England and its implications. *Proc. Inst. Civ. Engrs. Part 2* **71**, 1047-60.

Atkins (2013). *FD2628 Impact of Climate Change on Dams & Reservoirs. Final Guidance Report*. Defra, London

Austin BN, Cluckie ID, Collier CG and Harcaker PJ (1995) *Radar-based estimation of Probable Maximum Precipitation and Flood.* Met Office, Bracknell.

Babtie (2002). *Climate Change Impacts on the Safety of British Reservoirs.* Defra, London.

Clark, C *(1995).* New estimates of probable maximum precipitation in South West England. *Meteorol. Appl.,* **2**, 307–312.

Clark, C (2005). The Martinstown storm 50 years on. *Weather, 60 (9), pp 251-257.* Royal Meteorological Society, Reading, UK

Collier C (2009). *On the relationship between Probable Maximum Precipitation (PMP), risk analysis and the impacts of climate change to reservoir safety.* FREE Science Paper No. 2. NERC.

Defra/EA (2009). *Flood and Coastal Erosion Risk Management Research and Development Programme Reservoir Safety R&D Strategy*. Defra, London.

Easterling DR and Kunkel KR (2011). *Potential Impacts of Climate Change on Estimates of Probable Maximum Precipitation.* National Center for Atmospheric Research, USA.

Faulkner DS and Barber S (2009). Performance of the Revitalised Flood Hydrograph method. *J. Flood Risk Man.* **2**, 254–261.

FEMA (2013). *Selecting and Accommodating Inflow Design Floods for Dams.* Federal Emergency Management Agency, USA.

Hough MN and Hollis D (1997). Rare snowmelt estimation in the United Kingdom. *Meterol. Appl* **5**, 127-38.

H R Wallingford (2005) *Flooding in Boscastle and North Cornwall, August 2004. Phase 2 Studies.* Report to Environment Agency. H R Wallingford, Wallingford

ICE (2015) *Floods and Reservoir Safety. 4th edition.* ICE Publishing, London.

IH (1999) *Flood Estimation Handbook. 5 volumes.* Institute of Hydrology, Wallingford.

Jakob D, Smalley R, Meighen M, Xuereb K, and Taylor B (2009). *Climate Change and Probable Maximum Precipitation. Hydrology Report Series No. 12* – Water Division, Melbourne, Australia.

Kjeldsen TR, Stewart EJ, Packman JC, Folwell S and Bayliss AC (2005) *Revitalisation of the FSR/FEH rainfall-runoff method.* Defra R&D Technical Report FD1913/TR, CEH, Wallingford.

Kjeldsen T, Kim H, Jang C-H and Lee H (2016). Evidence and implications of nonlinear flood response in a small mountainous watershed. *J. Hydrologic Engineering, in press.*

Kunkel K, Karl TR, Easterling DR and Redmond K (2013). Probable maximum precipitation and climate change. *Geophysical Research Letters*, 40, 1402–1408.

MacDonald DE and Scott CW (2000). Design floods for UK reservoirs – a personal view of current issues. *Proceedings of British Dam Society conference, University of Bath pp 1-11.* Thomas Telford, London.

Mawdsley JA, Dixon AK and Adamson AC (1991). Extreme snow melt in the UK. In: *Proc. BHS Third National Hydrology Symposium, Southampton 1991.* 5-17-5.22. British Hydrological Society.

Nathan R and Weinmann E (2015). Very to extreme flood estimation. *Book 8 in Australian Rainfall and Runoff.* Institution of Engineers, Australia. Advanced draft, November 2015.

NERC (1975) *Flood Studies Report*. Natural Environment Research Council, London.

Reed DW and Anderson CW (1992). A statistical perspective on reservoir flood standards. In: *Water Resources and Reservoir Engineering. Proc. 7th Conference of the British Dam Society, Stirling*. 229-39. Thomas Telford, London.

Reed DW and Field EK (1992). *Reservoir flood estimation: another look. Report No. 114*, Institute of Hydrology, Wallingford.

Schaller N, Kay AL, Lamb R, Massey NR, van Oldenborgh GJ, Otto FEL, Sparrow SN, Vautard R, Yiou P, Ashpole I, Bowery A, Crooks SM, Haustein K, Huntingford C, Ingram WJ, Jones RG, Legg T, Miller J, Skeggs J, Wallom D, Weisheimer A, Wilson S, Stott PA and Allen MR (2016). Human influence on climate in the 2014 southern England winter floods and their impacts. *Nature Climate Change* doi:10.1038/nclimate2927

Stewart EJ, Jones DA, Svensson C, Morris DG, Dempsey P, Dent, JE, Collier CG, and Anderson CA (2013) *Reservoir Safety – Long Return Period Rainfall*. R&D Technical Report WS 194/2/39/TR, Defra.

WHS (2015). *The Revitalised Flood Hydrograph Model ReFH 2: Technical Guidance*. Available at http://files.hydrosolutions.co.uk/refh2/ReFH2_Technical_Report.pdf, accessed March 2015. Wallingford HydroSolutions, Wallingford

Wass P, Lindsay D and Faulkner DS (2008). Flash flood! A lucky escape for 10,000 bikers, *Tenth British Hydrological National Symposium, Sustainable Hydrology for the 21st Century*, September 2008, University of Exeter.

Dams – Benefits and Disbenefits; Assets or Liabilities?
ISBN 978-0-7277-6133-0

doi: 10.1680/dbdal.61330.101

Floods and Reservoir Safety 3rd and 4th Edition Flood Freeboard Estimates Compared

D A BRUGGEMANN, Atkins Ltd
J A RIBEIRO CORREIA, Atkins Ltd

SYNOPSIS The paper provides a short summary of some of the changes introduced in the 4th Edition such as the introduction of the Safety Check Flood in conjunction with the Design Flood and the move away from wave freeboard to a permissible wave overtopping discharge. The freeboard estimate in both editions is based on the significant wave height and the paper outlines the differences in approach after the significant wave height has been determined.

The variation of the flood freeboard estimates derived from both editions as a function of fetch and other parameters will be examined. Examples of the application of the two approaches to existing dams in the UK are provided.

INTRODUCTION

In the UK the primary source of guidance on the selection of appropriate inflow floods and freeboard provisions is Floods and Reservoir Safety published by the Institution of Civil Engineers, updated from the 3rd Edition (ICE, 1996) to the 4th Edition (ICE, 2015). The changes introduced in the new edition regarding the selection of the design inflow floods and the freeboard provisions are summarised as follows;

- Assessment based on a Design Flood in conjunction with a Safety Check Flood rather than a single design flood;
- Revised terminology to distinguish between "overtopping" and "overflowing".
- Acceptance of some wave overtopping through a permissible overtopping discharge rather than elimination;

RESERVOIR FLOOD INFLOW

The flood categories of the 3rd Edition have been retained in the 4th Edition but Table 2.1 (in Floods and Reservoir Safety) now includes standards for the Safety Check Flood and the Design Flood rather than a single design flood. The definition of these floods has followed ICOLD Bulletin 82 (1992) as set out below:

- Design Flood (DF) – the inflow that must be discharged under normal conditions with a safety margin provided by an accepted freeboard limit.
- Safety Check Flood (SCF) – the inflow beyond which the safety of the dam cannot be assured (i.e. key components exhibit marginally safe performance for this flood condition).

It is implicit in the definition of the SCF that some damage to the dam may occur, but it should be unlikely to result in dam failure. However, it should be recognised that in the case of a still water flood level above the crest of an embankment dam, it is difficult to predict whether or not the dam would fail.

FREEBOARD AND OVERTOPPING

The dam freeboard is defined as the vertical height from top water level to the top of the dam crest or wave wall. The freeboard should include the flood surcharge, wave surcharge, settlement allowance and any other pertinent factors.

The approach adopted in the 3rd Edition was to make sufficient allowance for wave surcharge to eliminate wave overtopping on embankment dams not designed specifically for such operation. The 4th Edition has moved away from this approach for these embankment dams by specifying an allowable wave overtopping discharge (Table 6.2, Floods and Reservoir Safety, 4th Edition (2015)).

The 4th Edition has also introduced definitions for "overflowing" and "overtopping" to bring reservoir wave terminology into line with coastal wave terminology as follows:

- "Overflowing" refers to (relatively) steady flows from flood rise;
- "Overtopping" refers to intermittent flow from wave overtopping.

Table 1 below summarises the allowable overtopping flows for the DF and SCF.

Floods and Reservoir Safety, 4th Edition, provides formulae for calculating the wave overtopping discharge and reference should be made to the document for these formulae. These formulae can be manipulated to arrive at the following explicit formula which yields the limiting / minimum freeboard requirement, R_c, for a sloping upstream face.

$$R_{C\_min} = ln\left(\frac{q_{max} \cdot \sqrt{tan\alpha}}{\sqrt{gH_s^3} \cdot \gamma_b \xi_{m-1,0} \cdot 0.067}\right)$$

Where q_{max} = allowable overtopping discharges (see Table 1 below), α = inclination of the upstream slope, g = acceleration due to gravity, H_s = significant wave height, γ_b = dimensionless berm factor, $\xi_{m-1,0}$ = wave breaking parameter.

Table 1 Allowable overtopping discharges

Design Flood	Safety Check Flood
≤ 0.001 litre/sec/metre (effectively no overtopping)	1 litre/sec/metre (good grass cover) 0.1 litre/sec/metre (poor grass cover)

Table 2.1 in the 4th Edition includes minimum freeboard requirements varying from 0.3m for a Flood Category D reservoir up to 0.6m for Flood Category A and B reservoirs.

FREEBOARD ESTIMATES FROM THE 3RD AND 4TH EDITIONS

The two approaches have been applied to an embankment with a 1V:3H upstream slope (no wave wall) and a smooth upstream slope. The results are shown in Figure 1 below.

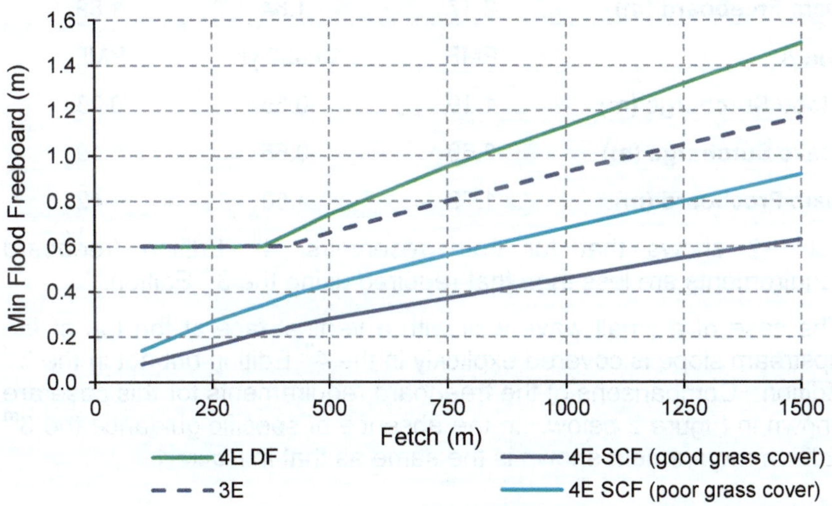

Figure 1 Sloping Upstream Face – No Wave Wall

Figure 1 suggests that the 4th Edition estimate of flood freeboard for the DF, all parameters being equal, is always greater than the estimate based on the 3rd Edition. However, interpretation of the Figure needs to take into account the flood surcharge for the DF and the SCF and the wave surcharge under these conditions. When the difference between the SCF and DF flood surcharges is less than the difference between the 4th Edition DF and the 3rd Edition DF wave surcharges, the 4th Edition will yield a larger freeboard requirement. In reservoirs with very wide spillways, the difference between the flood surcharge for the DF and SCF tends to be small and generally less than the wave surcharge and in these cases the 4th Edition may yield a larger freeboard requirement. Typically, UK reservoirs do not have very wide spillways, with the possible exception of flood storage reservoirs, so it would be expected that the 4th Edition requirements would be less than the 3rd Edition.

Both methods have been applied to two existing dams in the UK and the results are shown in Table 2 below.

Table 2 Freeboard Estimates (Smooth Slope, No Wave Wall)

Description	**3rd Edition**	**4th Edition**	
	DF	DF	SCF
Cat B	10,000 yrs	1,000 yrs	10,000 yrs
Flood Surcharge (m)	1.57	1.24	1.57
Wave Surcharge (m)	0.6	0.6	0.32
Dam Freeboard (m)	2.17	1.84	1.89
Cat A	PMF	10,000 yrs	PMF
Flood Surcharge (m)	1.19	0.85	0.23
Wave Surcharge (m)	0.60	0.65	1.19
Dam Freeboard (m)	1.79	1.50	1.42

Table 2 shows that for both reservoirs, 4th Edition freeboard requirements are less than that required using the 3rd Edition.

The case of a small wave wall with a vertical face at the top of the upstream slope is covered explicitly in the 4th Edition but not in the 3rd Edition. Comparisons of the freeboard requirements for this case are shown in Figure 2 below. In the absence of specific guidance the 3rd Edition requirement shown is the same as that in Figure 1

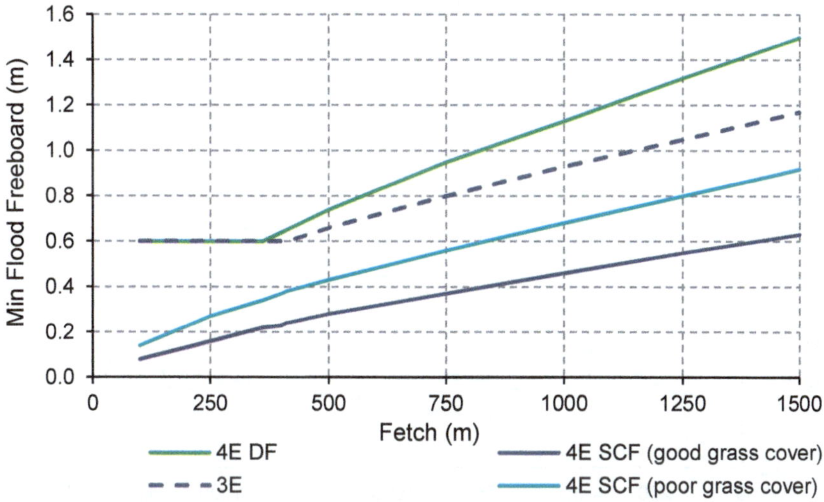

Figure 2 Sloping Upstream Face – Small Wave Wall (vertical face)

Figure 2 suggests that for the case of a small wave wall with vertical face, the 4th Edition yields to lower freeboard estimates than would be obtained from the 3rd Edition.

Table 3 shows an example calculation from an existing dam with a small wave wall.

Table 3 Freeboard Estimates (Smooth Slope with Small Wave Wall)

Description	**3rd Edition**	**4th Edition**	
	DF	DF	SCF
Cat B	10,000 yrs	1,000 yrs	10,000 yrs
Flood Surcharge (m)	1.24	0.98	1.24
Wave Surcharge (m)	0.96	0.79	0.32
Dam Freeboard (m)	2.20	1.77	1.56

Table 3 shows that for this example, the 4th Edition freeboard requirement is considerably less than that from the 3rd Edition.

A comparison has also been made for the case of a vertical face in deep water and this is shown in Figure 3 below.

The formulae in the 4th Edition for calculating the wave overtopping discharge for a vertical face in deep water have been also manipulated to arrive at the following explicit formula for the limiting / minimum freeboard requirement, R_c, for a vertical upstream face.

$$R_{C\_min} = \mathrm{N}\left(\frac{q\_max}{\sqrt{gH_s^3 \cdot 0.04}}\right) \cdot \frac{H_s}{-2.6}$$

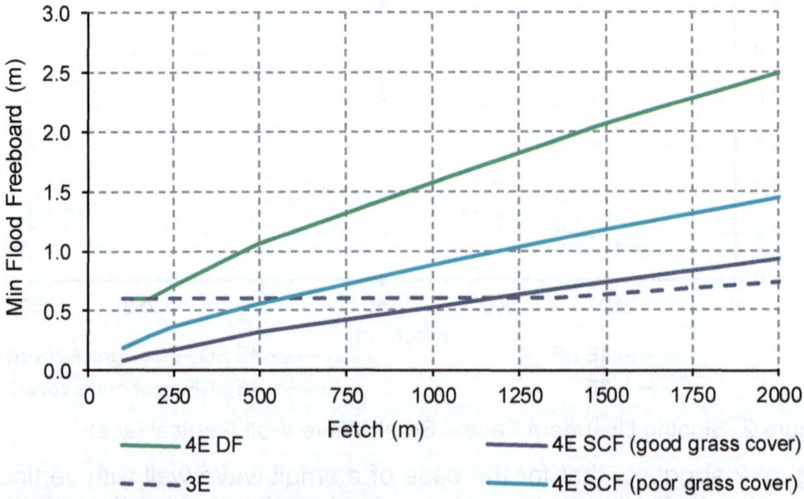

Figure 3 Vertical Face in Deep Water

The Figure shows that the 4th Edition freeboard estimates are significantly higher than those that would be obtained from the 3rd Edition. The freeboard also increases relatively rapidly with increasing fetch. A vertical face in deep water is typical of concrete dams, generally which can tolerate significant overflow. Figure 3 has been prepared using the criteria for embankment dams and therefore is not applicable directly to concrete dams.

Table 4 shows the results of a calculation for an existing dam with a vertical face in deep water.

Table 4 Freeboard Estimates (Vertical Face)

Description	3rd Edition	4th Edition	
	DF	DF	SCF
Cat A	PMF	10,000 yrs	PMF
Flood Surcharge (m)	1.40	0.87	1.24
Wave Surcharge (m)	0.60	0.95	0.27
Dam Freeboard (m)	2.00	1.82	1.67

Table 4 shows that for this example, 4th Edition freeboard requirement is slightly less than that derived from the 3rd Edition.

The influence of the upstream slope inclination has also been investigated for a smooth slope. The results are shown in Figure 4 below.

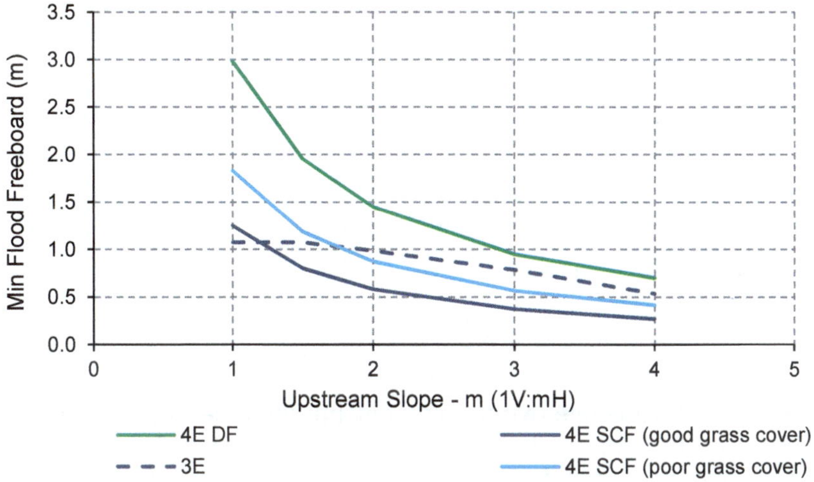

Figure 4. Influence of Slope Inclination Freeboard Estimates (smooth slope)

Figure 4 shows that the rate of increase of the estimates from the 4th Edition increases rapidly for slopes steeper the 1V:2H. While embankment dams with upstream slopes steeper than 1V:2H might be rare in the UK, the result suggests that increasing freeboard by crest raising and localised steepening of the upstream slope could be counterproductive. In such cases it could be better to select a vertical wave wall or focus efforts on strengthening the downstream face to resist erosion damage.

CONCLUSIONS

The key differences between Floods and Reservoir Safety 3rd and 4th Editions with respect to freeboard estimates have been highlighted. These differences are summarised as follows:

- Assessment based on a Design Flood in conjunction with a Safety Check Flood rather than a single design flood;
- Revised terminology to distinguish between “overtopping” and “overflowing”.
- Acceptance of some wave overtopping through a permissible overtopping discharge rather than elimination;

It has been pointed out that for the Safety Check Flood it can be difficult to predict whether or not failure will take place under sustained flow over the crest of an embankment dam. Caution should be exercised in the assessment of overflow conditions on embankment dams not designed specifically to resist overflow.

The equations given in the 4th Edition have been manipulated to arrive at the following explicit expressions for the limiting / minimum freeboard:

- Sloping upstream face

$$R_{C\_min} = ln\left(\frac{q_{max} \cdot \sqrt{tan\alpha}}{\sqrt{gH_s^3} \cdot \gamma_b \xi_{m-1,0} \cdot 0.067}\right)$$

- Vertical upstream face

$$R_{C\_min} = \mathrm{N}\left(\frac{q\_max}{\sqrt{gH_s^3} \cdot 0.04}\right) \cdot \frac{H_s}{-2.6}$$

Where q_{max} = allowable overtopping discharges (see Table 1 above), α = inclination of the upstream slope, g = acceleration due to gravity, H_s = significant wave height, γ_b = dimensionless berm factor, $\xi_{m-1,0}$ = wave breaking parameter.

The freeboard estimated based on Floods and Reservoir Safety, 3rd Edition (1996) and 4th Edition (2015) have been compared for a smooth, sloping upstream face without a wave wall, a smooth sloping upstream with a small wall with a vertical face, and for a vertical face in deep water. The results of these investigations suggest that the 4th Edition yields smaller freeboard requirements than those derived from the 3rd Edition. However there are some cases, like reservoirs with long fetches and with steep or vertical upstream slopes, where the freeboard requirements can be higher with the new edition.

The effect of the upstream slope inclination has also been examined and showed a rapid rate of increase in the freeboard required when the slope is steeper than 1V:2H. This result suggests that it might be more effective to provide a vertically faced wave wall or strengthen the downstream slope against erosion damage.

REFERENCES

ICE (1996), *Floods and Reservoir Safety, 3rd Edition*, Thomas Telford Ltd, London

ICE (2015), *Floods and Reservoir Safety, 4th Edition*, ICE Publishing, London.

ICOLD (1992), *Selection of Design Flood – Current Methods. Bulletin 82.* International Commission on Large Dams, Paris

SECTION 3:

GEOTECHNICAL ASPECTS OF EMBANKMENT DAMS

Dams – Benefits and Disbenefits; Assets or Liabilities?
ISBN 978-0-7277-6133-0

doi: 10.1680/dbdal.61330.111

Prevention of Internal Erosion in Homogeneous Dams - A Case Study

N BENNETT, Mott MacDonald Ltd, UK
Dr M EDMONDSON, Mott MacDonald Ltd, UK
Dr P BENNETT, Mott MacDonald Ltd, UK
A WOOD, Mott MacDonald Ltd, UK
P RIGBY, United Utilities, UK

SYNOPSIS Under the 1975 Reservoirs Act, United Utilities (UU) is the statutory undertaker for 140 impounding reservoirs (IRs) retained by earth-fill embankment dams.

Many of the dams were constructed around the turn of the 19th century, before the phase associated with the 'Pennine' style clay core. Therefore, owing to their early construction, the dams are substantially homogeneous, comprising locally available materials, including a mixture of peat, sand, gravel and clay.

UU has utilised a 'Portfolio Risk Assessment' (PRA) method, in combination with the seepage 'Toolbox' (Rigby *et al,* 2014), to provide quantification and ranking of various potential failure mechanisms in order to allow a targeted approach to risk reduction.

To provide the highest risk reduction possible and remove three such dams (Blackstone Edge, Whiteholme and Springs IRs) from the 'intolerable' zone, Mott MacDonald Bentley (MMB) was appointed to address the potential risk of internal erosion via a poorly compacted or high permeability zone, either within the embankment or surrounding an existing conduit.

This paper presents case studies of remedial works which have been undertaken at these sites to reduce the risk of failure to an acceptable level. The schemes comprised a variety of methods, including a granular filter combined with partial sheet pile cut-off wall and permeation grouting by both 'Tube-à-Manchette' (TaM) and end-of-case techniques. Grouting operations have included the innovative addition of dyed grouts to allow permeation to be traced.

PORTFOLIO RISK ASSESSMENT

In the AMP6 business plan United Utilities (UU) made a commitment to OFWAT to ensure that its impounding reservoirs have an annual probability of failure of less than 0.0001 (1 in 10,000 years); defined as the 'intolerable' threshold.

UU uses a Portfolio Risk Assessment (PRA) system, which is based on the University of New South Wales (UNSW) method. This enables ranking of the dams at a portfolio level and determines those which lie within the 'intolerable' region. This is based on an assessment of seepage, stability, flooding impacts and seismic risk.

When used in conjunction with UU's seepage 'Toolbox', internal erosion risks can be assessed from initiation, through continuous progression, to potential failure. The 'Toolbox' makes an assessment of the annual probability of failure of a reservoir embankment against 29 individual perceived embankment failure modes, termed 'Initiation Mechanisms' (IMs). This allows remedial solutions to be targeted to provide the required risk reduction. Table 1 defines the IMs which have been addressed as part of the remedial works covered within this paper.

Table 1: Initiation mechanism definitions

Initiation Mechanism	Definition
IM14	Poorly compacted or high permeability layer in the embankment.
IM16	Cracking in the crest due to desiccation by freezing.
IM18	Poorly compacted or high permeability layer around a conduit through the embankment.
IM19	Poorly compacted or high permeability layer around and along the conduit, with flow into the conduit through a crack or open joint.

HOMOGENEOUS EMBANKMENT DAMS

As a result of recent Portfolio Risk Assessment (PRA) scores, the three homogeneous embankment dams which are the subject of this paper were assessed using the 'Toolbox' as being within the 'Intolerable' zone.

Rigby *et al* (2014) observed that 'a homogeneous dam over its life is considered to be over five times more likely than a central clay core dam to fail by piping failure'.

Table 2 gives the background information for each of the dams. The following sections outline the remedial works undertaken at each site to bring the level of risk in relation to an identified potential failure mechanism from 'intolerable' to 'as low as reasonably practicable', 'ALARP' (i.e. <1:10,000) (Le Guen, 1999).

Table 2: Case study site comparison

	Blackstone Edge IR	**Whiteholme IR**	**Springs IR**
Construction (year)	1803	1816	1830
Reservoir Capacity (m^3)	772,000	1,601,000	609,000
Downstream Slope (v:h)	1:2.5	1:2.5	1:3
Dam height (m)	14	16.2	13.7
Dam length (m)	350	1300	786
Crest width (m)	3.2	3.5	4
Probability of failure	3.57×10^{-4}	1.35×10^{-2}	1.09×10^{-3}
Initiation Mechanism	IM18, IM19	IM16, IM18	IM14, IM18, IM19

All three reservoirs described in this case study are Large Raised Reservoirs under the Reservoirs Act 1975 (HMSO, 1975). As such, all remedial works have been constructed under the supervision of a Qualified Civil Engineer (QCE). All three have also been designated Category A reservoirs, as defined in Floods and Reservoir Safety (ICE, 2015).

BLACKSTONE EDGE EMBANKMENT DAM

Blackstone Edge reservoir is located approximately 6km to the north east of Littleborough, Rochdale. The earth embankment dam was constructed in 1803 as one of a series of 'Canal' reservoirs for the Rochdale Canal Company.

A seepage 'Toolbox' assessment was undertaken by UU for Blackstone Edge reservoir in 2011, which showed that the likelihood of failure as a result of wash-out of fines into the conduit through a crack or open joint (IM19, Table 1) was within the 'intolerable' zone.

Blackstone Edge Construction and Geology

British Geological Survey (BGS) maps show Blackstone Edge IR to be located on the Pennine Watershed, at an elevation of approximately 380m. Beneath a cover of peat overlying thin residual sandy soils, the area of the embankment and reservoir is underlain by the Lower Kinder Scout Grit Formation, comprising coarse, massive, well-jointed gritstones with occasional shale partings (Figure 1). The reservoir's earth fill embankment is of substantially homogeneous construction, comprising layers of sand and peat.

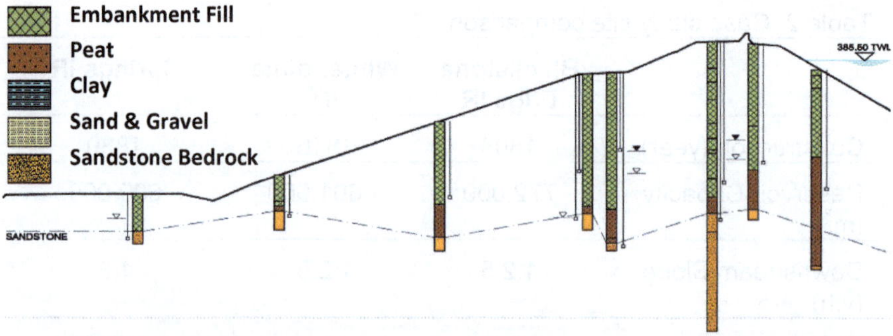

Figure 1: Geological build up at Blackstone Edge

There is a draw-off tunnel which extends part of the way beneath the embankment, housing the original draw-off pipework from the reservoir valve tower. This pipework was grouted up and abandoned in 1989.

<u>Blackstone Edge Remedial Works</u>
The scope of remedial works at Blackstone Edge included:

1. Permeation grouting around the redundant draw-off pipework.
2. Grouting to infill the culvert tunnel.

Permeation Grouting Works
The objective of the grouting works was to form a grout 'collar' around the redundant draw-off pipework to reduce the risk of loss of fines along a flow path down the outside of the pipe. The grout zone was informed by known levels from historical data and for pipe falls of between zero and 1 in 20.

The methodology for the grouting works (Figure 2a) was as follows:

- A grid of four boreholes at 1m spacing (two either side of the anticipated pipe location) was drilled from the embankment crest to a depth of 14m using rotary open hole drilling techniques.
- Falling head permeability tests were undertaken in the first borehole over the specified grout zone.
- Tube-à-Manchette (TaM) grout tubes were installed within the boreholes.

- A cement, Bentonite, Ground Granulated Blast-furnace Slag (GGBS) and plasticiser grout mix (12.5kg; 3.6kg; 25kg; 0.41litres respectively) was injected through an inflatable packer installed within each TaM port.

(a) Permeation grouting from crest (b) Tunnel infilling operations

Figure 2: Remedial works at Blackstone Edge reservoir

For each TaM port, grout takes were monitored until a limiting pressure (80% of overburden) was reached, to avoid the potential risk of hydrofracturing the embankment.

During grouting operations through the lower TaM grouting ports, injected grout was observed issuing into the culvert through defects in the gritstone culvert lining. This provided clear evidence of an existing flow-path and historical internal erosion behind the existing culvert, validating the original initiation mechanism.

On-site trials were undertaken with a 'thickened' grout mix, in an attempt to infill voids whilst limiting grout loss towards the tunnel portal. This was achieved by increasing Bentonite content and decreasing water content within the mix. However, the thicker mix could not be driven along the grouting lines without requiring excessive pressures.

Therefore, in order to create a 'firm' barrier against which to grout, the phasing of site operations was amended and the tunnel was infilled before completing TaM grouting.

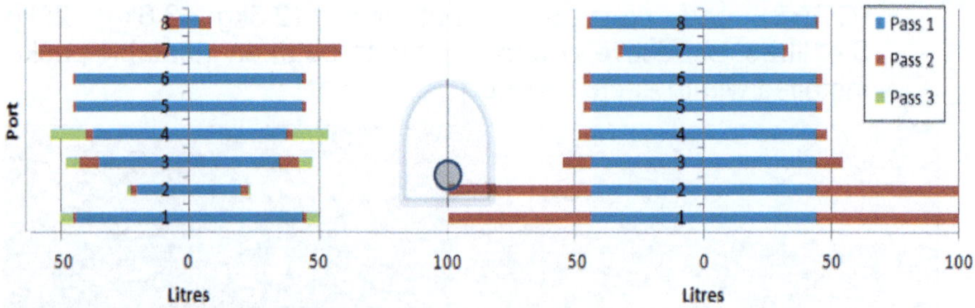

Figure 3: Grout takes surrounding the conduit

Figure 3 shows grout takes through each port for two of the boreholes. It was observed that grout takes in the lower ports (in the vicinity of the pipe main) were greater in the first borehole (right hand side) than the second (left hand side). This demonstrates that grout flow was achieved from the first borehole towards the second (right to left), thus limiting the ability to inject grout in the second borehole. A total grout volume of 5893 litres was injected.

Tunnel Infilling Works

Tunnel infilling was undertaken by constructing a blockwork headwall (Figure 2b) and injecting cementitious grout into the tunnel. Inlet and breather pipework was used during grouting to allow the tunnel to be filled in separate horizons (500mm; 500mm; 500mm; 150mm), over four days. This served to limit the active pressure on the tunnel headwall during injection and allowed continued observation within the tunnel. The final, smaller, horizon to the crown of the culvert utilised a horizontally installed TaM pipe, initially sealed in place with Bentonite slurry. The final phase of tunnel infilling involved injecting cementitious grout through the grout ports to displace the Bentonite slurry.

This methodology allowed a back pressure to be achieved, indicating successful filling of all voids within the tunnel. On completion, Bentonite slurry was observed issuing through the headwall. In addition, this allowed TaM grouting works around the conduit to be successfully completed.

WHITEHOLME EMBANKMENT DAM

Whiteholme IR is located approximately 1km to the north of Blackstone Edge, and was constructed shortly after (Figure 3a).

The reservoir is approximately rectangular in shape and is retained by an embankment dam which extends along the south-east and north-east sides (Figure 3b).

(a) Reservoir locations (b) Whiteholme embankment

Figure 3: Whiteholme & Blackstone Edge reservoir locations

A 'Toolbox' assessment completed in 2011 identified IM18 (a poorly compacted or high permeability layer around a conduit through the embankment) as requiring remedial works to reduce the perceived risk of failure to be 'ALARP'.

In addition, a Section 10 Inspection Report completed in 2012, which was based on the results of the 'Toolbox', gave the following recommendation 'In the Interest of Safety':

'Improvement works shall be constructed which will act to reduce the seepage through the dam and prevent wash-out of erodible material from the dam body'.

Whiteholme Construction and Geology

Due to its proximity to Blackstone Edge IR, the geology at Whiteholme is substantially similar. The Whiteholme dam is homogeneous, with no clay core, and appears to have been constructed predominantly from locally available materials, comprising sand and peat, with no discernible layering.

Historic boreholes (Figure 4) indicate that, in the location of the redundant draw-off pipework, the embankment is formed with mixed deposits of weathered sandstone and peat, overlying natural peat and Glacial Till. These deposits overlie Kinderscout Grit.

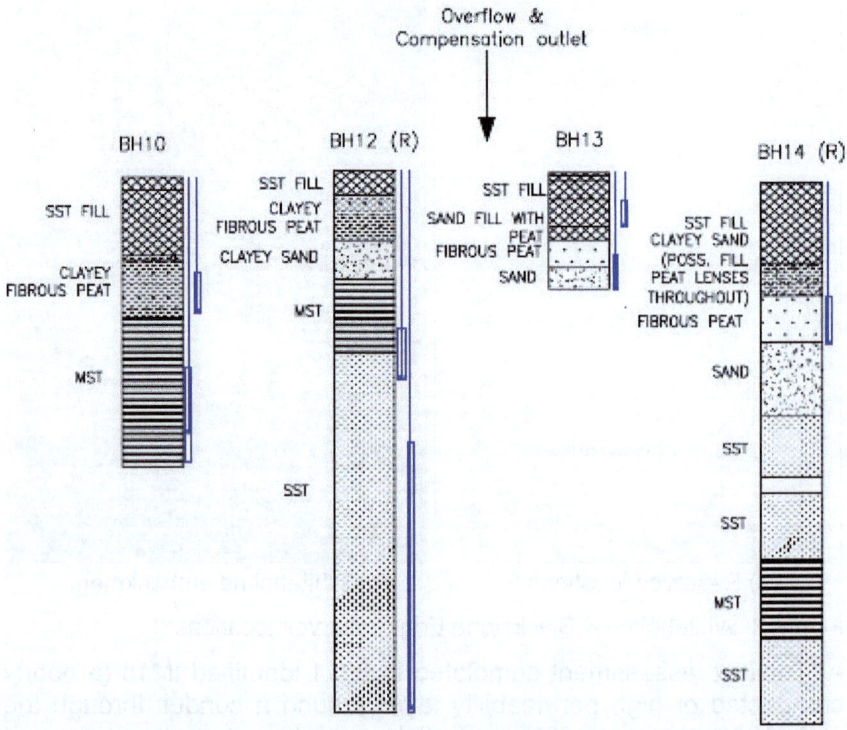

Figure 4: Historical boreholes surrounding Whiteholme outlet tunnel

Preliminary falling head permeability tests indicated low permeability strata over the proposed grouting zone (in the order of 10^{-7} to 10^{-8} m/s), representative of clay and peat.

Available information also indicated a historical burst of the draw-off pipework (prior to abandonment). This gave concerns over the existence of a pre-existing flow path connecting the grouting zone with the downstream tunnel portal, confirming the risk perceived using the 'Toolbox'.

<u>Whiteholme Remedial Works</u>

The scope of the remedial works at Whiteholme IR was to undertake permeation grouting from the embankment crest to form a collar around the redundant drawoff outlet, to prevent wash-out of fines along the outside of the conduit (IM18, Table 1). In addition, the weir was modified to permanently lower the top water level (TWL), in order to mitigate the risk of any flow through cracks within the crest (IM16, Table 1).

These works were undertaken following on from the Blackstone Edge IR grouting and so lessons learnt were used to refine the grouting technique. As a result, end-of-case grouting was employed, to enable grout thickening if required.

Four boreholes straddling the pipe location were drilled using open hole rotary drilling techniques to the top of the grouting zone. The hole was continued to final depth by auger drilling to allow recovery of disturbed samples. The grout design incorporated the addition of food grade colouring, utilising a different colour per borehole, to allow identification of grout travel between holes. This aimed to allow verification of grouting around the pipe and to provide a degree of confidence to the QCE that works had been successful (Figure 5).

(a) Injecting coloured grout (b) Coloured grout cube

Figure 5: Use of coloured grout at Whiteholme

Grouting was completed, with boreholes taking a total volume of grout of 3460 litres and with no evidence of grout loss in the downstream tunnel. A fifth borehole was drilled between the two boreholes with greatest grout take for validation purposes. This borehole was formed by cable percussive boring to allow continuous undisturbed U100 sampling. The recovered arisings provided evidence of blue pigmentation from the coloured grout, indicating successful grout travel between boreholes (upstream to downstream and across the conduit) and giving confidence that a grout 'collar' had been formed in the vicinity of the drawoff main. Coloured grout was also noted in one of the U100 samples.

SPRINGS EMBANKMENT DAM

Springs Reservoir lies to the northeast of the A675 road, between Bolton and Belmont. The dam was completed in 1830 and as such was one of the last dams to be constructed before the introduction of the 'Pennine type' dam. Although historic drawings indicated there to be a clay core, recent geotechnical investigations have shown no evidence of such. At the base of the Springs embankment lies Dingles Reservoir. This was constructed 15 years after Springs. The toe of Springs embankment is submerged by Dingles when the latter is at its TWL.

Springs Construction and Geology

BGS maps indicate that Springs reservoir is located over superficial Glacial Till deposits underlain by rock of the Millstone Grit series.

Historic boreholes show the embankment material to comprise very soft to firm variably sandy and gravelly clay, with the foundation comprising a distinct peat layer overlying Glacial Till deposits.

Springs Remedial Works

The main remedial works at Springs IR have included construction of a filter blanket on the downstream embankment face and associated toe drainage to address the perceived risk of a 'poorly compacted or high permeability layer in the embankment' (IM14).

Due to the challenging time frames associated with the construction of a full face blanket, budget considerations and safety concerns, MMB and UU collaboratively developed a solution to include a partial sheet pile cut off wall within the crest of Springs embankment. This has greatly reduced the volume of filter material required, resulting in significant cost savings, due to reduction in costly material required to comply with the filter envelope, as well as a significantly reduced placement duration associated with the smaller blanket area.

Sheet Piling Works

Steel sheet piles with an interlock sealant have been selected to provide a watertight barrier, adequate section modulus for drivability and sacrificial thickness to ensure a 100 year design life. Piles have been driven to depths of between 6m and 8m below Probable Maximum Flood (PMF) level.

Different installation methods were investigated for the sheet pile cut-off wall. Silent piling technology was considered versus the use of an excavator with a Movax vibrating hammer attachment. Due to the increased manoeuvrability and affordability of the Movax and the risk associated with recovering the silent piler in the case of a breakdown, the Movax was selected for the pile installation (Figure 6a).

However, site trials found that the vibrating hammer only drove the piles to within two metres of the design depth along the initial section of embankment. This was likely due to the presence of cobbles/ boulders or a stiffer layer. Subsequently, a Doosan air hammer was successfully used to drive the piles to the final required depth. Utilising the Movax for the majority of the piling operations had the advantage of low noise and so minimal disruption to local residents.

Piles will also be installed at the toe of Springs embankment to avoid the requirement to run the filter blanket into Dingles reservoir basin. This secondary sheet pile wall will address the potential failure mechanism arising from a poorly compacted layer surrounding the tunnel culvert (IM18, Table 1), thus removing the need for additional work to construct a filter collar in this area. This also removes the requirement for deep excavations and confined space entry.

Vibration monitoring has been ongoing throughout the works, as well as monitoring of any embankment and wave wall movement, to ensure the works are not adversely affecting embankment stability.

(a) Movax sheet pile installation (b) Nuclear density testing of filter

Figure 6: Sheet piling and filter blanket placement

Filter Blanket Works

The filter blanket comprises three layers; 250mm filter media, 250mm drainage media, 200mm filter media, overlain by topsoil. The filter drains into an integral toe drain. A stringent inspection and testing regime has been implemented on site to ensure that the filter material fits within the design grading envelope, including on site grading checks. Covered material storage bays were also utilised on site to avoid material cross-contamination.

Temporary formwork has been used in conjunction with GPS levelling to ensure that the layer thicknesses are controlled. In addition, on site nuclear density testing (Figure 6b) is being carried out to ensure sufficient compaction is achieved for each layer.

Works at Springs IR are ongoing and will also include slip-lining of the existing scour pipe within the tunnel and infilling of the tunnel to prevent wash-out of fines into the tunnel culvert (IM19, Table 1).

CONCLUSIONS

Remedial works have been undertaken at three sites to bring the risk categorisation of each dam into the 'ALARP' range.

At Blackstone Edge IR, permeation grouting using a TaM technique was used as well as infilling the tunnel culvert. Challenges resulted due to loss of grout into the tunnel portal caused by existing flow paths and the inability to adapt the technique for different grout viscosities, or increase grout pressure due to dam safety precautions. The sequence of works was subsequently amended to address this issue and allow completion of grouting.

At Whiteholme IR, following learning from Blackstone Edge, end-of-case grouting techniques were used with the introduction of coloured grout in combination with permanently lowering the TWL. A validation borehole showed evidence of blue grout confirming grout travel from the original injection point located upstream. However, other colours were less discernible. It is thought that this was due to the relatively 'impermeable' nature of the ground and the resultant low grout takes.

At Springs IR an embankment face filter blanket has been combined with partial sheet pile cut off walls at both the crest and toe. These will act as a crack resistant, physical barrier to prevent the passing of fines through the embankment and along the conduit. In combination with slip-lining and tunnel infilling works, the probability of failure will be reduced to within the 'ALARP' range. Challenges on site relating to the drivability of the piles to the design depth have been overcome with the use of a Doosan air hammer following initial installation with a Movax vibrating hammer attachment.

REFERENCES

Rigby PJ, Thompson AN and Jones DE (2014), 'The nearly perfect nineteenth century embankment dam' in *Maintaining the Safety of our Dams and Reservoirs, Proceedings of the 18th Biennial Conference of the British Dam Society*. ICE Publishing, London

Le Guen J (1999) *Reducing risks, protecting people.* Health and Safety Executive, London.

HMSO (1975). *Reservoirs Act 1975.* Her Majesty's Stationery Office, London

ICE (2015), *Floods and Reservoir Safety. 4th edition.* ICE Publishing, London.

Dams – Benefits and Disbenefits; Assets or Liabilities?
ISBN 978-0-7277-6133-0

doi: 10.1680/dbdal.61330.125

An investigation and assessment of embankment stability at Daer Reservoir

R MORRIN, Scottish Water
M SULLIVAN, Jacobs UK Limited
A MACDONALD, Jacobs UK Limited
C HOLT, Jacobs UK Limited

SYNOPSIS Daer Reservoir, constructed in the 1950s, is one of the largest earth embankment dams in Scotland at approximately 40m high with capacity of around 25.5Mm³. The dam has an articulated concrete core with bitumen seals provided for the movement of joints. It is classified a Category A dam as defined by Floods and Reservoirs Safety, 3rd Edition (ICE, 1996).

On Friday 13th December 2013 a slip was observed on the downstream face of the dam during heavy rainfall. The paper describes the short term actions taken to address stability, and the subsequent ground investigations and failure analysis, including assessing if other areas of the embankment were at risk.

HISTORY OF DAM

Daer reservoir was completed in 1954 for the Lanarkshire County Water Joint Committee to supply water to the industrial areas of Lanarkshire, with a population of around 500,000. It was constructed by direct labour to a design by Binnie, Deacon & Gourley. The reservoir is now a key component of Scottish Water's supply network to Lanarkshire and the area south east of Glasgow. (Figure 1)

The reservoir has a capacity of 25,460,000m³ and is retained by an embankment dam with a height of 42m and a crest length of 793m. The upstream slope varies between 1 in 3.25 and 1 in 2.5. The downstream slope varies between 1 in 2.5 and 1 in 2.25 with a berm just below the mid-point. The embankment is formed of Glacial Till with an articulated concrete core wall keyed a minimum of 0.90m into the underlying rock and a single line grout curtain. There is a rockfill drain at the base of the core wall with foundation finger drains leading from this to a drainage blanket below the outer portion of the downstream face.

Figure 1. Daer reservoir

Figure 2. Wet patch on face

An inspection under Section 10 of the Reservoirs Act 1975 (HMSO, 1975) was undertaken on 23rd August 2013 by Dr A K Hughes. The report noted that the embankment was generally dry with few signs of distress or instability. However concerns were expressed regarding the volume of water passing into the main seepage monitoring chamber from fissures in the rock in excavations that were open at that time. Measures in the interests of safety were included for this; for the re-establishment of seepage monitoring points on the embankment; and for installation of drainage in a wet patch first noted in 2008 towards the right hand end of the downstream face (Figure 2).

THE SLIP AND IMMEDIATE ACTIONS

Around 08:00 on Friday 13th December 2013, a contractor reported that a slip had occurred on the dam overnight. The team on site relayed the message to their team leader, who immediately called the Reservoir Engineer. There was heavy rain and strong winds at the time, and there had been heavy rain in the preceding days.

Scottish Water's Reservoir Engineer (SWRE) was on site by 10:15, at which time it became clear that although relative to the length of the dam the slip affected a small area, a large volume of material had moved, and that an All Reservoirs Panel Engineer (ARPE) was immediately required. Mr A Macdonald was called, and arrived on site by noon.

The slip measured around 20m x 25m in area (Figure 3), and appeared to be around 0.5-1.0m deep. The depth was difficult to assess because the movement had left a very uneven surface. The slip was located immediately above the berm, towards the right hand end of the dam and near to the wet patch identified in the inspection report. The displaced material had flowed across the berm, and engulfed the concrete mitre drainage channel.

Figure 3. Slip area

Figure 4. Rockfill berm

The SWRE liaised with the local Operations Manager and the internal delivery team to assess the availability of resources to assist as directed by the ARPE. As this was a Friday, close to Christmas, resources were not immediately easy to find from the larger framework contractors. However the owner of a local contractor made his way to site and confirmed that he could assist.

The ARPE advised that the slip was significant, and while it appeared to be a surface slip at the underside of the topsoil this could not be confirmed without further investigation. It was agreed that urgent action was required to prevent further deterioration of the situation. With all parties now on site a plan for the immediate requirements was developed. The key points of this were:

- Careful and safe removal of the displaced material from the lower mitre area and the berm
- Clearance and preparation of the slip "crater"
- Stabilisation of the embankment using rockfill (Figure 4)
- 24 hour working
- Drawdown of the reservoir using the reservoir scour provisions

The contractor mobilised a team to site with a variety of equipment; including two tracked excavators, two large capacity dump trucks, flood lighting, skilled operatives and operators. The benefits of a having a local contractor were shown as he was able to rapidly source not only plant and materials from his own yard but, most importantly, around 1400 tonnes of crushed rock from a local quarry.

The displaced material was removed from the mitre using an excavator transferring material into dumpers, and then to a stockpile area beyond the toe of the dam. The concrete mitre channel was cleared by hand, over a length of around 50m. The berm area was then cleared, allowing access to the slip area. The slip "crater" was then cleared and prepared; with any remaining slipped material being removed. Geotextile was placed, followed by the placing of a rockfill stabilising berm.

During the preparation works crushed stone had begun arriving, and was stockpiled adjacent to the toe of the dam. This was transferred by dumper up the mitre to the berm, where a working platform was created. This was formed to ensure that the excavator could reach the top of the slip to lay the stone fill. The working platform was built up in layers until it reached the bottom of the slip area. The fill was then placed within the slip area and at a shallower gradient than the downstream face of the dam to provide toe support and aid stability. After working through the night, the rockfill stabilising berm was completed at around 11:45 on Saturday 14 December.

Coincident with work on the slip, efforts were being made to drawdown the reservoir. However one of the scour lines could not be used as work was being undertaken on an in-line turbine and this, coupled with high catchment inflows, meant that drawdown was extremely slow. The 2013 inspection report had stated that even with all scour facilities operating and a small inflow it could take around 77 hours to draw the reservoir down by a metre. The ARPE was asked if emergency pumps should be mobilised. However given the size of the reservoir and the number of pumps required to make any significant impact, the decision was taken to concentrate on stabilising the embankment rather than improving the drawdown rate.

REVIEW OF RECORDS

In the immediate aftermath of the event a review began of available records of the dam construction. There were few relevant records digitised but some drawings were located within the Daer WTW boardroom. A request was put into Black and Veatch, successors to the original designers, to search their archives and they were able to provide all the as-built drawings. There were also several papers available for review. One issue these helped with was confirming that the clearly defined linear features that ran down the downstream face of the embankment were from a turf drain system designed to carry rainwater during the final stages of construction, to preserve the rest of the embankment after seeding.

There were instrumented sections on the embankment with standpipe piezometers. However there had been no requirement in recent Inspecting Engineers' reports to read these and so no records from them were available. Readings recommenced but were of little immediate use as the condition of the piezometers was unknown.

The record drawings showed that there were precast semi-circular channels behind each vertical joint in the core wall which acted as drains carrying any water down to the drainage layer at the base. On exposing these at the surface, water could clearly be heard flowing.

The embankment had six different seepage/leakage points, five of which had been read regularly, and the other (known as the mid-stream chamber) was in the process of being replaced. Monitoring data was reviewed and no concerns were evident.

GROUND INVESTIGATION

A ground investigation was designed and executed between March and May 2014 in order to determine the cause of the slip and to enable the risk of deeper seated failure mechanisms to be assessed. The aims of the ground investigation were:

- to investigate the ground conditions in and around the area of the slip including the nature and engineering properties of the embankment fill;
- to establish the general phreatic level and presence of perched water levels within the embankment; and
- to identify if there was leakage of water through the core wall.

Excavations were undertaken to expose the vertical drains behind the core wall in the area of the slip and at mid-bank for comparison. This facilitated drain extension to crest level, CCTV survey and future water level monitoring. Many of the drains contained significant depths of water, limiting the depth of the CCTV surveys

Rotary boreholes were drilled on the crest and downstream face, extending through the embankment fill and into the underlying bedrock. In planning the investigation, it had been intended to recover high quality (Class 1) soil cores of cohesive embankment fill for laboratory testing. However the fill was found to be predominantly granular. An open-hole technique was adopted to progress the boreholes to full depth. In situ SPTs were undertaken to determine shear strength parameters for the embankment fill.

Standpipe piezometers were installed within boreholes through the crest and face of the embankment. An inclinometer was installed in one of the crest boreholes located above the slip.

Hand pits were excavated behind the concrete mitre drain to the right of the area of slip to determine the presence of any water flows. Only one of these showed a moderate ingress of water, the remainder being dry or with only minor water seepage. Trial pits and trenches were excavated by hand on the face of the embankment to investigate the interface between topsoil and embankment. The depth of excavations was shallow, not more than 1.0m. Hand digging proved difficult due to the coarse granular fill.

FINDINGS AND GEOTECHNICAL ANALYSIS

Drainage

Evidence was found of leakage through the core wall near the area of the slip, with significant water inflow noted through vertical joints in the core wall in two vertical drains. Inflow was also recorded within other drains in the area of the slip but the rate was significantly less and appeared to enter through pipe joints. Minor inflow to drains was also noted at mid-bank locations remote from the area of the slip.

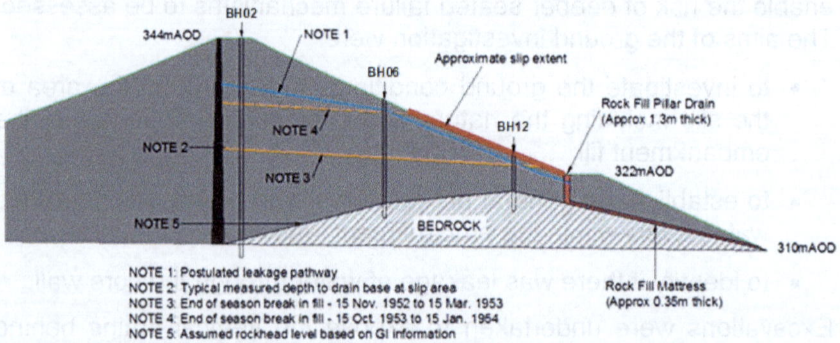

Figure 5. Simplified cross section and postulated leakage pathway through embankment at area of slip

The depth to the base of the drains was measured and found to be significantly shallower than expected on the right hand side of the embankment in the area of the slip. This may be an indicator of widespread blockage of vertical drains along the right hand side of the dam embankment. The cross section presented in Figure 5 suggests that the measured base of the vertical drains in the area of the slip coincides with an end of season break in placing embankment fill. Such blockages may have occurred during construction. Depths were generally as expected towards mid-bank. The vertical drains were initially monitored on a weekly basis, revised to quarterly once a pattern of dip level against reservoir water level was established.

Figure 6 illustrates that the water level within the vertical drains above the slip zone responds significantly to relatively small changes in reservoir water level. Away from the area of the slip, water levels within the drains remain responsive to changing reservoir level but the magnitude of such changes is less, significantly less at mid-bank.

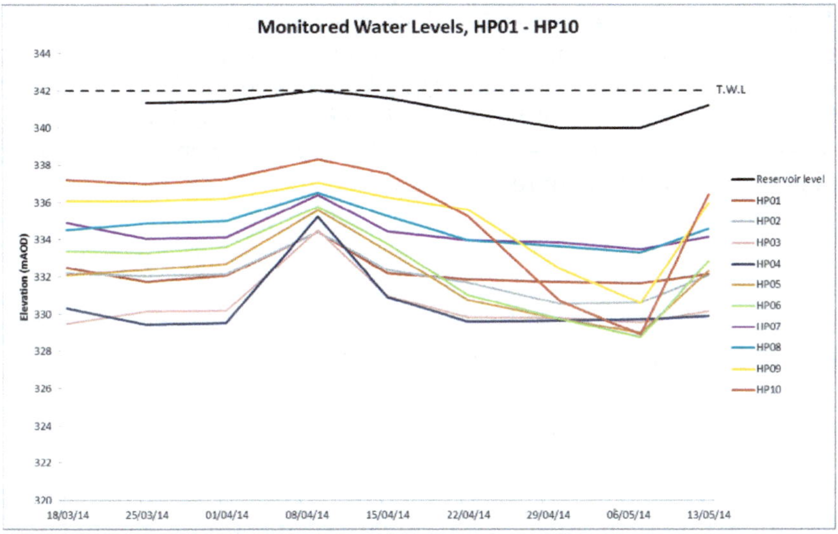

Figure 6. Water levels in drains within the area of the slip

The sensitivity of water levels in the vertical drains to changing reservoir water levels is not restricted to drains where core wall leakage was identified. Whether this is due to further leakage at greater depth or connectivity at the level of basal drainage could not be identified. There may be some connectivity between the vertical drains themselves.

Attempts were made to pump water from a number of the vertical drains to enable a CCTV survey of the full drain depth. It was not possible to complete these surveys as the recharge rate was too fast, again suggesting there is connectivity between the vertical drains.

<u>Embankment</u>

The boreholes and excavations undertaken on the embankment face confirmed the presence of a significant topsoil layer of between 0.2m and 0.7m thick. This layer was of greater thickness higher up the slope and could act as a sponge to retain water.

The fill material was found to be reasonably consistent over the full height of the embankment, typically described as medium dense or dense grey, sandy gravel with cobbles and boulders. Discrete layers within the embankment fill were recorded to be silty. There did not appear to be any depth correlation of these layers.

SPTs typically recorded N-values ranging between 6 and 48 with only three recording N-values less than 20. The results of SPTs indicate the embankment fill to be typically medium dense to dense. Twenty six SPTs recorded N-values greater than 50, suggesting a very dense material or the presence of obstructions. The results were used to derive shear strength properties, using the relationship between N-value and φ' (Peck *et al*, 1967). Figure 7 and Figure 8 present the range of values for φ' for the fill beneath the crest and the face of the embankment respectively.

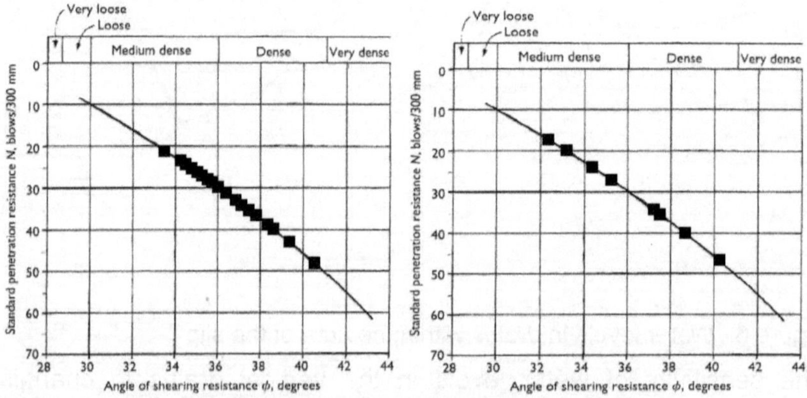

Figures 7 and 8. Relationship between SPT N value and φ'

Based on the correlations it was considered reasonable to adopt φ'=34^0 as a fairly conservative parameter to input into slope stability models. The lowest value adopted for sensitivity testing was φ'=32^0. Given the granular nature of the embankment fill, the effective cohesion (c') was assumed to be zero.

The investigation confirmed the phreatic level to be very low, with groundwater generally not encountered until rockhead or just above. However perched water was recorded within the embankment.

The embankment fill was found to be reasonably permeable. Permeability testing was undertaken within boreholes located on the crest returning a permeability ranging from 10^{-5} to 10^{-7} m/s. The boreholes on the downstream face were drilled using water flush and on suspension of drilling the monitored water levels were recorded to fall relatively quickly, suggesting a free-draining material. In contrast to the above, the trial pits undertaken on the face of the embankment recorded fill material immediately above the berm to be wet with logs describing the embankment fill as "very wet" or "saturated". This saturated surface zone above the berm was also identified by visual observation during the investigation, the top of the "wet" horizon being at approximately 332mAOD tying in with the top of the slip.

The investigation indicated that rockhead varies across the dam in the area of the slip as illustrated by Figure 5, despite construction records suggesting a fairly uniform rockhead profile. While a certain degree of natural variation in rockhead should be expected, there is a significant increase from approximately 311mAOD beneath the crest to 318mAOD in the area beneath the slip. This may have implications for the effectiveness of the drainage system.

Bedrock was described as medium strong to very strong grey fine to medium grained sandstone. The top 1.5m of bedrock was recorded to be non-intact or mostly broken.

Postulated leakage pathway

The likely pathway for water flow is shown in Figure 5. Construction season breaks in fill are also represented. The postulated leakage pathway was interpreted on the basis of:

- the shallowest recorded water level within the core wall drains and water level monitoring at BH06;
- from BH06 water was considered to be present reasonably close to the surface of the embankment face (within 1m) until it intercepts the French drain or pillar drain on the berm;
- the top of the slip was at approximately the same level as the "wet horizon", below which the embankment face is soft;
- the top of the slip also appeared to tie in with an end of season break in fill between October 1953 and January 1954;
- the measured depth to the base of the core wall drains appeared to match another end of season break in fill, between November 1952 and March 1953.

Slope stability

Slope stability analysis was undertaken to assess the risk of further instability to the embankment. Four cases were considered:

1. A completely dry downstream shoulder (a best case scenario) representing a well-drained dam embankment;
2. A high phreatic surface (a worst case scenario);
3. The likely ground and groundwater conditions with $\varphi'=32^{\circ}$;
4. The likely ground and groundwater conditions with $\varphi'=34^{\circ}$.

The minimum Safety Margin for each analysis is presented in Table 1. As well as the minimum Safety Margin the corresponding likely slip depth is also stated (e.g. shallow, deep, etc).

The analysis undertaken for a dry embankment suggests a very shallow slip surface and safety margin of the order of 1.4, even with a cautious value for φ'. It should be noted that using a slightly higher (and more realistic) value of φ' would result in a higher safety margin. This indicates that the likelihood of slope instability on a dry embankment face is negligible.

Table 1. Summary table of slope stability analysis

Case No.	Scenario	Shear Strength c'	Shear Strength Φ'	Slip depth	Minimum Safety Margin
1	Dry embankment	0kPa	32	Very shallow	1.40
2	High phreatic surface, saturated embankment	0kPa	32	Deep	0.99
3	Postulated failure condition	0kPa	32	Shallow	0.93
4	Postulated failure condition	0kPa	34	Shallow	1.01

A model with a high phreatic surface within the embankment fill was used to force a deep-seated slip although the investigation had confirmed that the phreatic surface is in fact very low. However it was modelled as a comparison to the best case scenario. The model used a conservative value of φ' and the resulting safety margin is therefore considered the lowest possible. The slip surface is at a maximum depth of approximately 8m and controlled by rockhead. The resulting safety margin is in the order of 0.99. However, the likelihood of a deep slip is considered to be low due to the conservatism in the model.

Cases 3 and 4 used the postulated leakage pathway as the basis for a ground model. The leakage pathway is represented by perched water within the embankment fill. Accordingly the embankment fill was modelled as two separate layers – a shallow, saturated layer around 2m thick under the influence of perched water and a deeper, well drained layer holding no water. The same shear strength parameters were applied to both layers. Both analyses suggest a failure similar to that which occurred in December 2013.

The analyses indicated that the risk to the embankment of a deep-seated slope failure is low. However there is a risk of further shallow slips, considered to be dependent on four main components:

- source of water, for example leakage through the core wall;
- insufficient drainage capacity;
- existence of a sub-horizontal pathway for water flow to the face of the embankment; and
- a thick topsoil layer acting as a sponge to retain water.

The influence of rainfall is also likely to have an effect on embankment stability. A period of heavy rainfall, as occurred at the time of the slip, could saturate the topsoil layer and increase the likelihood of a shallow failure along the topsoil/fill interface.

WILLOWSTICK AND SECTION 10 INSPECTION

A Willowstick survey was undertaken in June 2014 to see if this would highlight particular zones of seepage not only at the location of the slip but elsewhere along the embankment. It identified two primary zones of seepage, one in the area of the slip and one around the location of the old river channel, one secondary zone near to the left abutment, and one tertiary zone around 215m from the left abutment. All of these were noted as being at depths from the crest which were at or below foundation level.

Scottish Water decided that a further Section 10 inspection should be carried out with a view to determining future actions and the timescale for these. This was undertaken by Mr A Macdonald on 7 November 2014. The report contained seven measures in the interests of safety related to the slip and the subsequent investigations and analysis. These included further assessment of drawdown capacity, additional embankment drainage, improvements to seepage monitoring including a new downstream chamber in the area of the old river channel, mapping of rock exposures on the right abutment to assess the implications for flow through foundation strata, and further investigations into the seepage flows under and through the dam core wall including remedial works to reduce seepage if deemed necessary by an ARPE.

ROCK MAPPING

Desk study indicated that the bedrock underlying the dam is greywacke sandstone belonging to the Gala Group and of Silurian Age. The rock is recorded to be fractured and affected by faulting, and is within an area affected by the Moniave Shear Zone, that the line of the Sandhead Fault potentially underlies the dam foundation along its entire length or in part, and that the underlying bedrock is generally impermeable with groundwater confined to near surface within fractures and joints.

Inspection of the exposure at the right hand end of the dam was made with rock found to comprise light grey fine to medium grained sandstone (greywacke), observed to be strong to very strong. The rock face was recorded to dip steeply at angles of between 69^0 and 90^0, with dip directions ranging from 199^0 to 251^0. Persistence of block joints ranged from less than 1m to around 6m, with apertures of up to 60mm. Joints were typically clean without infill, or infilled with decomposed / disintegrated rock.

Groundwater flow in this bedrock tends to be discontinuity driven and the persistence and maximum aperture of 60mm suggests the potential for significant water flow through joints. Intersecting discontinuity sets also highlighted the potential for water flow. Potential fault lines identified in the rock exposure and the published geology make it likely that the bedrock beneath the dam is faulted.

DRAINAGE IMPROVEMENTS

It was recognised that improvements to the collection and monitoring of seepage water in the old river channel downstream of the dam were required. Between the toe of the dam and an existing cut-off wall across the old river channel, a herring bone drain system has been constructed leading to a new mid-stream monitoring chamber. At the time of writing this paper the flow recorded during dry weather conditions on the ultrasonic monitor has been in the range 14-18l/s.

Drainage has also been installed to two areas on the downstream face which were soft underfoot and showing signs of historic movement. This involved constructing filter drains to just below the base of the topsoil which was up to 600mm thick in places.

CONCLUSIONS

The slip at Daer appears to have been caused by seepage through and under the core, heavy rainfall and a high reservoir level which, combined with a relatively impermeable construction horizon and a thick topsoil layer, resulted in saturation of the face and a slip on the topsoil/general fill interface. The immediate actions following the slip showed the benefit of having local contractors able to respond to an unexpected event in a short period of time. Larger framework contractors are not always in a position to address such urgent maintenance work. The slip location also allowed relatively easy access for plant and materials. Being able to locate high quality construction records enabled a good understanding to be obtained of the design and the possible implications of construction seasons on the cause of the slip.

Further investigations are required into the seepage through and under the core wall. Improved monitoring will allow the extent of seepage and its relationship with reservoir level to be better understood. Other work to be carried out in the short term will include further drainage on the downstream face and assessment of the effectiveness of the pillar drains in carrying water from the berm to the foundation drainage blanket.

REFERENCES

HMSO (1975). *Reservoirs Act, 1975.* Her Majesty's Stationery Office, London, UK

ICE (1996). *Floods and Reservoir Safety, 3rd Edition.* Thomas Telford, London, UK.

Peck, RB, Hanson, WE and Thornburn, TH (1967). *Foundation Engineering, 2nd Edition.* John Wiley, New York.

Dams – Benefits and Disbenefits; Assets or Liabilities?
ISBN 978-0-7277-6133-0

doi: 10.1680/dbdal.61330.139

Slurry trench cut-off wall and permeation grouting of Chapel House Embankment Dam, Cumbria.

C F PAILING, United Utilities, Warrington, UK
D E JONES, United Utilities, Warrington, UK
S WORTHINGTON, Keller Geotechnique, UK

SYNOPSIS United Utilities Chapel House embankment dam is located 1.5km southeast of Uldale, Cumbria and was a significant early construction project for the firm John Laing in 1902 for the Aspatria and Silloth Joint Water Board.

In 2008 improvements were required, to reduce the probability of dam failure associated with internal erosion. The Portfolio Risk Assessment (PRA) had already assessed overall failure of the dam but the United Utilities seepage "Toolbox" workshops identified the main risks of failure associated with internal erosion. An impermeable cut-off through the embankment, around the spillway and into the rock was identified as the appropriate technique to control the risk. United Utilities design, developed with Keller Geotechnique, comprised a slurry trench cut-off wall along the length of the embankment into the underlying foundation soils along with permeation grouting adjacent to the spillway, conduits and below the slurry trench into the rock. This paper details the design, construction and monitoring of the ground engineering with considerations including stability of the slurry trench; the interface of the permeation grouting to the slurry trench; and, crucially, demonstrable benefit in terms of reduced seepage and risk of internal erosion.

HISTORY AND CONSTRUCTION OF CHAPEL HOUSE EMBANKMENT

Chapel House embankment in Cumbria is one of 170 embankment dams owned by United Utilities (UU) and although many are larger dams this reservoir is an important asset holding nearly 100 million litres of water which is a vital part of the drinking water supply system for communities such as Wigton, Siloth and Aspatria. The dam has the historical distinction as it was the first significant civil/ground engineering contract for the firm John Laing in 1902.

It is believed that John Laing was about 23 years old at the time of construction with reference to the reservoir in his biography 'The Good Builder; the John Laing Story' (Ritchie, 1997). Despite best efforts to locate the diary/log book referred to, it could not be obtained and comparatively little information was available regarding the construction details of this 125m long and approximately 8m high dam. It is believed to be constructed with locally derived Alluvial deposits and boulder clay placed by horse and cart to form an earthfill embankment with a central puddle clay core. The regional geological mapping shows the embankment to be founded on Alluvial deposits in the valley floor with Glacial Till (boulder clay) and outcropping Eycott Volcanic Group, generally comprising andesite, lapilli tuff and volcaniclastic sandstones on the abutments. The stilling basin constructed as part of additional central spillway works in 1982 exposed water bearing gravel considered to be part of the Alluvial deposits.

The general arrangement of the embankment is shown in Figure 1. The reservoir has a drawoff shaft which is connected by an iron bridge to the crest of the dam, Figure 1. The shaft joins a 275mm cast iron supply main through the body of the dam and a 610mm diameter scour pipe controlled by a valve on the upstream end on the valve shaft and another at the toe. Anecdotally it was suggested that the pipes were housed within a concrete lined tunnel through the embankment but there is little evidence to support this theory.

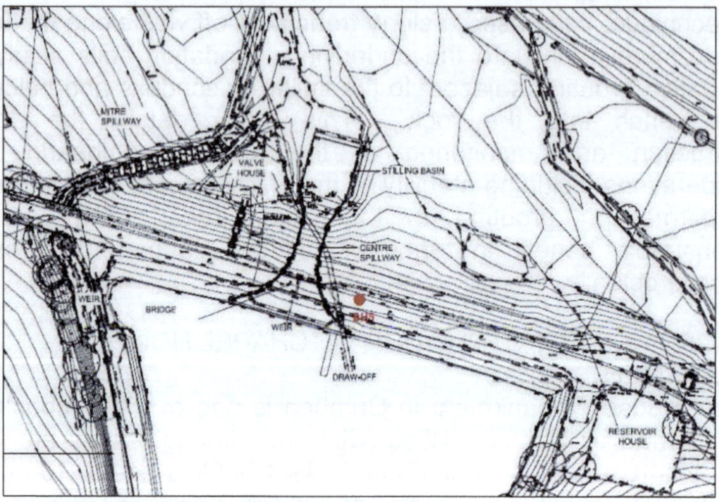

Figure 1. Chapel House Embankment Layout

HISTORIC SEEPAGE AND INSPECTION RECOMMENDATIONS

Localised wet patches and extensive reed growth along the toe of the embankment and around the stilling basin had been noted during the Section 12 inspection and a total of ten drains/pipes had been installed over the years to monitor seepage on the downstream face. Between 1988 and 1998 increased leakage at the toe of the slope below the central overflow was recorded and resulted in limited works to stem the leakage. Material alongside the sheet pile either side of the central overflow was excavated and replaced with puddle clay.

In July 2008 the Section 10 inspection identified a number of ITIOS requirements, one being 'works are carried out to reduce the probability of failure associated with internal erosion to bring the dam within the "acceptable" or "tolerable" regime' (HSE, 2001)'. Following on from the inspection a desk study of the dam was carried out and the findings presented to a Risk Estimating Team as part of the first stage, workshop 1, of the UU "Toolbox". The toolbox is used to identify potential flow and/or failure paths and their probability of failure by internal erosion and piping using an event tree approach to review the process of internal erosion from initiation, through continuation and progression and finally breach (USBR, 2008).

GROUND INVESTIGATION WORKS

An initial review of information and assessment of possible failure modes highlighted the need and objectives for a ground investigation and to develop the ground model of the dam. The ground investigation work presented challenges of its own, such as access to the crest (requiring road closure) and to the downstream shoulder where a suitable scaffold working platform had to be constructed off the crest.

The ground investigation works was carried out by UU framework contractor Environmental Services Group Ltd. A total of six trenches were excavated along the crest to determine the location and depth to the clay core followed by seven cable percussive boreholes, some with rotary follow on, along with three on the downstream slope and toe of the embankment. During the investigation a number of issues were reported. During the rotary coring within the foundation material a dense medium to coarse subangular to subrounded gravel was encountered from 10.30m bgl within BH5 (refer to Figure 1.) and at 12.30m bgl the borehole was terminated due to water bubbling up against the back of the wing wall on the upstream side of the central spillway. This was believed to be the result of a water flush being used to enable rotary coring to progress through the dense gravel, and it indicated connectivity between the foundation and the upstream embankment. Rotary drilling within the very dense gravel and the underlying very strong volcanic Tuff proved difficult, with slow progress and significant wear and tear of the drill bits, which required replacing during the works.

Although no 'puddle clay' was encountered during the investigation the pits and boreholes on the 4m wide crest indicated a selective cohesive material had been placed in the central section of the embankment generally described as slightly sandy, gravelly organic Clay with frequent pockets of peat. The shoulder material was generally described as slightly sandy, gravelly clay with cobbles.

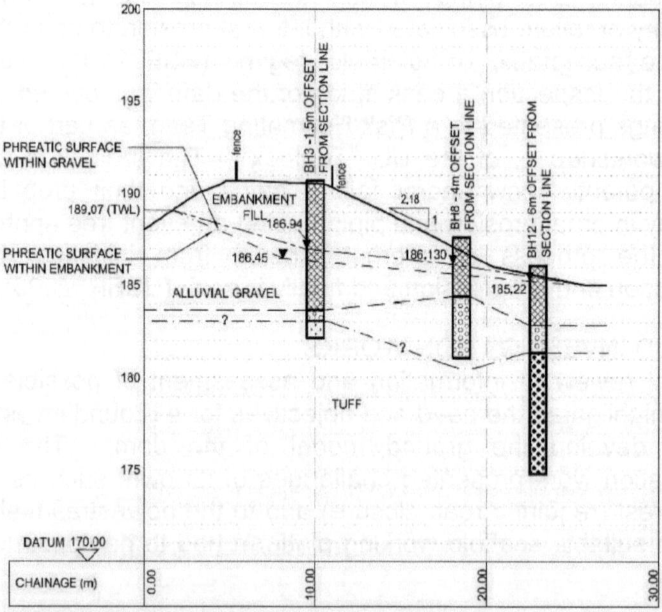

Figure 2. Embankment Cross Section

Piezometers were installed within the embankment fill and underlying gravel and were monitored as part of this initial investigation between September 2010 and June 2011. Figure 2 shows the phreatic surface based on the average groundwater level data within the underlying gravel foundation and within the embankment. Due to their similarity in level a hydraulic connection between the embankment and foundation was indicated. The determined phreatic surface downstream of the crest is lower than would be anticipated for a homogenous dam and it seems likely that the selected lower permeability materials are performing to some extent as a core, albeit of low efficiency.

Non-intrusive ground penetrating radar and electrical resistivity was carried out in September 2010 to investigate the location of the 275mm diameter cast iron supply pipe and the 610mm diameter scour pipe. Although the geophysical survey was unable to locate the pipework it was able to confirm that there was no indication of a tunnel through the embankment.

DETERMINATION OF REMEDIAL WORKS

The results of the ground investigation were presented to the Risk Estimating Team as part of the internal erosion Toolbox workshop 2 and the resulting assessment identified seven failure paths within the intolerable range of probability of failure, as presented in Table 1.

Table 1. Failure path descriptions within the intolerable range of probability of failure

Failure Path	Description
FP14	Poorly compacted or high permeability layer in the embankment
FP15	Poorly compacted or high permeability layer on the core-foundation contact
FP18	Poorly compacted or high permeability layer around a conduit through the embankment
FP20	(New Spillway) Poorly compacted or high permeability zone associated with a spillway or abutment wall
FP21	(New Spillway) Crack/gap adjacent to a spillway or abutment wall
FP24	Backward erosion in the soil foundations
FP25	Suffusion in soil foundations

The optimum solution identified to address the failure paths for Chapel House was an impermeable cut-off through the embankment into the rock. The challenge was to develop such a solution whilst addressing a number of constraints and considerations including pollution prevention, stability and maintenance of reservoir storage and supply. UU Engineering undertook an optioneering process to develop an initial design which comprised a slurry trench cut-off along the embankment into the underlying foundation gravels, along with permeation grouting adjacent to the spillway, around the pipework and between the slurry trench base and into the rock. The solution and the reduction in the probability of failure demonstrated through the toolbox analysis was presented to and accepted by the QCE and the Project Team.

The specification adopted for the slurry trench was based on the ICE Specification for the Construction of Slurry Trench Cut-off Walls (ICE, 1999) and the permeation grouting was a performance based specification based on the British Standard for grouting works (BSI, 2000); this was in order to allow the Specialist Ground Engineering Contractors to evaluate the ground conditions and provide the opportunity to match the optimum permeation drilling and grouting techniques with their proposed grout mix design.

DETAILED DESIGN

The contract to install the cut-off was awarded to specialist contractor Keller Geotechnique in June 2012 with a programme for the slurry trench cut-off wall and associated permeation grouting extending until March 2013. The specification required the contractor to produce a Construction Quality Assurance Plan detailing the method of installation, the specification and properties of the materials to be used and how performance would be validated during installation through monitoring, testing and reporting. In addition there was a requirement for stability risk assessments for the construction plant and activities and an environmental control plan including flood contingency.

The ground investigation completed for the toolbox assessment highlighted that greater certainty of the actual cross valley profile would be required for design. One of the first objectives was therefore to carry out ground investigation to identify the rock head valley profile; likely to be deeper than some of the initial ground investigation holes. A variety of drilling systems was required to advance the holes through the gravel and then the underlying rock. Rotary percussive drilling techniques were employed to drill the holes to the required depths, utilising a down-the-hole hammer and rock roller in conjunction with temporary casing. Where fractured ground was encountered the Symmetrix drill system was used to allow the temporary casing to be extended behind the drill bit until competent stable rock was reached. The investigation confirmed that the embankment fill extended up to a maximum of 9.60m bgl with significant thickness of sandy gravel and cobbles, up to 12.2m. Below the gravel and cobbles a highly fractured tuff was encountered between 9.4m and 18.8m bgl with 'intact' rockhead between 10.5m and 22.3m bgl.

Keller Geotechnique proposed to install a 600mm minimum wide slurry trench cut-off wall with a target permeability of 1.0 x 10^{-8} m/s. This would extend along the length of embankment from crest level to the maximum excavated depth of 11m bgl or 500mm below the base of the gravel layer using a Komatsu 24 tonne long reach excavator.

Based on the ground investigation results it was apparent that the thickness of ground to be treated by permeation grouting between the base of the slurry cut-off wall and below the gravel was greater than indicated by the initial ground investigation and the technique selected was Tube-à-Manchettes (TaMs). Below the central spillway the grouting technique was also TaMs. TaM pipes were installed in holes along the dam at regular centres. The TaMs have grout ports at regular centres which allows grout to be injected at specific levels, and pressures, across the dam and allows for revisiting particular grout ports if required. To prevent grout migration into the watercourse and to treat larger voids End of Case (EoC) grouting was proposed along the downstream side of the proposed slurry trench cut-off prior to the construction of the slurry trench. EoC grouting was also proposed around the vicinity of the existing supply, scour and draw off pipes to minimise the amount of pressure applied and reduce the risk to existing assets.

The permeation grouting had a target permeability of between 1.0 x 10^{-6} m/s and 1.0 x 10^{-7} m/s. Both grouting techniques were spaced at 1.2m centres with TaM grouting injections at 0.5m vertical intervals and EoC at 1.0m vertical intervals.

A high strength grout was not required; the more important design characteristic was to have a low enough viscosity to allow penetration of seepage paths within the embankment, granular material and fractured rock. Following on from on-site grout mix trials a cement bentonite grout mix was used for the EoC grouting and TaM sleeve grout and a microfine cement grout mix was used for TaM grout injection. Table 2 shows the performance parameters that were adopted for the grout mix.

Table 2. Grout Mix Performance Parameters

Mix	Bleed %	Viscosity secs	Specific Gravity
EoC & TaM sleeve grout mix	20	45	1.32
TaM Grout mix	20	42	1.54

The aim of the grout is to reduce the permeability of the existing material and infill any existing seepage paths. If the grout injection is too high, new pathways can be created by fracturing the existing material. Therefore it was important to establish a limiting pressure which was less than the overburden pressure and this was set at 0.5bar/metre. In addition to a limiting pressure a stop criteria was determined for TaM and EoC grouting. For the EoC the stop criteria was defined as injection refusal at the specified pressure, or surfacing of grout. Once the stop criteria was reached the casing was retracted by 1m and process repeated. If there was no injection refusal, secondary and tertiary holes were completed as required. For TaM grouting an estimate of anticipated grout take at each injection point was calculated as 148 litres, again the stop criteria was defined as injection refusal at the specified pressure or surfacing of the grout. Where volume limit was reached grouting was terminated and the sleeve position was revisited on a subsequent day following the same procedure.

It was proposed to construct the slurry trench cut-off in two sections either side of the central spillway. The initial slurry trench mix was based on Keller Geotechnnique slurry trench cut-off design for around a landfill site in Burnham-on-Sea. The slurry trench wall was constructed using a cement, Ground Granulated Blast furnace Slag (GGBS) and bentonite mix, with a 28 day unconfined compressive strength greater than 100kN/m^2.

CONSTRUCTION AND MONITORING

The general sequence of work was to carry out the EoC holes along the downstream side of the crest prior to installing the 119m length slurry trench cut-off wall either side of the central spillway. The slurry wall was constructed using a long reach 360° excavator. Slurry was continually supplied to the trench to maintain a slurry level 500mm below ground level during excavation to prevent trench collapse. To ensure no cold joints if there was an unavoidable pause in construction the excavation was taken back into the previously formed slurry.

The TaM grouting was carried out on completion of the slurry trench cut-off to reduce the permeability of the ground from the base of the slurry wall to below the gravel. Below the central spillway, where the gravel/cobble layer and fractured rock is at the deepest, it was agreed with the QCE to extend the grout to at least a depth of 1.5 times the height of the embankment.

Figure 3. Installation of slurry trench cut-off wall with vacuum excavator removing arisings from skip

The spillway TaM grouting installation was achieved using an excavator fitted with a front mounted rotary percussive drill mast. It was a requirement that no load be imposed on the spillway bridge and the rig was therefore sited on the crest road and only the mast placed on the spillway as shown in Figure 4.

Figure 4. Installation of TaM holes over central spillway showing reservoir water level maintained during the works

A rigorous monitoring regime was in place prior to and during the works to monitor the integrity of the embankment and maintain water quality of the reservoir and River Ellen, a salmonid river. In addition to daily water quality monitoring of the reservoir and river a total of 22 temporary level pins were installed at various locations on the embankment and five settlement/deflection monitoring points on the concrete spillway and bridge. These were installed and monitored prior to the works to establish a baseline and to set trigger levels relative to the original baseline. Monitoring was carried out twice daily from a remote monitoring observation point on an adjacent hillside. Trigger levels were agreed that could have resulted in a cessation of works pending an on-site review.

RESULTS AND VALIDATION

The geotechnical works were completed on 13 March 2013. The slurry trench provided a cut-off area of 837.4m² with a total volume of slurry mix as 502.44m³. A total of 91 EoC holes and 109 TaM holes were drilled with over 3000 grout injections. The volume of grout used in the treatment of the embankment and foundation are summarised in Table 3, below.

Table 3. Total Volume of Grout used (m³)

Type	Volume
EoC Cement Bentonite Grout	221.73
TaM Installation Cement Bentonite Grout	62.46
TaM Treatment Microfine Grout Injection	242.47

The successful validation of the works was reliant on a combination of records, testing and monitoring which all formed part of the Construction Quality Assurance Validation report. To ensure the quality and integrity of the grout mix, specific gravity (mud balance), viscosity (marsh funnel) and bleed testing was conducted daily. In addition five in-situ permeability tests were carried within the permeation grouting areas and indicated a successful average in-situ permeability of 3.09×10^{-7} m/s. Validation of the slurry trench cut-off wall was achieved by 90 day triaxial permeability laboratory testing which showed the that the target permeability of $>1.0 \times 10^{-8}$ m/s was achieved with an average permeability of 2.71×10^{-10} m/s. Grout takes of the EoC and TaM holes were recorded and presented graphically, Figure 5.

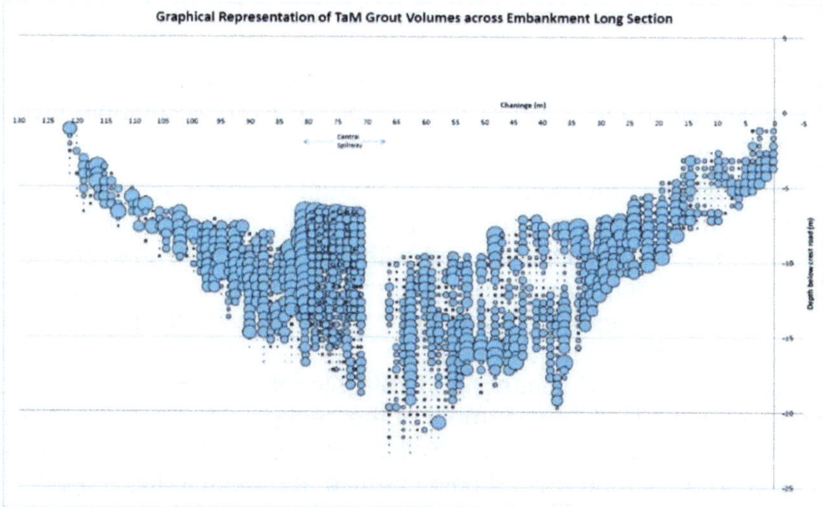

Figure 5. Graphical Representation of TaM Grout Volumes across the Embankment Long Section (note: between chainage 65 – 70 EoC grouting used to fill area around pipework through the embankment)

As part of the validation post monitoring of piezometers were carried out and showed the impact the cut-off has had on the recorded groundwater levels upstream and downstream as shown on Figure 6.

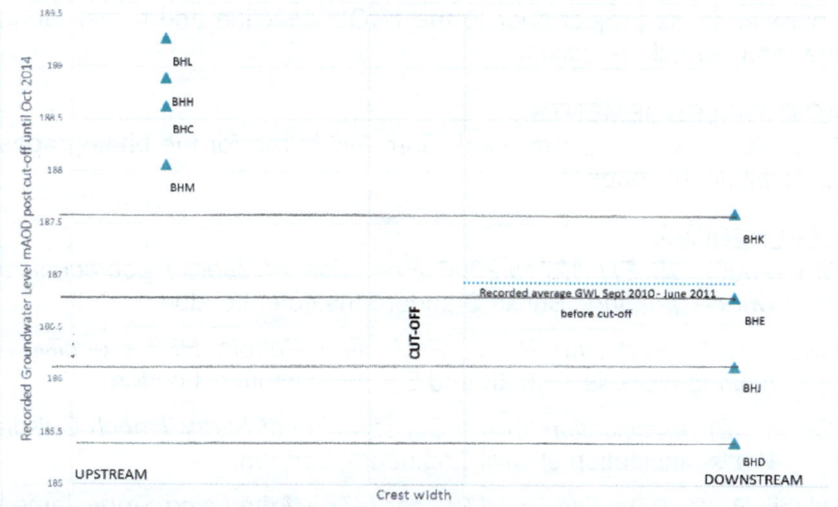

Figure 6. Highest recorded GWL upstream and downstream of cut-off (July 2013 until October 2014)

CONCLUSION

Although no major works had been required to John Laing's first ground engineering project for 110 years some potential risks were identified by regular inspection and later quantified by applying UU Engineering's process. The innovative UU Engineering Toolbox allowed early identification and optimisation of solutions and crucially demonstrated the reduced probability of failure to acceptable levels for the solution prior to committing to expenditure.

UU Engineering developed the conceptual design based on slurry trench and permeation grouting and developed a comprehensive performance specification agreed with the QCE. UU identified the significance of the ground engineering elements of the work and considered that early engagement of a specialist geotechnical contractor and the requirement for a Geotechnical Adviser (UK Register Ground Engineering Professional RoGEP) would be important in the joint development of the design. Associated with this decision the appointment of Keller as the Principal Contractor assisted in promoting a good collaborative approach to the project. A collaborative relationship was maintained throughout the construction phase into testing validation and monitoring. The UU Geotechnical Engineer worked closely with the contractor to develop the design and processes to deal with expected variation to the ground model, such as the depth of cut-off in the valley section. Both parties worked to ensure good Construction Quality Assurance records were maintained, which was key to the successful completion and approval of the project prior to the ITIOS deadline and to delivering the required risk reduction.

ACKNOWLEDGEMENTS

The authors would like to thank Sam Fishburne for the photographs used within this paper.

REFERENCES

BSI (2000) *BS EN 12715:2000 Execution of special geotechnical work – grouting.* British Standard Institute, London

HSE (2001) *Reducing Risks, Protecting People HSE's decision-making process.* Health and Safety Executive, London

ICE (1999) *Specification for the Construction of Slurry Trench Cut-off Walls.* Institution of Civil Engineers, London.

Ritchie B (1997). *The Good Builder – The John Laing Story.* James & James, London

USBR (2008). *Unified Method for Estimating Probabilities of Failure of Embankment Dams by Internal Erosion and Piping Guidance Document.* The University of New South Wales and URS.

Dams – Benefits and Disbenefits; Assets or Liabilities?
ISBN 978-0-7277-6133-0

doi: 10.1680/dbdal.61330.151

Some particular issues in the application of Eurocode 7 to embankment dams

M W HUGHES, Atkins

SYNOPSIS Eurocode 7 (BSI, 1997a) sets out the "principles and requirements for safety and serviceability" and "sets out the basis of design" to be applied to the geotechnical aspects of buildings and civil engineering works.

Whilst there are several guidance documents and text books which discuss the application of Eurocodes to infrastructure embankments, there is very little guidance on how to deal with the assessment of embankment dams.

Using British Research Establishment Report BR 363 (Johnston *et al*, 1999) as a benchmark, and making particular reference to flood attenuation embankments, this paper summarises available guidance on the subject of limit state (GEO) design of the embankments (excluding rapid drawdown). It uses worked examples to test the use of partial factor modifiers to increase the level of safety for embankment dams and includes various approaches for dealing with variable water levels/pressures.

LIMIT EQUILIBRIUM WITH GLOBAL FACTOR OF SAFETY

The use of the limit equilibrium approach to determine the stability of a slope, with results compared against a prescribed "global" factor of safety, has been the standard approach in engineering practice for decades.

In the United Kingdom no prescribed factors of safety are available for embankment dam slope stability. In the absence of guidance, designers use the typical values presented in the British Research Establishment Report "An engineering guide to the safety of embankment dams in the United Kingdom" (Johnston *et al*, 1999). The values are repeated in Table 1 below.

Table 1. Extract from BRE Report BR 363

Loading Condition	Typical minimum acceptable factors of safety
End of Construction	1.3 to 1.5
Steady seepage with reservoir full	1.5
Rapid drawdown	1.2

LIMIT STATE WITH PARTIAL FACTORS

Since publication of the Second Edition of BR 363 in 1999, the application of Eurocode 7 - EN 1997 (BSI, 1997b) and the principals of limit state design have largely become mandatory for the design of geotechnical aspects of buildings and civil engineering works.

In terms of determining the stability of a slope, EN 1997 requires a check to be made to verify that the relevant limit state (GEO), relating to failure or excessive movement of the ground, is not exceeded.

In the United Kingdom the verification of the overall stability of slopes (under EN 1997) should be checked against Design Approach 1 which comprises the application of two separate combinations:

- Combination 1 (DA1-1): Actions govern stability (capturing the interaction between applied loads (their actions) and the ground).
- Combination 2 (DA1-2): Ground strength governs stability.

Partial factors are directly applied to combinations of material strengths, actions (or effects of actions) and resistances. The analysis must show that the design effect of actions (E_d) is less than or equal to the corresponding design resistance (R_d). The result is sometimes reported as $U = E_d/ R_d$, where U = degree of utilization (%) which must be less than or equal to 100%.

Despite this change in approach, the application of limit state design to earthworks still requires the designer to use the same limit equilibrium methods of analysis to determine safety of a slope (with the corresponding target factor of safety being ≥ 1.0).

> EN 1997 states that its provisions apply to "embankments for small dams and infrastructure". What constitutes a "small dam" is not defined. The technical publication "Design of Small Dams", published by the United States Department of the Interior defines small dams as "structures with heights above streambed not exceeding 50ft…" This tallies with the ICOLD corresponding definition for "large dams" of >15m high above its foundation.

The introduction of partial factors has led to some frustration as:

- Their application can change the location of the critical surface relative to an analysis completed using characteristic values.
- Pore pressures are difficult to include as an independent variable especially when considering the division between permanent and variable elements.

These difficulties have caused some confusion and mistrust when undertaking design of slopes to EN 1997.

APPLICATION OF EUROCODE 7 TO THE DESIGN OF LEVEES

In 2013 CIRIA published the "International Levee Handbook", C731 (CIRIA, 2013) - an all-embracing guide on the safety assessment, management, design and construction of levees (flood embankments). Whilst C731 provided extensive commentary on the application of EN 1997 and the various limit states, it did not provide specific guidance on the application of Design Approach 1 in the United Kingdom.

In 2014 CIRIA published a guide on the "Application of Eurocode 7 to the design of flood embankments", C749 (Pickles *et al*, 2014). Its aim is "to improve clarity on key issues relating to the design of flood embankments, which are not addressed in detail by the current EN 1997". In terms of understanding limit state (GEO), the key findings are:

Partial Factor Multipliers

Partial factor multipliers are proposed which, depending on consequence for loss of human life and economic, social or environmental damage, are to be applied to basic EN 1997 partial factors as shown in Table 2 below.

Table 2. Partial Factor Multipliers

Description	Consequence/ Reliability Class	Partial factor multiplier
Low consequence for loss of human life *and* economic, social or environmental consequences small or negligible	CC1/RC1	0.95
Medium consequence for loss of human life, economic, social or environmental consequences considerable	CC2/RC2	1.00
High consequence for loss of human life *or* economic, social or environmental consequences very great	CC3/RC3	1.05 *

* Most likely to apply to levees

Design pore water pressures

C749 recommends that design water levels are best derived by adding a safety margin to the characteristic water level instead of factoring water pressures.

- Characteristic water level = 50% probability of exceedance level
- Design water level = 1% probability of exceedance level
- Upper limit = crest level + overtopping margin
- Characteristic water level = Design water level for embankments that are designed to overtop.

For a design life of 100 years, water levels should be determined for 150 year (50%) and 10,000 year (1%) return period events.

Sufficient Margin

C749 instructs designers to consider whether the direct application of design water levels provides sufficient margin of safety (compared to previous design practice).

C749 does not discuss the corresponding global factor of safety used as a benchmark when developing their guidance.

If the designer determines that the margin is insufficient then applying partial factors to water pressures should be considered.

Permanent/variable partial factors

Box 4.1 of C749 states: "if the 'reference period' quoted in the definition of a permanent action is taken as the duration of the design situation (for example the duration of a flood) then the water levels and pressures will generally rise to a maximum and then reduce. This is consistent with monotonic variation as used in the definition of a permanent action, and it may be reasonable to take water as a permanent action."

TESTING CIRIA C749 APPROACH TO LEVEES

An exercise has been undertaken to test the approach set out in C749 to understand the benefits and limitations of the approach. Particular emphasis has been placed on determining how likely a suitable margin can be achieved and what may constitute a reasonable margin between design and characteristic water levels.

Software

The August 2015 release version of Geostudio (GEO-SLOPE International Ltd) was used to analysis the example sections. This version includes in Slope/W the ability to apply partial factors.

SLOPE/W does not apply partial factors to characteristic water pressures. It is the responsibility of the designer to satisfy the requirements of EN 1997 by defining appropriate pore pressure conditions to the model.

Design Approach

By inspection, it is apparent that DA1-2 will govern the design of these slopes (there are no structures or applied loads, whilst water actions will be determined using suitable margins rather than application of partial factors). On this basis, as is allowed in EN 1997, no calculations have been carried out for DA1-1.

Benchmark

Historically, geotechnical structures were designed to high factors of safety as a means of indirectly controlling movement of structures (i.e. mobilisation of passive resistance or settlement of footings). Using the phraseology of limit state, the high factors of safety controlled both SLS and ULS.

The design of slopes differed as global factors of safety were significantly lower (the analysis considers ULS only). This allows direct comparison of the output from the limit state approach with global factors of safety from previous design practice.

Prior to the application of Eurocode 7 there was no code of practice that applied directly to levees. Designers tended to adapt related good practice guidance such as BR 363. The design water level would be determined following discussions with the designer's hydrologist – the level typically being fixed at crest level.

This test makes use of available contemporary guidance on suitable global factors of safety, that is, State of California's Urban Levee Design Criteria (ULDC) (DWR, 2012). The ULDC provides recommendations for minimum landside slope stability based on "design water surface elevation" (taken to correspond to characteristic water level). Minimum factors of safety of 1.4 and 1.5 are recommended for intermittently and frequently loaded levees respectively.

Three metre high embankment with 600mm freeboard

Three design sections, each 3m high and with a crest width of 4m, were tested (see Figures 1 to 3 below). A typical freeboard (river defence) of 600mm was applied to the 100 year return period water level (Q_{100}). Design water levels were established by applying the following safety margins to the characteristic water level:

1. No safety margin added. It was assumed that the characteristic water level (Q_{150}) is 100mm above the Q_{100} level (-500mm).
2. Safety margin added to characteristic water level. The 10,000 year flood level was set at crest level (zero mm).
3. Safety margin added to characteristic water level. The 10,000 year flood level included overtopping flow margin set at 50% of the freeboard (+300mm).

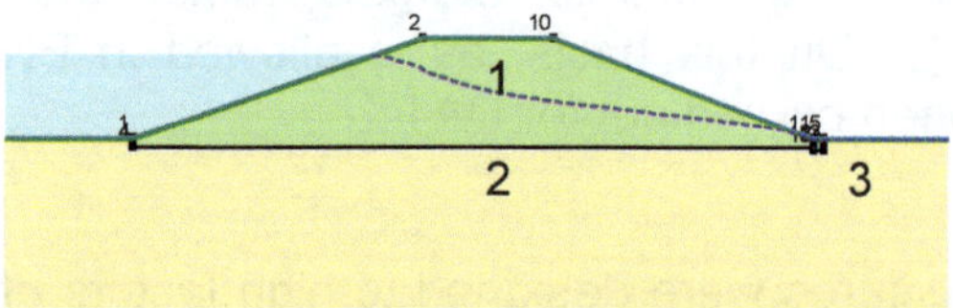

1. Clay Fill (ϕ' = 30°; c' = zero; k = 10^{-8} m/sec)
2. Silt Formation (ϕ' = 27°; c' = zero; k = 10^{-6} m/sec)
3. Toe drain

Figure 1. Homogeneous Embankment (Steady State)

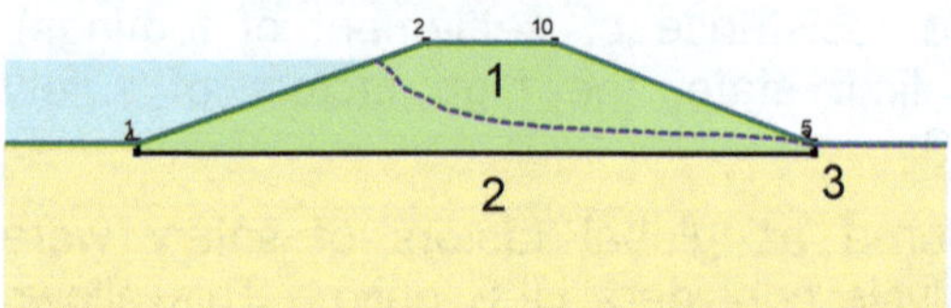

1. Clay Fill (as above)
2. Silt Formation (as above)
3. Toe drain removed – transient analysis undertaken to justify.

Figure 2. Homogeneous Embankment (Transient)

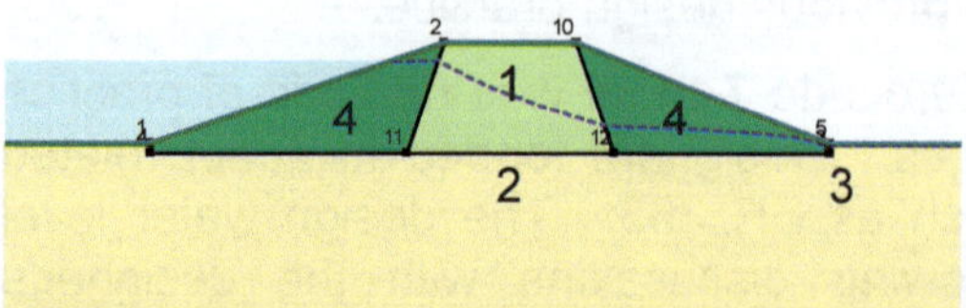

1. Clay Fill (as above)
2. Silt Formation (as above)
3. No toe drain.
4. General Fill (ϕ' = 30°; c' = zero; k = 6 x 10^{-6} m/sec)

Figure 3. Zoned Embankment (Steady State)

The landside design slope profile was established using the partial factor method including a partial factor multiplier of 1.05.

The equivalent factor of safety, using characteristic material properties and water level, was then calculated for comparison against the recommendations in the ULDC.

Table 3. Results for three metre high embankment with 600mm freeboard

	Design Slope			Equivalent FoS		
Design water level (+/- Crest)	**-500 mm**	**Zero mm**	**+300 mm**	**-500 mm**	**Zero mm**	**+300 mm**
Steady State	2.65	3.00	3.25	1.31	1.40	1.48
Transient	2.65	3.00	3.25	1.31	1.39	1.45
Zoned (SS)	2.55	2.90	3.10	1.31	1.39	1.42

Fixed Design Water Level

A check has been undertaken on the sensitivity of the approach to variations in embankment height and characteristic water level (the design water level being fixed at crest level).

The homogeneous steady state model was used with the landside slope set at 1V:3H. Three embankment heights where checked (up to 4m). The crest width was fixed at 4m (the 2m high embankment was also tested with a 2m wide crest).

Unfactored factors of safety were calculated as the characteristic water level was varied to represent various freeboard allowances (reported as design water level minus characteristic water level).

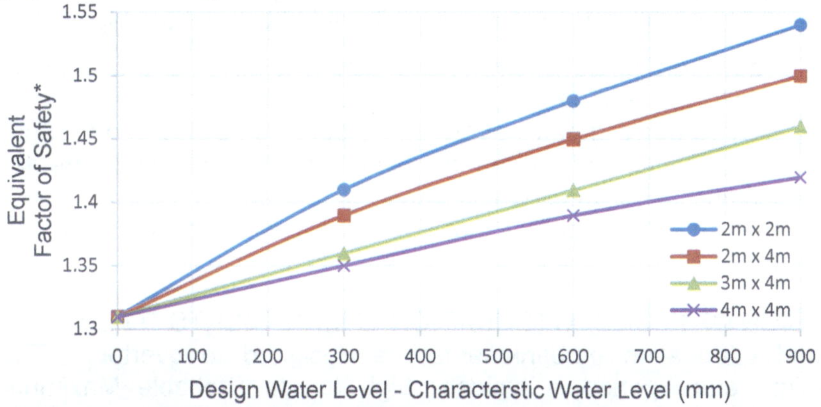

Figure 4. Variation in Unfactored FoS with Fixed Design Water Level

* Reported for design including partial factor multiplier of 1.05

Discussion

Table 3 shows some apparent consistency emerging using the C749 approach of applying margins to characteristic water levels. The results were similar to the Urban Levee Design Criteria for intermittently loaded levees (with a partial factor multiplier of 1.05).

Figure 4 illustrates that margin of safety is influenced by the scale of embankment in relation to flood depth and the difference between characteristic and design water levels. As a result, it is far from clear how the C749 approach provides assurance to the designer that sufficient margin of safety can be consistently achieved "compared to previous design practice" since:

- Using determined design water levels relies on the design level being sufficiently greater than the characteristic level to introduce an appropriate margin of safety.

- Freeboard is calculated using a significant number of variables which are largely independent of the scale of the embankment (e.g. catchment characteristics, and superelevation).
- The 10,000 year water level is not routinely analysed for flood defence purposes. Return period events up to 1,000 year can be difficult to stabilise whilst the 10,000 year level can only be guest at by extrapolating from lower return period levels.
- There are obviously issues with extrapolating the 10,000 year water level to estimate overtopping depth from lower return periods. Overtopping volumes would be a best estimate – they will not be accurate as it is likely that the hydraulic characteristics would be quite different at the higher stage.

  The overtopping allowance is also sensitive to the channel profile and spill length. Small defences in steep sided valleys may return a greater overtopping depth across a short length than a wide shallow channel overtopping across much greater length. The latter would result in a lower margin of safety despite the possibility that the flood consequences might be greater.

DESIGN OF FLOOD ATTENUATION EMBANKMENTS TO EN 1997

Flood attenuation embankments are designed to overtop. The design return period is typically high (up to Probable Maximum Flood) and the design water level will be equal to the characteristic water level.

In this case, C749 suggests that alternative design pore pressures may be calculated by applying partial factors to the characteristic pore pressure.

By inspection, it is again apparent that DA1-2 will govern the design of these slopes. Given that flood attenuation embankments rarely impound water, all water actions within the embankment structure are likely to be variable. On this basis, it is not possible to increase the margin if all characteristic water actions are treated as permanent actions. Therefore it is clear that a partial factor must be applied to introduce sufficient margin of safety.

<u>Method</u>

Three design sections, similar to those given in Figures 1 to 3 above, were analysed against an overtopping depth of 1000mm. Both 3m and 6m embankment heights were tested (crest width of 4m).

In addition, two quite different real-life design sections were tested (see Figures 5 & 6 below). The design water levels differ insofar as the steady state conditions were determined based on the design water level being at crest level.

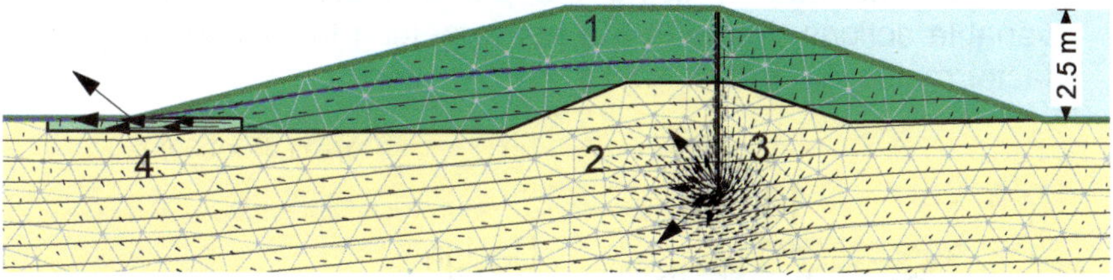

1. General Fill
2. Silty Formation
3. Sheet pile cut off
4. Drainage blanket

Figure 5. Embankment with sheet pile cut off

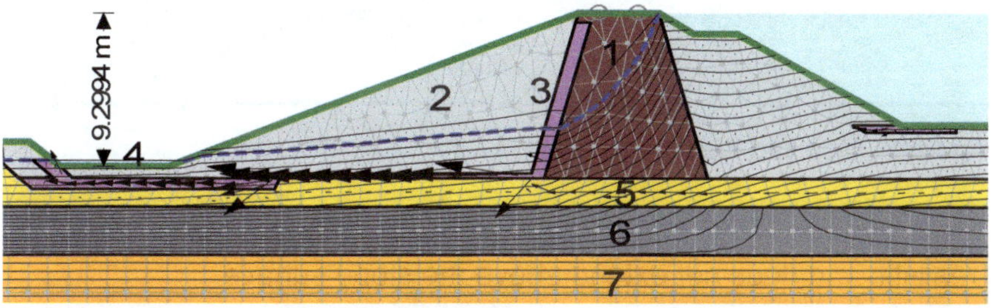

1. Clay Core
2. General Fill
3. Chimney Drain
4. Drainage Blanket
5. Alluvial Deposits
6. Glaciolacustrine Deposits
7. Glacial Sands and Gravels

Figure 6. Zoned embankment with drainage blanket

A partial factor multiplier of 1.05 was applied to partial factors greater than 1.0 for both actions and material properties.

The following procedure was applied:

1. Solve Seep/W to obtain the characteristic water profile. Use "Sketch Polylines" to trace the resulting piezometric line. Note the corresponding minimum factor of safety (A) from Slope/W (characteristic values). *For the purposes of this exercise, each model was amended until (A) = 1.50 (from BR 363).*

2. Clone the Slope/W analysis (no parent) and change PWP conditions to "Piezometric Line". In Define View, draw the piezometric line along the traced sketch line. Alternatively, for transient models, select a time step that best fits the sketch line.

3. Solve and note the corresponding minimum factor of safety (B) using characteristic values. If (B) is reported for the same slip surface as (A), and is within only a few percent of the value, then the piezometric line can be considered sufficiently representative.

4. Create a clone of this Slope/W analysis and, as shown in Figure 7, modify the profile of the piezometric line (using the toe level as the datum) using the partial factor for unfavourable variable actions (from DA1-2) not forgetting to include the partial factor multiplier, that is, 1.30 x 1.05 = 1.365.

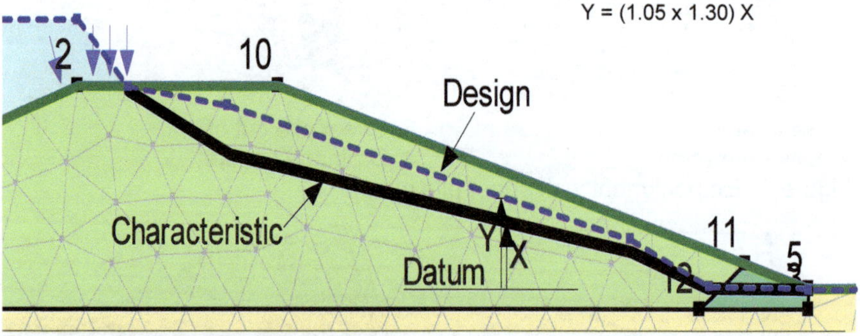

Figure 7. Modified Piezometric Line

> The designer must ensure that the profile of the modified line is realistic and does not extend above the slope surface.

5. Solve, with modified partial factors applied to material strengths for DA1-2, and note the corresponding factor of safety (C).

> The designer must review the results to ensure that the nature of the slip (location, shape & scale) is similar to (not necessarily the same) the critical slip from the characteristic analysis (A).

6. Adjust (C) to account for loss in definition of pore pressures where, (D) = (C) x (B)/(A). The degree of utilization can be reported as the reciprocal to the factor of safety, that is, U = (1/(D)) x 100%.

Results

Table 4. Results for flood attenuation embankments

	Height	Factors of Safety			Degree of Utilization
		B	C	D	
Steady State	3.0m	1.51	1.01	1.00	100%
	6.0m	1.49	0.98	0.99	101%
Transient	3.0m	1.50	1.01	-	99%
	6.0m	1.50	1.01	-	100%
Zoned (SS)	3.0m	1.54	1.02	0.99	101%
	6.0m	1.52	1.02	1.01	99%
Embankment with sheet pile cut off	2.5m	1.51	1.02	1.01	99%
Zoned embankment with drainage blanket	9.3m	1.92	-	-	-

Discussion

The results provided in Table 4 show that consistency was achieved, when applying partial factors to the characteristic pore pressure, with very similar degrees of utilization (100%±1%).

Overall, the application of partial factors to the characteristic pore pressure (at Step 4) did not prove to be too onerous. To clarify this point, the change in factor of safety at Step 4 is presented below:

- Steps 1 to 3: Analysis using characteristic material properties and water levels (A/B) FoS ≅ 1.50
- Step 4: Analysis using characteristic material properties with piezometric line uplifted by a factor of 1.365 FoS ≅ 1.31
- Steps 5 & 6: Analysis to DA1-2 with modified partial factors (C/D) FoS ≅ 1.00

The results show that applying a "variable" partial factor to water pressures (in DA1-2) creates a margin of safety which compares well with previous design practice (BR 363).

The pore water pressure conditions for the zoned embankment with drainage blanket were complex as they were strongly influenced by artesian conditions in the underlying glacial sands and gravels. This meant that the result of the analysis, using the traced piezometric line, was not representative and could not be progressed.

CONCLUSIONS

1. C749's recommended approach of deriving design water levels by adding a safety margin to the characteristic water level has the benefit of being relatively simple. However, the window of application appears narrow and is strongly influence by the scale of the levee in relation to freeboard and overtopping depths.
2. The onus is placed on the designer to be satisfied that an acceptable margin of safety is achieved. In the absence of guidance, it is suggested that UDLC is used for comparison for levees and BR 363 for dams.
3. For DA1-2, where design pore pressures are calculated by applying partial factors to the characteristic pore pressure, it is not possible to increase the margin if all characteristic water actions are treated as permanent actions.
4. For DA1-2, although C749 considers that applying a variable partial factor to water is too onerous, it is clear from the results that the variable partial factor must be applied (along with a partial factor multiplier of 1.05) to ensure the margin of safety is sufficient (compared against BR 363).
5. There are limitations to the suggested procedure for applying partial factors to pore pressure. It is suggested that the suitability can be established during Step 3, where the result from the analysis using the piezometric line will identify any complications.
6. It is important to understand how the software used applies the partial factor approach in order to be sure that it is compatible (allows application of partial factors applying to characteristic water pressures).

REFERENCES

BSI (1997a). *BS EN 1997-1:2004 Eurocode 7. Geotechnical design – Part 1: General rules.* British Standard Institute, London

BSI (1997b). *NA to BS EN 1997-1:2004:2007 UK National Annex to Eurocode 7: Geotechnical design – Part 1: General rules.* British Standard Institute, London

Johnston TA, Millmore JP, Charles JA and Tedd P (1999). *An engineering guide to the safety of embankment dams in the United Kingdom. Second Edition, BR 363*, CRC Ltd, London (ISBN 1 86081 272 4)

CIRIA (2013) *The International Levee Handbook C731*, CIRIA, London, UK (ISBN: 978-0-86017-731-4).

Pickles A, Sandham S, Simpson B and Bond A (2014). *Application of Eurocode 7 to the design of flood embankments, C749.* CIRIA, London, UK (ISBN: 978-0-86017-754-8)

DWR (2012). *Urban Levee Design Criteria.* Department of Water Resources, The Natural Resources Agency, State of California.

Dams – Benefits and Disbenefits; Assets or Liabilities?
ISBN 978-0-7277-6133-0

doi: 10.1680/dbdal.61330.165

Retrofit of Fibre Optics for Permanent Monitoring of Leakage and Detection of Internal Erosion

J DORNSTÄDTER, GTC Kappelmeyer, Karlsruhe, Germany
D DUTTON, GTC Kappelmeyer, Watford, UK

SYNOPSIS The experience of the last 20 years has shown the great advantage of the use of fibre optics for the efficient monitoring of dams. More than 100 large dams have been equipped during construction or major refurbishment. But old existing dams have been excluded until now from the use of this valuable technique.

In 2014 GTC developed a new fibre optic cable with optimized fibres which could be inserted into small diameter tubes. The fibres in these cables form an internal loop, allowing light to travel in both directions in the same cable. The well-established GTC's temperature sounding method is used to install high grade steel probes into the earth fill dam along its axis, and down into the foundation if required. The new cables are inserted and connecting cables form a "light pass" from one end of the dam to the other. Thereby a two-dimensional view of the temperature distribution within the dam is obtained which can be monitored remotely and in real time.

In 2015 three dams were equipped with this new technique, two in England and one in France; all three are monitored permanently. The cables have been installed to a maximum depth of 30m and a crest length of 430m. At one site the recorded data shows the successful sealing of a leak in the dam by a new slurry trench cut-off wall.

The new technique is described, the installation process is shown and results from permanent monitoring are demonstrated.

INTRODUCTION

Internal erosion is one of the most frequent causes of failure and deterioration of embankment dams. Internal erosion is strongly influenced by construction properties (e.g. filter and drain design, grain and pore sizes) and hydrodynamic conditions within the dam. Whilst construction properties are usually known, information is rarely available on the local hydrodynamic situation within the embankment. The most critical hydrodynamic parameter inducing internal erosion (material transportation phenomena) is the pore velocity of the seeping water. The onset of internal erosion starts at low pore velocities. Thus the detection of seepage zones of low pore velocities within a dam can prevent the development of damage and possible failure of the structure.

The existence of reliable methods for the detection of internal erosion is indispensable to anticipate the failure of embankment dams. The development and cross-tests of different geophysical and hydrogeological methods in the last 30 years have proved the high reliability of ground temperature measurements for the detection of leaks in embankments and foundations at reasonable cost.

The use of seepage water temperature as a tracer, applied to dams first in 1953 by Kappelmeyer (1953), has been shown to be a reliable method to detect and monitor in-situ the seepage flow conditions, even at extremely low velocities, and for detecting internal erosion at an early stage of development. With the ability to record temperatures over a period of time the technique can also now be used to estimate the leakage rate.

The development of this technique started in the 1950s with temperature measurements in boreholes and piezometer stand pipes. In 1992 GTC Kappelmeyer introduced greater accuracy and reliability by measuring in-situ ground temperatures with an array of purpose-designed small diameter temperature probes which were rammed vertically into the crest of a dam at regular intervals (Dornstädter, 1997). As an alternative measurement technique, since 1995, optical fibres have been incorporated into dams and into foundations but only during construction or major refurbishment (Aufleger *et al*, 1998). They can provide a continuous record of temperatures and can be remotely monitored.

Fibre optic temperature sensing operates by sending a short laser pulse (<10ns) into an optical fibre. The backscattered light is analysed with Raman spectroscopy, providing Stokes and anti-Stokes intensities. The ratio of Stokes to anti-Stokes intensities is proportional to the temperature at the point of reflection (the measuring point). The distance of the measuring point along the fibre is calculated from the returning time interval of the backscattered light and the velocity of light. The method provides a temperature profile distributed along the entire optical fibre.

NEW DEVELOPMENT

In 2014 new fibres were developed which facilitate the combination of temperature probes and fibre optics, thus providing the advantage of easy installation of fibre optic cables into existing dams and allowing a retrofit of two-dimensional seepage monitoring based on fibre optic (Patent DE19621797, 2011).

The key to the new solution are bend-optimized fibres, which can be bent to a very small radius without too much attenuation of light intensity when a laser pulse travels through them. The cable, with a typical outer diameter of 4 to 6mm including armouring and watertight protection, has a minimum of two fibres inside it. At the far end of such a cable one of the internal fibres is bent through 180° and welded to a second fibre by fusion splicing. The splice and bend (or 'optic loop') are then protected against mechanical damage by a cover with a typical outer diameter of 8mm. This cable with fibre optic loop at the far end is inserted into the small diameter tube of a temperature probe that has been previously installed into the dam.

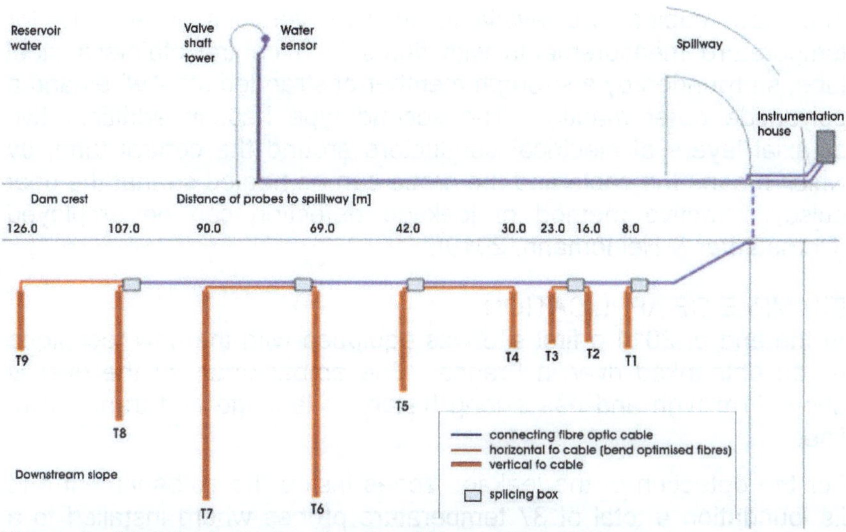

Figure 1. Sketch of a typical layout of fibre optic cabling

For the tube installation the well-established temperature sounding method is used, by which high grade steel probes of less than 25mm external diameter are vibrated into the earth fill dam and foundation along the dam's axis. The maximum depth ever reached is 45m. Individual cables are inserted in each probe and, in a shallow cable trench, a connecting cable runs from probe to probe and finally to the instrumentation house/cabinet. The fibres of each individual probe are spliced to the fibres of the connecting cable in a way that allows the laser pulse from the Distributed Temperature Sensing (DTS) instrument to travel along the connecting cable, running down and up each probe from one end of the dam to the other as shown in Figure 1. For high precision measurements and calibration reasons it is recommended that a complete "light pass" loop is formed starting and ending at the DTS instrument. With this configuration of cables so called double-end measurements can be carried out. If less precision is acceptable it is sufficient to carry out single-end measurements with an open ending of the light pass.

Since the attenuation loss of light on each bend is not negligible this must be taken into account when selecting the "laser power" of the DTS instrument. If many probes are to be monitored along a dam, there is always the option of creating several "light passes" by using a connecting cable with a large number of fibres with each pair of fibres connected to a group of probes forming a "light pass". Typically 8 to 10 probes can be put together in one pass. More "light passes" require multi-channels in the DTS system and are measured one after another by multiplexing.

The new cables are available in two versions, one only for temperature measurements with fibres in a central stainless steel tube, surrounded by a strength member of stranded steel wires and a polyamide outer mantle. The second type has, in addition, two co-axial layers of electrical conductors around the central tube, by which means the cable and the probe can be heated so that the heat pulse, or active method of leakage detection can be employed (Dornstädter & Heinemann, 2010).

EXAMPLE OF APPLICATION

At the end of 2014 a first site was equipped with the new technique on an embanked river in France. The embankment of the river is about 11m high and has a long history of leakage and transport of fines.

For the detection of the leakage zones inside the embankment and its foundation a total of 37 temperature probes where installed to a depth of 16m along a 430m long section of the dam. The profile follows the downstream edge of the embankment's crest (Figure 2).

Fibre optic cables were installed in all 37 probes and connected by a cable in a small trench. Four light passes were installed, each ending at the instrumentation house (Figure 3) situated on the downstream berm. Additional to the probes along the crest a fibre optic cable was installed from the instrumentation house in a trench along the downstream toe close to the open wide drain. The system is operated by solar power with a diesel power generator inside the instrumentation house as a back-up power supply for the batteries.

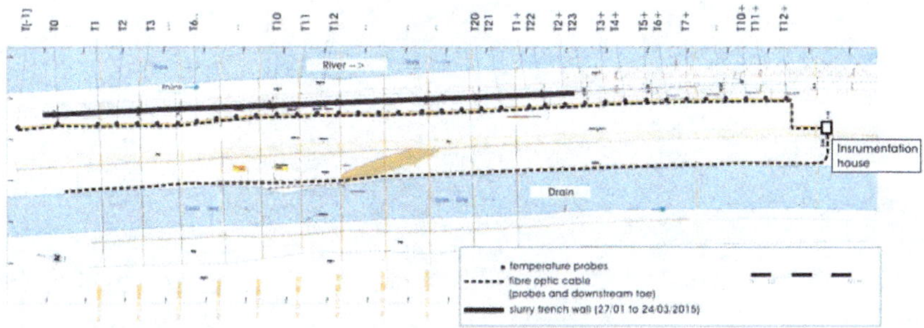

Figure 2. Site plan – 37 temperature probes

Figure 3. Instrumentation house on embankment berm

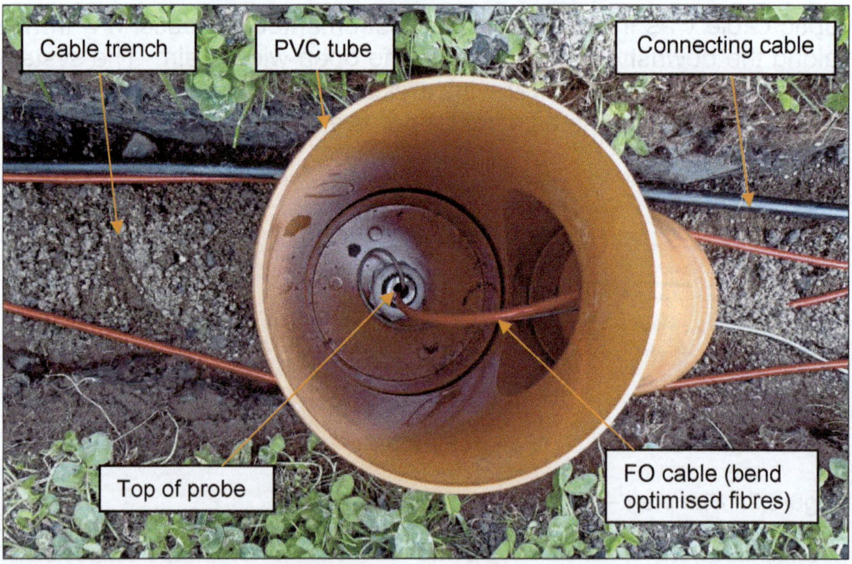

Figure 4. Installation situation

Figure 4 shows the position of one vertical probe. The small red cable with bend optimised fibres is inserted into the high grade steel probe, protected by a large diameter PVC tube. In the upper part of the figure the black connecting cable connecting all probes in a small 30 cm deep trench is visible.

Since the site was having leakage problems combined with transport of fines the client decided to build a slurry trench cut-off wall in Spring 2015. The remotely operated fibre optic monitoring system was installed in Autumn 2014. The two-dimensional temperature distribution of 4 January 2015, about three weeks before the start of construction of the new cut-off wall, is shown in Figure 5. In the centre a strong temperature anomaly is indicated by the blue colour corresponding to the low water temperature of the river in winter. The strongly percolated area extends from probe T6 to probe T21 from 7m to 14m depth below crest level, showing severe leakage flow through the lower part of the dam and through its foundation.

Two minor percolated areas were detected at T2/T3 and from T5+ to T7+ at the interface between the embankment and its foundation.

The temperature evolution before, whilst and after the construction of the cut-off wall was remotely monitored with automatic data analysis. During the construction procedure the client followed the success of the construction by the automatic temperature monitoring system. Figure 6 shows the temperature distribution some months after completion of the cut-off wall which was constructed along the array of probes between T0 and T2+.

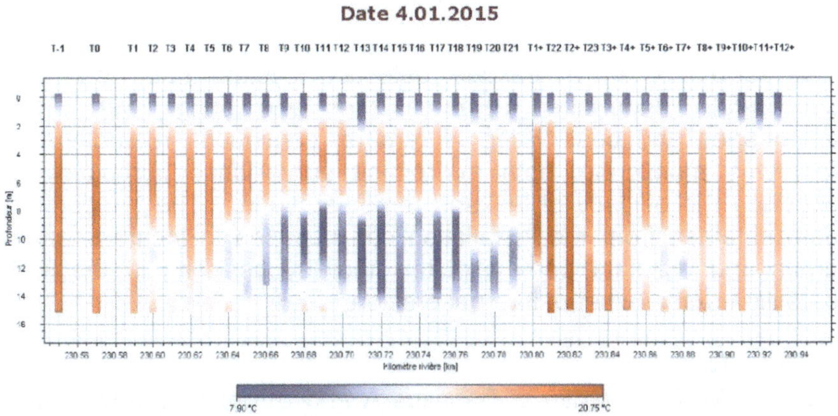

Figure 5. Temperature distribution before construction of cut-off wall

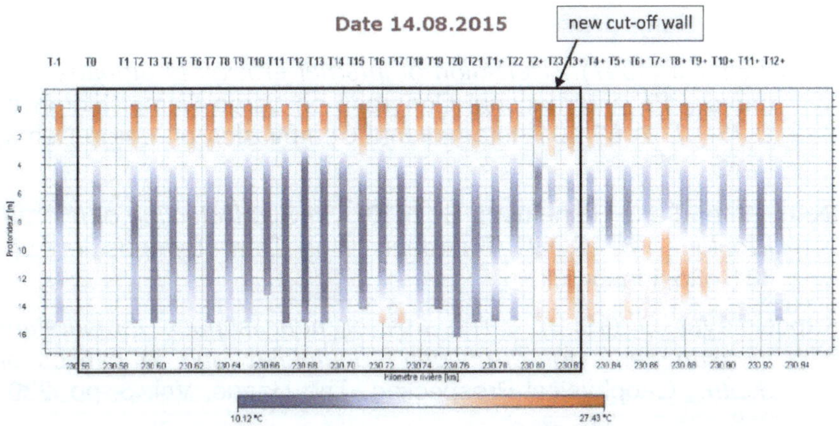

Figure 6. Temperature distribution after construction of cut-off wall

The result shows the appearance of an anomaly at the downstream end of the cut-off wall which indicates a lateral change in the direction of the seepage flow. Furthermore the previous minor leakage flow in the vicinity of probe T6+ has increased since the leak between probes T6 and T21 has been stopped. By stopping the leakage flow the hydraulic pressure upstream of the cut-off wall has increased and this has caused the stronger flow around the downstream end of the wall.

CONCLUSION

Recent developments in the manufacture of fibre optic cables have enabled two well-proven ground temperature leak detection techniques to be combined, providing a means whereby fibre optics can be readily installed deep within the body of an existing dam.

This retrofitting of a leakage detection system has now been successfully undertaken at three sites and brings with it the advantages and possibilities of remote, continuous recording and real time monitoring, both for routine surveillance purposes and checking the efficacy of leakage remedial works. The temperature data obtained from the installation can be used to locate seepage areas and estimate pore velocities thereby assisting in the early detection of the onset of internal erosion.

REFERENCES

Aufleger M, Dornstädter J, Fabritius A and Strobl Th (1998). *Fibre optic temperature measurements for leakage detection - applications in the reconstruction of dams.* 66th ICOLD annual meeting, New Delhi, pp. 181-189, International Commission on Large Dams, Paris.

Dornstädter J (1997). *Detection of internal erosion in embankment dams.* 19th International Congress on Large Dams, Florence. Q.73R.7. 2: 87-101, International Commission on Large Dams, Paris.

Dornstädter J and Heinemann B (2010). *In Situ Detection of Internal Erosion.* 8th ICOLD European Club Dam Symposium, pp. 481-485, Innsbruck.

Kappelmeyer O (1953). *The use of near surface temperature measurements for discovering anomalies due to causes at depth.* Geophysical Prospecting, The Hague, Vol. 3, pp. 239-258.

Patent DE19621797 (2011). *Verfahren und Vorrichtung zur Leckageüberwachung an Objekten und Bauwerken.* Deutsches Patent- und Markenamt, Munich.

SECTION 4:

DISCONTINUANCE OF RESERVOIRS AND THE ENVIRONMENTAL BENEFITS

Dams – Benefits and Disbenefits; Assets or Liabilities?
ISBN 978-0-7277-6133-0

doi: 10.1680/dbdal.61330.175

Haweswater Reservoir: an environmental asset or an environmental liability?

P J RIGBY, United Utilities, Warrington, UK
A THOMPSON, United Utilities, Warrington, UK
J BUTCHER, United Utilities, Warrington, UK
D E JONES, United Utilities, Warrington, UK

SYNOPSIS Haweswater Reservoir, owned and operated by United Utilities (UU), is the third largest reservoir within England behind Kielder Water and Rutland Water. Constructed in 1940 it holds approximately 85,000 million litres of water and supplies Manchester via 72 miles of aqueduct.

The dam itself was considered to be a major feat of engineering at the time and is a rare example of a buttress concrete dam. The reservoir location is viewed as one the most beautiful parts of the Lake District and is a vital for the local economy, tourism and diverse flora and fauna. But this was not always the case. At the time of its design and construction there was huge opposition to its construction from the local community. Marland, a village at the heart of the valley, was flooded and villagers were relocated, losing their homes and livelihood. The impact on the local flora and fauna in the short term was also significant.

This paper will look at the social and environmental impacts resulting from the construction of the reservoir through to its present day operation. It will also discuss the impacts associated with returning such reservoirs back to their natural environmental as part of discontinuance works and the challenges faced.

INTRODUCTION

Haweswater reservoir is situated in the Lake District National Park 4km to the west of Shap Village in Cumbria. It occupies the Mardale valley and is some 6.7km long and 900m wide making it the third largest reservoir in England behind Kielder Water and Rutland Water. The reservoir is retained by a 470m long, 33m high concrete dam which commenced construction in 1929 and was finally commissioned in 1941 (Figure 1). Water from Haweswater Reservoir is transported through approximately 72 miles of aqueduct to supply the urban conurbation of Manchester, terminating at Audenshaw.

Figure 1: Haweswater Dam and Reservoir

THE DEMAND FOR WATER

The 19th Century saw the increasing migration of the population from the country to the towns as part of the urban explosion driven by the Industrial Revolution. During the early part of the century the population of Manchester and Salford jointly grew by the order of 47% (Binnie, 1976). As a consequence, the demand for fresh water increased exponentially, driven by population and industry growth as well as the introduction of water borne sanitation.

Whilst a large number of reservoirs existed around Manchester following the prolific embankment dam building period from the 1850s to 1890s spearheaded by Bateman and Hawksley, these could not match the growing demand. In 1878 Manchester Corporation turned its attention to the Lake District with a view to enlarging and impounding Thirlmere (Sheail, 2002). Following the further demand for increased resources a Bill was submitted in 1919 to gain parliamentary agreement for the impounding and raising of the level of Haweswater.

The Haweswater Act of 1919 was passed which gave Manchester Corporation permission to acquire the Mardale valley and adjacent catchment areas in Westmorland to construct a new dam and impounding reservoir. The controversial proposal to impound Haweswater led to a challenge by Lancashire County and a request that such an upland catchment be treated in the character of a national trust (Sheail, 2002). However it was not until 9 May 1951 that this area gained its designation as a National Park. Interestingly however, some environmental conditions were included in the Haweswater Act to protect the existing fish population. This required a third of the stream water to be passed forward as compensation flow with a proportion released in the form of "freshets" or "spates" to assist with salmon and river trout spawning.

HAWESWATER RESERVOIR AND DAM

The area around Haweswater had long been considered by the locals to be a remote and quiet place away from the tourists who had invaded the Lake District following the construction of the railways. This remote idyll comprised a small natural lake which provided a peaceful and unviolated sanctuary amongst the hills (The Sphinx, 1869). The natural lake at Haweswater was approximately 4km in length split by a tongue of land at Measand. The area was populated by the farming villages of Measand and Mardale Green. These villages and the adjacent land would be lost following construction of the dam and flooding of the valley, which would raise the water level by the order of 95ft.

Construction of the dam commenced in 1929 despite the ongoing public outcry. Due to the high ground pressures created by this form of construction the foundations were required to extend to strong rock. Rock at the site was proved at depths of between 16ft to approximately 35ft below existing ground level and comprised the Birker Fell Andesite formation, a volcanic extrusive rock (Figure 2). Grouting was required at the base of the cutoff trench in order to improve the nature of the rock which was locally fractured due to faulting in the area.

Figure 2: Initial excavations for foundations underway (1935)

Construction works continued until 1931 when the lack of available funds during the Great Depression halted further works. During this time life in the villages of Mardale and Measand returned to normal with only a small skeleton crew left to oil the machinery at the dam site (www.mardalegreen) (Figure 3). It was some four years later in 1935 that works continued on the dam with its final completion in 1941.

The proposed dam was considered to be a major feat of engineering for its time. This was the first buttress dam of its type built in the UK and formed the template for several other dams built in Scotland after 1945 (Kennard, 1996). The dam was designed and constructed using direct labour under the supervision of the engineer for the scheme, Mr Holme Lewis (Davies, 1940).

For its time the dam was considered to be innovative, offering economies in construction resulting from reduced uplift pressures and efficient use of materials. The dam comprises 44 buttresses extending to a final length of 470m and height of 33m. The use of this type of buttress arrangement resulted in an estimated 27% saving when compared to a comparable gravity section dam (Kennard, 1996). In total the dam used an estimated 140,000 cubic yards of concrete, and over 30,000 tons of cement (www.mardalegreen). 190,000 tons of stone was sourced from local quarries including Shap with an early example of recycling demonstrated by the incorporation of stone from Mardale village church into the dam (Figure 4).

Figure 3: Excavations continue using direct labour

Consideration was given during the design to the visual aesthetics to ensure that, once completed, it would harmonise with the area as far as practical to appease the local community and blend into the local landscape. Its design utilised simplicity of form and line with any elaborations treated with disfavour. An example of this was the attempt to improve "unsightly" horizontal joints. Timber fillets were fixed to the shuttering to produce chamfers at the top and bottom of each lift. This was achieved at relatively little expense (Davies, 1940).

Figure 4: Dam under construction, concrete mixing plant in foreground

Figure 5: Visual impact of dam on landscape

The impact and success of the final construction was perhaps best summarised by Alfred Wainwright writing in his Pictorial Guide to the Lakeland fells (2005) (Figure 5):-

"If we can accept as absolutely necessary the conversion of Haweswater [to a reservoir], then it must be conceded that Manchester have done the job as unobtrusively as possible. Mardale is still a noble valley. But man works with such clumsy hands! Gone forever are the quiet wooded bays and shingly shores that nature had fashioned so sweetly in the Haweswater of old; how aggressively ugly is the tidemark of the new Haweswater!"

PRESENT AND FUTURE CONSIDERATIONS

The Lake District attracts over 16 million visitors a year with tourism contributing over £1,140 million to the local economy. Part of the attraction to the area is around the dramatic landscapes and the amenities provided by the "Lakes". Few people who visit the area would appreciate that Haweswater Reservoir was not a natural feature, having being part of the established landscape for many decades. The flora and fauna around the reservoir have re-established and flourished following the loss of land from impounding and the area is home to the only remaining Golden Eagle in England.

The impact on the environment from tourism is being felt with the increase in visitor numbers to the area. United Utilities (UU) is managing this impact at an annual cost of £40,000 and working with the RSPB on its SCamP initiative (Sustainable Catchment Management Programme). Some of its aims are to improve water quality, ensure a sustainable future for UU's tenant farmers and permit the habitat to start to become more resistant to long term climate change.

The impact of climate change is an important consideration for how reservoirs such as Haweswater are managed. The summers of 1984 and 1995 saw droughts in the region with water levels in Haweswater down to only 11% full, revealing the foundations of Mardale Village. Conversely the floods of 2009 and the recent floods of 2016 resulted in major flooding in the region (Low Carbon, 2012).

There could be opportunities around the management of water stocks within Haweswater reservoir to assist with the impact of climate change. The Environment Agency is exploring the opportunities for the reservoirs such as Haweswater to be held below current top water level to provide storage to capture water from extreme flood events, preventing downstream flooding. Conversely the water companies are being driven to ensure resilience against future droughts which would require reservoirs to be at full capacity. All of which comes with associated costs, risks and liabilities, not least from the customer.

Over the last few years UU has been reviewing its current stock of reservoirs balancing the need for the resources against the cost of ongoing maintenance and operation. This has led to a number of reservoirs, where water was not required for supply, to be discontinued and the areas returned back to their original landscape (Figure 6 and 7).

Figure 6: Hurst IR before discontinuance

Figure 7: Discontinuance of Hurst returning it back to the original landscape

Baystones and Hurst IR are two noticeable successes. However, this can often lead to opposition from locals who see this as a loss of an amenity. One such challenge presented itself with the discussion around Cogra IR near Cockermouth. The option to discontinue was considered against the costs associated with the required improvements, given that the water is not utilised for supply. An action group was started by locals concerned about the potential loss of the reservoir. However, would these same customers be prepared to pay additional costs on their water bills to preserve these facilities?

DISCUSSION

So has the view on Haweswater changed since its conception in 1919 when it was considered to be taking away the existing beauty of the lower parts of the Mardale valley (Nicolson, 2015)? A recent talk by the RSPB in 2014 described it as one of "Cumbria's iconic lakes" with diverse flora and fauna and this appears to reflect the views of the visitors to the Lake District. A Lake District without this particular "Lake" would be unthinkable even if the water was not a required resource.

REFERENCES

Anon (1967). *Wet Sleddale works of the Haweswater scheme of the Manchester Corporation Waterworks undertaking.* Water Engineering, vol 71, June, pp 223-226.

Binnie GM (1976). *The evolution of British dams.* 12th Dickinson Memorial Lecture. Transactions of Newcomen Society, vol 47, pp 207-224.

Binnie GM (1987). *Masonry and concrete dams 1880-1941*. Industrial Archaeology Review, vol 10, no 1, pp 41-58.

Davies DG (1940). *The Haweswater dam.* Transactions of Liverpool Engineering Society, session 66, vol 61, pp 79-132.

Davis DG (1940). *The Haweswater dam.* Water and Water Engineering, vol 42, no 521, pp 62-68 & no 522, pp 151-153.

Hopkins JK and Wickham DB (1988). *Repairs to downstream face of Haweswater dam. Reservoir Renovation.* Proceedings of BNCOLD Conference, Manchester, technical note 3.

Kennard MF, Owens CL and Reader RA (1996). *Engineering guide to the safety of concrete and masonry dam structures in the United Kingdom.* Report 148, CIRIA, London.

Low Carbon Lake District (2012) *Adapting to Climate Change in the Lake District National Park: Initial Assessment of Risks, Opportunities and Actions.* www.lake district.gov.uk

Nicholson N (2015) *The Dambuilders. The Destruction of Mardale and the building of Haweswater Dam.* Bookcase ISBN 9781904147893

Schofield L (2014) *Haweswater Wildlife, An Illustrated Talk.* RSPB lvcpnews

Sheail J (2002). *An environmental History of Twentieth Century Britain.* New York Palgrave.

Taylor GE (1951). *The Haweswater reservoir.* Journal of Institution of Water Engineers, vol 5, no 4, pp 355-392.

The Sphinx (1869). *Over the Nan Bield Pass to Haweswater.* Vol 11 No 59

http://www.mardale.green.talktalk.net/haweswater_dam.htm

Dams – Benefits and Disbenefits; Assets or Liabilities?
ISBN 978-0-7277-6133-0

doi: 10.1680/dbdal.61330.185

Environmental Benefits of Reservoir Discontinuance – Hurst Reservoir Case Study

H J BEEDEN, United Utilities
C D PARKS, United Utilities

SYNOPSIS Hurst Impounding Reservoir was situated east of Glossop, in Derbyshire, but after decommissioning of Hurst water treatment works, the 167Ml reservoir supplied only compensation flow to Hurst Brook. The design was a homogeneous embankment with an upstream puddle clay blanket. Investigations identified defects in the embankment and several options were considered to address these.

United Utilities' (UU) sustainable long-term solution was achieved through complete removal of the dam and restoring Hurst Brook to a natural watercourse. The project reused all excavated materials with the ambitious intention of zero waste removed off-site; an achievement fundamental to the project subsequently winning the Sustainability Category of the Ground Engineering Awards in 2015.

Ecological benefits of the project included extension of acidic grassland and moorland habitats within the Peak District National Park (PDNP), the creation of bird breeding areas and new aquatic habitats. A short construction programme ensured only one "game & native bird" nesting season was disrupted, alleviating stakeholder concerns.

Environmental Impact Assessment (EIA) (Halcrow, 2012) conclusions were incorporated in the planning consent. The project promoted good stewardship of the land in the Dark Peak and managed the operational risks whilst also achieving excellent consultation and stakeholder management.

INTRODUCTION

This paper describes the environmental considerations and benefits of the discontinuance of a reservoir embankment and focuses on the sustainable methods that can be employed to achieve complete restoration. The example used is the discontinuance of Hurst Reservoir, near Glossop. This was a difficult and complex project requiring the full engagement of the many different stakeholders. The main objective to achieve a successful outcome was to permanently return the valley to as close to its pre-1839 state as possible by creating ecological habitats which are as natural as possible and re-using 100% of the soils recovered from the 150m long embankment, reservoir sediment, and material from the structures within the landscaping. The associated commercial driver was to provide the works below the budget required for improvements that would be needed to allow continued operation of the reservoir under the Act.

Successful completion of this project was measured by achievement of a sustainable outcome assessed as mitigation of the technical solutions that would have been required had the reservoir continued in service. The design process included input by MWH advising on hydrogeology and flood risk assessment that, in turn, formed part of the Halcrow project EIA which enabled the UU Geotechnical Engineering outline design to be completed. The construction partner was Cheetham Hill Construction Ltd who was advised, particularly in the detailed geotechnical design and materials management, by White Young Green (WYG). The project to successfully remove the dam and restore Hurst Valley was completed in 2014.

HISTORY OF HURST IMPOUNDING RESERVOIR

Hurst Impounding Reservoir was situated 1km east of Glossop, in Derbyshire, nestled in a narrow valley (Figure 1) on the northern edge of Hurst Moor, immediately south of the A57, Snake Pass.

Figure 1. The former reservoir from the Inlet structure, looking west.

The reservoir embankment, at an elevation of 217mOD, was constructed over Hurst Brook and completed in 1839. The town of Glossop was a thriving mill town in the early 1800s. However, the supply of water to feed the mills and the population was a major issue. In 1831 a group of 50 wealthy local gentlemen, who became known as the "Glossop Commissioners" obtained an Act of Parliament to build "the Glossop reservoirs", the first of which was Hurst Reservoir, to improve supply to their mills and bring water to the community. The scheme engineer was John Ashworth and his assistant was a young civil engineer named John Frederick LaTrobe Bateman, who would later become the most renowned of the Victorian dam engineers.

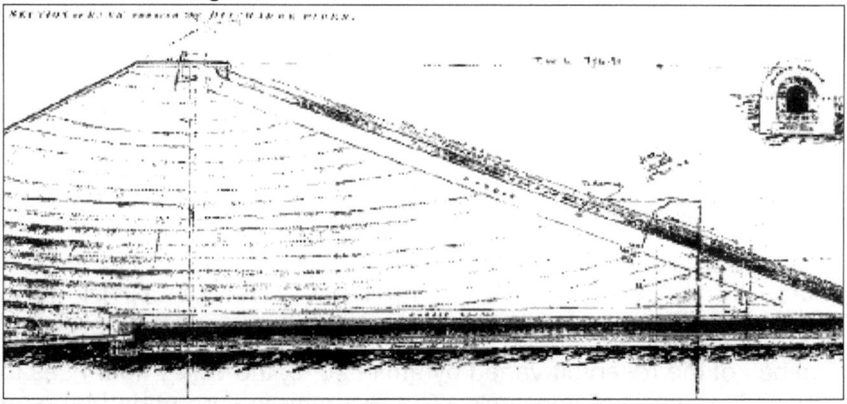

Figure 2. Design section of the upstream face of the embankment, showing the puddle clay blanket.

Hurst Reservoir embankment was designed as a homogeneous embankment with an upstream puddle clay blanket (Figure 2). The dam captures flow from below the confluence of Hurst Brook and Span Clough. A by-wash channel drained from the penstock, along the northern boundary of the reservoir. The reservoir fed the Hurst Water Treatment Works (WTW), situated below the dam. The reservoir had a capacity of 167Ml but after decommissioning of Hurst WTW in 1998, the reservoir supplied only compensation flow via Hurst Brook to Glossop Brook and the Upper Mersey catchment.

The dam operated with outlets at both the upstream and downstream ends of the reservoir. The embankment was subject to several phases of alteration, including raising of the crest prior to 1933, twice repositioning the by-pass channel, alterations to the overflow arrangement, construction of a new, and closure of an existing, overflow prior to the early 1960s, along with plugging the original outlet tunnel and construction of a siphon in 1962.

ENVIRONMENTAL BACKGROUND

Environmental setting

In 1951 the Peak District was designated the very first National Park encompassing the reservoir, adjacent to the later designated 'Right to Roam' heath and moorland. The land is owned by United Utilities and has been receiving flows and drainage from the surrounding catchment for over 170 years

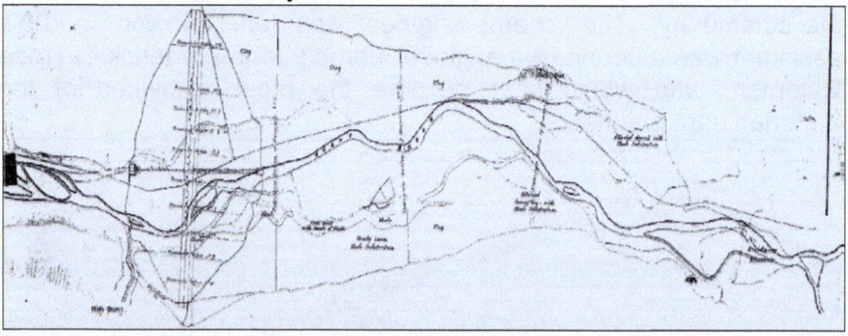

Figure 3. Design plan of the dam within the Hurst Valley

A survey of the valley (Figure 3) illustrated Hurst Brook's natural course in 1835, giving a reference for landscape and geomorphological designs.

The bed of the reservoir varied by 9m in 2013; the valley floor rises to 220mOD at the inlet penstock. The reservoir is surrounded by peat moorland, collecting regional drainage, but was not in direct hydraulic continuity with the Dark Peak Site of Special Scientific Interest (SSSI), adjacent to the south (MWH, 2011b).

Dark Peak SSSI is also designated as a Special Area of Conservation (SAC) and Special Protection Area (SPA) due to upland dwarf shrub heath habitat, although in "unfavourable recovering" condition in 2008. The project Ecology Report considered that, regarding the proposals, it was "highly unlikely that the engineering works will adversely affect the habitats and their associated fauna and flora within the SSSI through direct disturbance".

Hurst Brook was classified by the Environment Agency as of poor ecological status in 2012, based upon poor quality fish population and a moderate quality macro-invertebrate ecosystem. According to EIA ecological surveys, the reservoir did not contain a significant aquatic ecosystem or fish population as the embankment had provided a barrier to upstream migration, there being no fish pass installed, and the reservoir was never stocked.

Geological setting

British Geological Survey mapping (BGS, 2006 & 1981) shows Hurst valley to be underlain by bedrock comprising Shale Grit and Hebden Formation (Kinderscoutian). The dip of the rock strata within the valley is not shown, although the geological structure shows a dip to the West. The Hebden Formation, massive sandstone, has recently been interpreted as a "delta slope, turbidite channel deposit" (BGS, 2013). Superficial deposits are indicated to be absent and it is assumed that glacial deposits were excavated to construct the dam.

A series of record drawings of the dam construction, dated 1836, indicate a thin covering of clay above the "Shale Grit" bedrock along the valley sides with alluvium shown along the valley floor.

The substantial artificial deposits were designated as "Reservoir Sediments" and "Embankment Materials" to correctly reflect the heterogeneous particle size (up to large boulders) laid down since 1840, as sediments collected in the basin.

Figure 4. Hurst Reservoir embankment before discontinuance.

Historical Ground investigations and observations

Investigations were undertaken on the crest and downstream face of the embankment (Figure 4) in 1999 by Strata Surveys Ltd. Piezometers were installed in each hole and these were monitored along with the embankment toe drain on a regular basis for several months. In 2006 the panel engineer recommended that the reservoir level be held down by 3m, while a leak adjacent to the siphon chamber was investigated. Monitoring of the piezometers was resumed in the mid-2000s and it was established that a substantial toe drainage flow was closely linked to reservoir level, demonstrating that leakage was causing the increasing flows.

DECISION TO DISCONTINUE

Information obtained during the construction of the siphon in 1962 was confirmed by subsequent investigations of the embankment in 2008. It was determined that there was no discernible upstream clay blanket and it appeared that the embankment was constructed as a simple homogeneous earth dam. The investigation proved that the majority of the embankment material encountered in the boreholes was either very soft, or loose and typically wet or saturated. The investigation holes were grouted using tube-a-manchette techniques. Some of the holes took in excess of 1000 litres of grout to backfill, part of the data indicating that the embankment material was not in an acceptable condition.

Defects had been recorded during an inspection in 2005 (Reilly, 2006) and several internal defects had been recorded during the ground investigation. These included confirmed leakage paths, several possible leakage paths, potential voids and other defects in the embankment material and likely internal erosion of the foundation that, without rectification, would represent an increased risk, to the population and to property downstream, of possible failure of the dam.

In 2008 a statutory inspection of Hurst Reservoir under the Reservoirs Act 1975 (Carter, 2008), resulted in an observation by the Panel Engineer that, "In the interest of safety" either the reservoir must be discontinued or several defects addressed, including improvements to discharge capacity; strengthening the wave wall and making the reservoir watertight to remove all concentrated leakage paths,.

Several options were considered to rectify defects identified by the investigations including: improving the overflow, addressing leakage and improving the wave-wall. However, the optimum solution was determined to be breach of the embankment and to discontinue the reservoir. The client made the decision to progress this as a sustainable project, deciding to remove reservoir capacity and provide benefits to the environment, while demonstrating 40% lower costs to discontinue than to improve the asset, along with future savings in relation to maintenance and associated regulatory obligations.

Stakeholder impacts

The committee representing the adjacent Golf Club was concerned about disturbance of the club programme, use of the joint access road, playability of the first tee and degree of environmental disturbance affecting their eastern boundary.

The Moorland Owners Group maintain shooting rights and sheep grazing on the land to the north, and voiced concerns about loss of habitat and disturbance of ground nesting birds during shooting seasons.

Regulatory stakeholders including the PDNP and Natural England were also consulted on the proposals and made valuable contributions to the project.

PROJECT DRIVERS AND SUSTAINABILITY

United Utilities is required to provide value for money for its customers while maintaining customer service and minimising social, economic and environmental impacts of its operations and construction activities. UU is also required to follow guidance from regulators to implement the Water Framework Directive (WFD), to remove engineered structures (weirs and dams) which block natural watercourses and for artificial water bodies to achieve good ecological potential by 2015.

Project drivers formalised or generated by the EIA included:

- Creation of a mosaic of terrestrial habitats that, in the longer term, offer greater biodiversity than the current reservoir;
- Providing long-term landscape management under a 15 year funded plan;
- Meeting all safety issues determined in the Statutory Inspection Report;
- Removal of future maintenance obligations under the Reservoirs Act;
- No waste soil generation;
- Minimising traffic movements and impacts; and,
- Providing new terrestrial and river habitat for birds, fish populations and invertebrates, e.g. stone and caddis flies.

Whilst:

- Maintaining access for the local Fire Operations Group's (FOG) helicopter to draw water from an on-site water body to tackle moorland fires;

- Maintaining water quality standards in Hurst Brook;
- Protecting existing local species and upland dwarf shrub heathland habitat and the adjacent SAC/ SSSI; and,
- Addressing local stakeholder requirements for access.

This project was based upon the "Three Pillars of Sustainability": addressing the social, environmental and economic impacts of the project.

Social Factors

There were many stakeholder impacts that needed to be addressed as part of the discontinuance.

Stakeholder management

Complex stakeholder involvement and buy-in was seen as key to completing this project, making consideration of social impacts central to the scheme design and carried through the construction period. The views of all stakeholders were captured and moulded the project.

Stakeholders were invited to comment on the proposed development via a public meeting and subsequently at a public exhibition. The affected parties were consulted formally as part of the planning process by the PDNP, which made the submitted EIA documentation available to view on its website. The EIA application went to planning committee where the Moorland Owners Group lodged objections to the proposals, which were subsequently resolved by mutual agreement. Throughout construction on-going liaison with affected stakeholders, including the Golf Club, Moorland owners and FOG group, ensured buy-in at all levels.

Flood Risk Assessment

Detailed assessment of flood risk (MWH, 2011a) within the valley and downstream confirmed that reinstatement of the valley slopes would ensure the system coped with expected rainfall patterns, including summer storms and predicted climate change. Furthermore it was established that the design met the Environment Agency's requirements for flood attenuation and compensation flow to Hurst Brook.

Traffic

Traffic congestion within Glossop and on the A57 and A635 was a major concern to local stakeholders and businesses. The option for discontinuance provided for the lowest level of traffic impacts possible on the adjacent infrastructure, removing more than 12,000 wagon movements to and from site to dispose of excavated material.

Public Access
The land around Hurst Reservoir is private land. However, it had a history of use by the public, including dog walkers and pedestrians, and the frequency of access to the reservoir edge increased following the 3m water level reduction in 2008.

Environmental Factors

Landscape and Visual Impact Assessment (LVIA)
The proposed change from reservoir to valley was visually assessed, by landscape architects for the EIA. The design had to be in keeping with the five surrounding local character areas designated by the PDNP. Hurst IR fell within National Character Area 51: Dark Peak, as a landscape of large-scale sweeping moorlands, contrasting with the more urban Manchester Pennine Fringe. Subjectively, the visual amenity is now improved by the removal of the embankment and restoration of the valley.

Economic Factors
Various options for the reservoir site were assessed including retention of the asset. Consideration was given to current operational spend, potential capital investment in both substantial maintenance and the cost of the proposed project. The discontinuance scheme was assessed in terms of its future whole life costs.

Sustainable Solution
Discontinuance was determined to be the most economic solution. This could have been achieved in a number of ways, however, only complete removal of the embankment and reuse of the material in the reservoir basin achieved all of the social and environmental requirements. Primarily due to the thorough understanding of the geotechnical materials' properties and the geoenvironmental risk assessment to determine material reuse on the valley sides, the cost of disposal to landfill was avoided and the complete reuse of material offered a 30% lower cost than partial discontinuance and 40% lower than removal for waste disposal.

DESIGN OF THE DISCONTINUANCE

Outline Design

Considerable historical research and ground investigation information was used by the client to assess the materials comprising the embankment and basin sediments. Detailed geotechnical design was required to determine how the soils could be reused to infill former borrow areas and provide landscaping to the original valley profile. Early contractor involvement in the design of stable slopes following the 19th Century landform allowed a collaborative approach.

A relatively small quantity of hydrocarbon impacted embankment fill was identified during investigation. The client's risk assessment allowed re-use of hydrocarbon tainted soils within the landform design. By placement in Hurst Quarry the vertical and lateral migration pathways were extended, which decreased risk of impact to Hurst Brook to acceptable levels.

Detailed Design

The detailed earthworks design by the contractor used site-won materials, incorporating reservoir sediments with poor geotechnical engineering properties, rip-rap, masonry and embankment fill to reprofile the valley sides and partially fill the adjacent quarry.

Construction

Figure 5: Weather impacts hampering reservoir drawdown.

The major issue facing the contractor was the logistics of removal of the 17m high embankment and structures, whilst generating no waste materials and maintaining flow in Hurst Brook, using only a single access track with a 3T weight limit. The weather also had a major impact in determining how sequence of works and was physically managed by the contractor during the scheme (Figure 5).

Completion was achieved in eight months, with a total of 51,700m³ of embankment materials and 51,300m³ reservoir sediments being re-used in landscaping (Figure 6) following the CL:AIRE Code of Practice (CL:AIRE, 2011) for re-use of uncontaminated soils. Structures were crushed and used in the stream bed, quarry and by-wash channel, under an agreed Materials Management Plan. Only the railings, pipes and iron valves were taken off site for recycling.

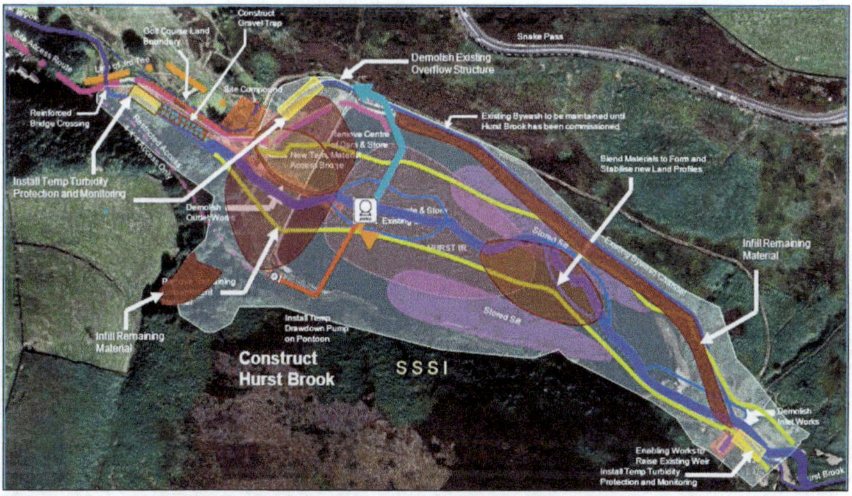

Figure 6: Materials Management Plan (MMP) pictogram (WYG).

PROJECT BENEFITS

Landscape Management Plan

The landscape management plan will run for 15 years following the one year period. This will support the long-term sustainable solution using locally-sourced heather brash and native Pennine wildflowers and grasses.

Post-earthworks construction, a specialist landscape contractor was employed specifically to deliver the habitat requirements of the landscape scheme, in keeping with the surrounding Local and National Character Areas. Landscape management costs have been assigned as project capital expenditure with the intention to 'transfer' this sum as ring-fenced operational expenditure.

Habitat Creation / Ecological Restoration

The major ecological benefits of the project are the extension of the acidic grassland and moorland habitats within the PDNP, the creation of breeding areas for Little Ring Plover, Common Sandpiper and Oystercatcher and new aquatic habitats for fish and invertebrates. The creation of 600m of watercourse, removing the physical restriction to fish migration routes and creating new aquatic environments; riffles and bog pool scrapes, ensures the project is compliant with the WFD.

Letting Nature take its course

There have been some minor concerns that the course of the brook was not as intended as there has been marked erosion in some areas of the landscaped stream bed and deposition in others (Figure 7). This erosion also revealed that the contractor had used geotextile in some areas to contain reservoir sediment in the landscaping, which has subsequently been removed where exposed.

Figure 7: Deposition (right foreground) and erosion (right bank distant) in the recreated Hurst Brook channel.

CONCLUSIONS

The experience gained on this project will significantly improve the team's ability to provide regulators and stakeholders with confidence that these scheme designs can be successfully constructed, at minimum cost, whilst realising sustainable benefits (Figures 8 and 9).

The work on Hurst Reservoir represents a development, building on the experience of an earlier, smaller scheme at Baystone Bank, completed in 2010. This increase in the knowledge and experience of ground engineering, environmental and construction professionals underlines that, properly executed, discontinuance is a credible, economic and sustainable approach to management of reservoir assets.

Stakeholder buy-in, resulting in only limited customer complaints, and the short programme ensured only one "game and native bird" nesting season was disrupted, alleviating the neighbouring moorland owner's concerns and minimising disturbance via the Golf Course access. This engineering project, restricted by significant environmental constraints, had minimal impact on the local population and Hurst Brook and reused all of the 103,000m³ excavated materials, which was a significant achievement when considering the limited information available on material properties at the commencement of the project.

Figure 8: View of Hurst Reservoir before Discontinuance.

Figure 9: Composite view of Hurst Brook valley post Discontinuance.

The 'Restoration of Hurst Brook' project provides UK panel and supervising engineers with a framework to seek advice on land contamination, integrated waste management along with visual and landscape impact assessment as part of the discontinuance process.

The major environmental and sustainable achievement of this project is the habitat creation and ecological restoration of the valley of the Hurst Brook. This will be further assessed in the future by evaluation of the effectiveness of the landscape management plan, using locally sourced seed and vegetation establishment on the upland areas. Information on the resulting local biodiversity and visual amenity value will be publicly shared.

In a paper Dr Andy Hughes (Hughes *et al*, 2008) stated that "the discontinuance of dams is often not a cheap or easy option". While it is challenging to demonstrate that discontinuance is a cheap process, the work at Hurst was certainly demonstrably the most economic solution and, although not easy, was designed and programmed to minimise impact and enhance the environment.

Discontinuance of an impounding reservoir, including embankment removal to regenerate the reservoir footprint, has rarely been undertaken in the UK.

The social, environmental and ecological benefits of restoring the Hurst Brook were economically delivered in terms of capital investment in the discontinuance project and future whole life cost.

REFERENCES

BGS (2006 & 1981). *Digital mapping & 1:50,000 scale, Sheet No.86 Glossop (Solid and Drift edition).* British Geological Survey, Nottingham

BGS (2013). *GeoRef website: Hebden Formation, Kinderscoutian.* British Geological Survey, Nottingham

Carter I (2008) *Report of an Inspection under Section 10 of the Reservoirs Act 1975, MWH Ltd Report No. 41514800.* MWH, High Wycombe, UK

CL:AIRE (2011). Code of Practice *'The Definition Of Waste: Development Industry Code of Practice 2011 Version 2.* CL:AIRE, London

Halcrow (2012). *EIA Environmental Statement for United Utilities Discontinuation of Hurst Reservoir and Restoration of Hurst Brook.* Halcrow, Swindon, UK

Hughes AK, Evans A and Galloway R (2008) 'The Discontinuance of Dams – it's not easy and it can be expensive!' in *Ensuring reservoir safety into the future. Proceedings of the 15th Biennial Conference of the British Dam Society.* Thomas Telford Ltd, London

MWH (2011a). *Geomorphology Desktop Assessment, Ref. 80019277.* MWH, High Wycombe, UK

MWH (2011b). *Hydrogeological Assessment Desk Study.* MWH, High Wycombe, UK

Reilly N (2006) *Report of an Inspection under Section 10 of the Reservoirs Act 1975, Report No E22311.* Pell Frischmann Ltd, London

Dams – Benefits and Disbenefits; Assets or Liabilities?
ISBN 978-0-7277-6133-0

doi: 10.1680/dbdal.61330.201

The Discontinuance of Sunnyhurst Hey and Improvements to Earnsdale Reservoirs, Darwen, Lancashire

S TENNANT, GHD Livigunn
Dr C D PARKS, United Utilities

SYNOPSIS Sunnyhurst Hey and Earnsdale Reservoirs are adjacent reservoirs situated on the north-western edge of Darwen Moor, some 1.5km west of Darwen town centre. Earnsdale Reservoir was constructed in 1863 and impounds 433Ml with a 300m long, 24m high embankment. A history of seepage and settlement and a more recent stability assessment, that indicated an insufficient factor of safety (FoS) on global stability, required mitigation measures. As a result a unique solution was formulated that incorporated a geogrid reinforced earth berm with a basket/rock facing, constructed on a geogrid reinforced load transfer platform that spanned between a number of deep soil mixed foundation cells and a filter that covers the downstream face of the dam. These were substantially completed in 2015 and form the main focus of the paper.

Sunnyhurst Hey reservoir was constructed in 1875 and formerly impounded 436Ml. The embankment, of 855m in length, suffered from a history of seepage, damp areas and soft ground at the toe. During investigations in 2008 the embankment was found to have no "core" and no "cut-off", contrary to historical information. Given the extent of the mitigation measures that would be required to bring the reservoir to an acceptable safety standard it was decided to permanently discontinue the reservoir.

INTRODUCTION AND OBJECTIVES

Earnsdale Impounding Reservoir is owned by United Utilities (UU) (the client) and regulated under the Reservoirs Act, 1975 (HMSO, 1975) and provides drinking water to about 100,000 customers in the Blackburn area. Following an investigation UU determined that the embankment did not meet a minimum criterion of 1:10,000 annual probability of failure; thus placing it in the 'intolerable' category on

UU's Portfolio Risk Assessment. Detailed geotechnical assessment in accordance with the 'Risk Analysis for Dam Safety' (UU, 2008) confirmed that remedial works were necessary to reduce, to an acceptable level, the risk of failure due to slope instability and/or internal erosion. The UU concept design was a toe berm to improve stability and a weighted filter, designed in accordance with the UU design guide (UU, 2012) to control seepage and prevent internal erosion.

Sunnyhurst Hey Reservoir is owned by UU and regulated under the Reservoirs Act 1975 and in tandem with Earnsdale Reservoir supported the supply of drinking water to customers in the Blackburn area. Investigations of the embankment confirmed concerns over stability and possible leakage through the embankment and foundation. Following the same geotechnical assessment method, it was determined that remedial works would be necessary to reduce the assessed risk of failure. For Sunnyhurst Hey the remedial measures required to maintain the reservoir in service were such that discontinuance was determined to be the preferred option.

In spring 2014 the client awarded a single contract, on a design and construct basis, to Askam Civil Engineering (the contractor) who engaged GHD Livigunn (the designer) to undertake the detailed design of the improvement works for both Sunnyhurst Hey and Earnsdale Reservoirs and to perform the role of Contractor's Geotechnical Advisor and be responsible for coordinating the various specialist design elements and providing on-site geotechnical supervision. The client appointed Mr Nicholas Reilly as Qualified Civil Engineer (All Reservoirs Panel) to oversee the design and construction of the improvement measures in accordance with the Reservoir Act 1975.

SUNNYHURST HEY IR DISCONTINUANCE

The discontinuance works at Sunnyhurst Hey Reservoir comprised the creation of a notch (Figure 1) in the western arm of the embankment in order to reduce its capacity such that it would no longer fall within the Reservoirs Act 1975. A proportion of the embankment fill material was reused at Earnsdale Reservoir as general fill behind the reinforced earth berm. A new pipe was constructed between the two reservoirs to permit the transfer of flows from the former Sunnyhurst Hey Reservoir to Earnsdale Reservoir. Sunnyhurst Hey Reservoir Basin has been engineered to form a new wetland habitat which will develop over the coming decades. Hughes *et al* (2008) state "...that the discontinuance of dams is often not a cheap or easy option", however, in this case it was determined to be the most economic solution.

Figure 1. Cutting the notch at Sunnyhurst Hey Reservoir

Geological setting

The 1:50,000 scale British Geological Survey sheet No.75 (Preston) shows the Sunnyhurst Hey reservoir to be underlain by bedrock comprising the Rough Rock sandstone of the upper Millstone Grit series (Namurian) to the north and west and by Pennine Lower Coal Measures siltstone and mudstone to the south and east. At the boundary between these rocks lies the Six Inch Mine coal seam and in close proximity but slightly lower in the sequence, the Sand Rock Mine coal seam. These are indicated to run in a south west to north east direction directly below the centre of the reservoir. While coal seams shown at higher elevations on Darwen Hill are indicated to have been worked, there was no indication of coal having been extracted from either the Six Inch or Sand Rock seams, although it is considered likely that drift mining of the seams has occurred. Superficial deposits are shown to be absent and there are no indicated faults below the footprint of the reservoir.

Ground investigations and studies

Willowstick Aquatrack and seepage and stability surveys were undertaken in 2007 which confirmed concerns with stability and leakage through the embankment and foundation. Soil Mechanics Ltd undertook the first known intrusive ground investigation of the embankment dam at Sunnyhurst Hey in 2008. The investigation entailed twelve cable percussion/rotary cored boreholes, six trial pits and six slit trenches. Boreholes were located in order to establish the construction of the embankment and downstream embankment shoulder at four cross sections. These were sunk through the embankment crest, through the downstream embankment shoulder, and through the foundation. Slit trenches were excavated across the embankment crest, to establish the presence and location of a puddle clay core and trial pits were excavated through the embankment toe and into the foundation to verify the location of the interface between embankment fill and natural founding material.

The results of the investigations indicated that the dam had no discernible core and no cut-off was evident. The geotechnical analysis that followed concluded that remedial works would be necessary to reduce the risk of failure due to slope instability and internal erosion to an acceptable level. Following further assessment the discontinuance of Sunnyhurst Hey Reservoir was deemed to be the preferred option.

A second investigation was then undertaken by Geotechnics Ltd in 2013 to establish the detailed ground conditions within the area of the embankment proposed for removal or notching. The investigation comprised three cable percussive boreholes through the embankment and five Terrier rig window sample boreholes located along the route of the proposed outfall pipe.

Sunnyhurst Hey discontinuance works design and construction

The design of the discontinuance works included construction of a number of new structures including:

- New overflow structure at the location of the breach approximately 8m below the original crest level of the reservoir,
- New pipeline connecting the overflow from the new weir level into Earnsdale Reservoir and
- New inlet chamber including screen and overflow structure that diverts normal inlet flows directly to Fishmoor Water Treatment Works with overflows falling into the empty Sunnyhurst Reservoir basin and ultimately over the new weir into Earnsdale Reservoir.

The design of the excavation works necessary to remove the notch of embankment material was also included in this package of work.

The excavation of the notch took place during the summer and autumn of 2014 and included the installation of the new pipeline and weir structure (Figure 2). Excavated soil was stockpiled for re-use at Earnsdale Reservoir as general backfill to the reinforced earth berm construction; the construction of the inlet structure and screen followed during 2015 with the Certificate of Discontinuance subsequently received from the AR Panel Engineer. Following drawdown and discontinuance, the remaining reservoir basin was levelled and landscaped to form a wetland area that will be allowed to develop as a wildlife habitat in the future.

Figure 2. Completed notch cut through Sunnyhurst Hey embankment (centre right) with new concrete inlet screen structure (background, centre left) and new concrete outlet weir (centre bottom).

EARNSDALE IR IMPROVEMENTS

Geological setting

The 1:50,000 scale British Geological Survey sheet No.75 (Preston) shows the Earnsdale site to be underlain by Glacial Till overlying rocks of the upper Millstone Grit series (Namurian). A sub-crop of the Holcombe Brook Coal seam is shown to have been proved along the westernmost bank of the valley, while the crop of the Bottom Brook is shown along the eastern bank. In view of the steepness of the valley sides and the indicated dip direction, it is considered that the Bottom Brook Coal seam may not extend below the embankment. An unnamed geological fault is shown to pass beneath the eastern end of the embankment. The fault is shown to down throw to the east and to form a boundary between the Holcombe Brook Grit and rocks of the younger Lower Haslingden Flags formation. By reference to the generalised vertical section, the magnitude of the throw is of the order of 80m.

Ground investigations and studies

Prior to undertaking ground investigations in 1975 it had been assumed that the embankment was constructed as a Pennine dam, i.e. with a puddle clay core and, possibly, with a concrete or clay infilled cut off trench. The 1975 investigation established that the dam was of homogeneous construction, with fill comprising silty clay, combined with gravel, cobbles, boulders, and small proportions of deleterious material. Borehole records also indicated that the embankment was constructed directly onto the original ground surface, without the prior removal of topsoil and vegetation.

Site investigations were carried out in 1997, 2009 and 2013 which further refined the ground model and established that the embankment fill material was a soft to firm sandy gravelly clay, with occasional pockets of yellow brown sand. The material is consistent with reworked glacial till and is considered likely to have been won from excavations within the reservoir basin. The embankment fill materials were underlain by soft to firm becoming stiff glacial till (clay). The thickness of glacial till was found to vary between 8m and 20.9m. Rockhead was proven in ten boreholes and comprised Coal Measures sandstone, mudstone and siltstone.

Slope stability analysis

A slope stability analysis was been carried out using slope stability software Geoslope Slope W (v.7.16). Eight cross sections were analysed and included a transect through the highest and steepest part of the embankment coincident with the centre of the embankment and the draw-off channel structure. The geotechnical parameters used in the analysis are given in the Table 1, below:

Table 1. Soil parameters used in slope stability analysis

Material Type	**ϕ'(°)**	**c' (kN/m²)**	**γ(kN/m³)**
Embankment Fill	31	0	20
Glacial Till	25	0	21.2
Glacial Sand and Gravel	36	0	18
Coal Measure Bedrock	45	0	22
Reinforced Earth Wall	45	30	19
Filter Drainage Layer	36	0	18

The piezometric surface modelled in the analyses was based on average water levels recorded between 1997 and 2009 together with more readings taken by the Headworks Controller throughout 2013. For the purpose of the stability analyses the piezometric surfaces modelled in each cross section were modelled passing through the embankment at top water level and falling linearly through the embankment materials, corresponding with monitored water levels observed in borehole piezometers closest to each section.

The analyses considered the 'critical' failure surface to be the first deep seated (>5m) slip affecting the integrity of the embankment crest close to, or upstream of, the wave wall and therefore resulting in a slip that would compromise embankment safety (Figure 3). The stability analysis indicates that, as could be expected, the highest areas of the embankment in the central third have the lowest FoS with the analysis indicating a FoS of 1.03 for a slip surface passing through the foundation and exiting through the base of the draw-off channel structure, which was clearly inadequate.

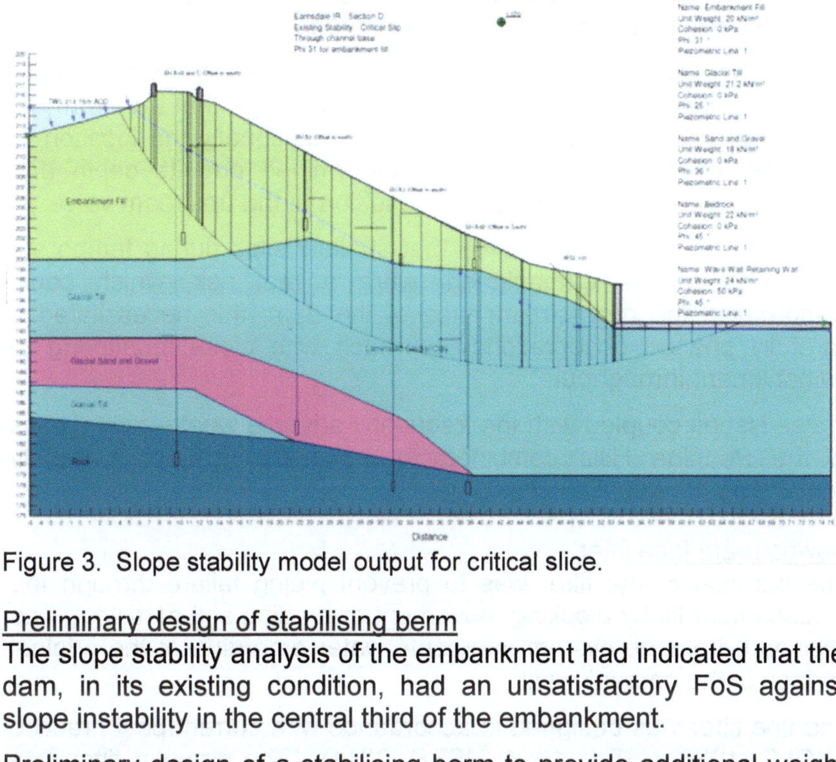

Figure 3. Slope stability model output for critical slice.

<u>Preliminary design of stabilising berm</u>

The slope stability analysis of the embankment had indicated that the dam, in its existing condition, had an unsatisfactory FoS against slope instability in the central third of the embankment.

Preliminary design of a stabilising berm to provide additional weight at the toe of the embankment was undertaken which indicated that, following installation of the berm, the FoS could be enhanced to 1.48 or greater. However there were a number of constraints that needed to be overcome in the design and build phase.

<u>Earnsdale design and construction</u>

The contractor was responsible for the detailed design of the stabilising berm and associated works which, at face value, seemed to be a relatively straightforward undertaking. However, investigations of the area around the toe of the embankment had revealed a difficult geotechnical situation requiring a challenging and complex geotechnical design solution.

Constraints

Access to the works was very restricted; at the base of the embankment access was only possible from the eastern flank and wide enough for plant to track in one direction only, furthermore an assessment of the existing walls deemed they had limited structural capacity and they could not support any surcharge or imposed loads in either the temporary or permanent condition. Loading to the crest access track was also restricted to light construction plant given the low FoS within the central third of the embankment.

The existing scour main runs through the middle of the works and it was a requirement of the permanent works design that the berm did not impart additional load on the main. Additionally the location of the main could not be accurately determined due its depth and difficulties with isolation due to the condition of the upstream valve.

Geotechnical analysis indicated that shallow slips during temporary excavation of the embankment were a real risk, which could compromise the downstream face of the dam; this necessitated a carefully phased observational approach with close monitoring of embankment throughout.

These issues coupled with the frequently adverse weather conditions in the Pennine Hills combined to make this project a hugely demanding construction challenge.

Downstream face filter

The purpose of the filter was to prevent piping failure through the embankment fill by blocking movement of eroding soil particles. The filter and drainage system intercepts water emerging at the original surface of the embankment.

The fine filter was designed in accordance with current best practice (ICOLD, 1986; USDA, 1994; USBR, 2011). The granular filter was designed to be placed directly onto the embankment fill following a topsoil strip.

The filter comprises primary and secondary filter layers, the latter also functioning as the primary drainage layer. The design of the secondary filter layer follows the same procedure, utilising the particle sizes of the primary filter as the base soil; thereby preventing erosion of the primary filter into the secondary filter layer. The relatively high permeability of the secondary filter layer encourages water to flow within the layer to the toe drainage. The filter and drainage layers were specified as a minimum thickness of 300mm (normal to the slope) and compacted to an end-product compaction specification using a long reach excavator with validation by nuclear density testing.

3D ground model

The designer utilised a 3D model (Figure 4) as a coordination and communication tool. The use of the 3D model enabled all organisations involved in the design and construction to appreciate the complexity and phasing issues associated with the construction. This supported the method statement and contingency plans so the works could be executed safely and also allowed the detail to be understood by non-specialists.

The model, created from the topographical survey using AutoCAD Civil3D and Keynetix Holebase to import geotechnical data, enabled rapid and accurate calculation of earthworks volumes which informed design decisions.

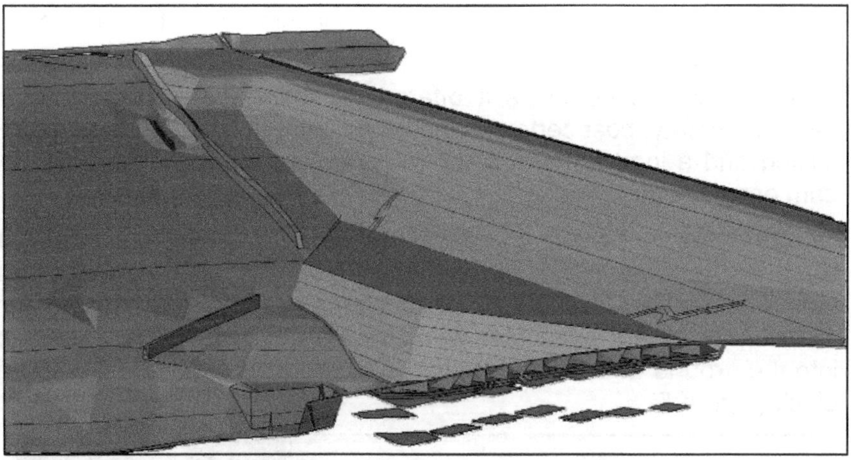

Figure 4. Rendered image from 3D earthworks model

Additional investigations

The contractor undertook additional ground investigation in the form of trial pits, window samples and dynamic probing. These confirmed earlier results that the ground conditions at the toe comprised a significant depth of normally consolidated cohesive made ground with insufficient bearing capacity to support the stabilising berm, and enabled these to be delimited. Static load testing and analysis demonstrated that under the maximum berm loading of 200kPa, settlements of more than 1000mm could be anticipated which were well in excess of the berm serviceability limit of 320mm.

Berm foundation

Following the geotechnical assessment of the ground model of the fill beneath the footprint of the berm, this information was inputted into the 3D model and an optioneering exercise undertaken. The options described in Table 2, below, were considered.

Table 2. Eco Wall Foundation Options

Ref	Description	Comment
1	Excavate and replace with granular fill	Significant temporary works required to safely excavate to depths of up to 6m.
2	Surcharge by phased construction.	Uncertainty over duration and prolonged exposure to weather; the predicted settlements would affect berm serviceability.
3	In situ ground improvement using deep soil mixing	Unproven for this application; large volume of treated material (1500m³) and associated cost.
4	Hit and miss deep soil mixing cells with load transfer platform	Volume of treated material reduced (600m³) weighted filter material incorporated into load transfer platform. Transfer platform enables bridging of the scour main.

Option 4 was selected as it offered the best solution in terms of reservoir safety, cost certainty and programme. The use of deep soil mixing and a load transfer platform represented another first for UK dam engineering.

Deep soil mixing is a commonly-used technique in Japan, the Far East, Scandinavian countries and more lately central Europe. It was introduced to the UK in 2012. The process uses an "Allu" mixing arm mounted on a 35T excavator which mixes cement binder and water into the ground at depths of up to 5m to produce a cement stabilised block (Figure 5).

Figure 5. Illustration of Deep Soil Mixing Plant and photograph of cell construction

Specialist sub-contractor Deep Soil Mixing was engaged by the contractor to undertake site trials and undrained triaxial tests. Three different mixes were tested using 10%, 15% and 20% CEM1 binder and tested at 7 days and 14 days. The tests demonstrated that a 10% binder mix would attain the required undrained shear strength of 110kPa at 14 days.

The load transfer platform (Figure 6) was designed in accordance with BS8006:2010 using the methods ibn Cl.8.4 "reinforced embankments over areas prone to subsidence", this method conservatively treats the area of soil between the DSM cells as a complete void. The platform comprised 2 layers of PET 1000 high strength geotextile incorporated within the 6T/6U weighted filter material. The strain within the geotextile was limited to 6% to satisfy the serviceability limit state of the berm.

Figure 6. Section through DSM cells showing transfer platform and scour main and photograph of installation.

<u>Stabilising berm</u>

The reinforced earth stabilising berm comprised a "TenCate EcoWall" retaining system reinforced with their Miragrid GX80/30 and GX55/30 geogrids, the front face of the wall slopes at 20° from the vertical; has a galvanised steel mesh front and is filled with dry stone (Figure 7). The overall length of the wall is 140m and it has a maximum retained height of 9.6m. A reinforced earthworks structure was used because a traditional toe berm would have provided insufficient mass to stabilise the embankment due the limited space available.

It is believed to be the first use of reinforced earthworks on a geotextile reinforced load transfer platform over DSM cells to improve the stability of an embankment dam in the UK.

Figure 7 Eco wall nearing completion

The following factors were adopted in the design:

- Long term creep factor of 64% of short term tensile strength (Pult)
- Manufacture and Extrapolation (fm) = 1.25
- Installation Damage (fd) = 1.05
- Environmental (fe) = 1.00

The wall was analysed in accordance with BS EN1997-1:2004 and BS8006-1:2010, Design Approach 1 and combinations 1&2 (Figure 8).

A major consideration was the stability of the existing embankment in the temporary condition given that the FoS of the central section was close to unity. This was tackled in 2 ways firstly the use of a good quality 6I granular material with a phi value of 41 degrees, and secondly to address the global sliding condition by extending the geogrid tails up the rear cut face. These measures reduced the base of the of the reinforced wall by 1.5m from 7.0m to 5.5m, this also reduced the extent of the berm foundation which is discussed separately in this paper.

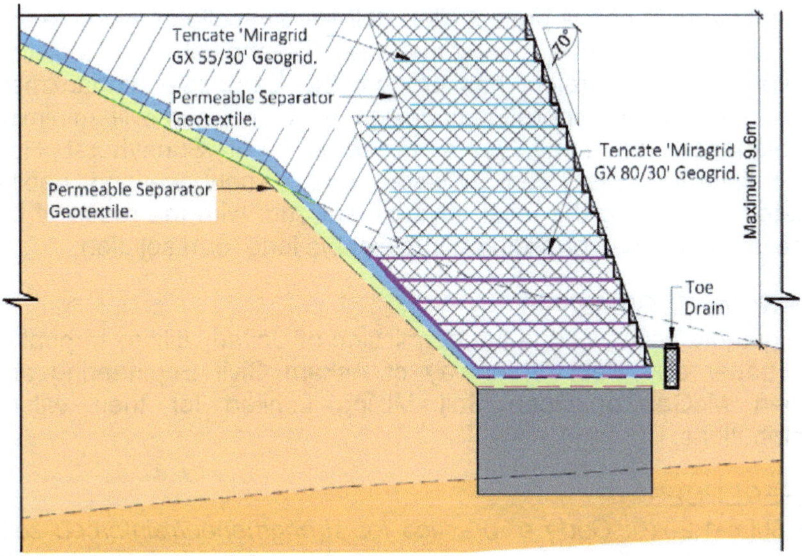

Figure 8 Schematic of Eco wall construction

CONCLUSIONS

March 2016 saw the successful completion of improvement works at Earnsdale and Sunnyhurst Hey. The work at Earnsdale included several ground-breaking technical innovations for UK dam engineering, specifically the first use of a reinforced earth wall, coupled with the use of deep soil mixing and a geotextile reinforced load transfer platform to improve stability of an earth dam (Figure 9). The reservoir safety improvements were implemented using the latest geotechnical standards and supported with a 3D earthworks model which coordinated and articulated the design issues and conveyed the phasing of works.

Figure 9. Eco Wall nearing completion

The many parts of this project were delivered in collaborative manner and its delivery has overcome many technical and logistical challenges. The project has brought the Earnsdale embankment within the desired operational design envelope and will minimise maintenance for many years to come. At Sunnyhurst Hey, discontinuance has been achieved and a new wetland habitat created, enhancing the local environment and, with the reuse of the excavated material, providing a sustainable long term solution.

ACKNOWLEDGEMENTS
The authors acknowledge the permission of United Utilities to publish this paper and Paul Greenway of Askam Civil Engineering and Robert McGall of Deep Soil Mixing Limited for their valued contributions.

REFERENCES

BS 8006-1:2010: *Code of practice for strengthened/reinforced soils and other fills*

BS EN 1997-1:2004: *Geotechnical Design: General Rules*

Hughes AK *et al* (2008) "The Discontinuance of Dams – it's not easy and it can be expensive!" in *Ensuring reservoir safety into the future. Proceedings of the 15th Biennial Conference of the British Dam Society, pp345-360.* . Thomas Telford, London

HMSO (1975). *The Reservoirs Act 1975.* HMSO, London

ICOLD (1986). *Bulletin No. 55. Geotextiles as Filters and Transitions in Fill Dams.* ICOLD, Paris

USDA (1994) *Part 633 National Engineering Handbook, Chapter 26 - Graduation Design of Sand and Gravel Filters.* United States Department of Agriculture Natural Resources Conservation Service, Washington, USA

UU (2012). *Weighted Filter Design Guide* (unpublished). United Utilities Water plc, Warrington, UK

UU (2008). *Risk Analysis for Dam Safety – A unified method for estimating probabilities of failure of embankment dams by internal erosion and piping guidance document, modified from the USBR, USACE, UNSW and URS manual by United Utilities for UU Embankment Dams* (unpublished). United Utilities Water plc, Warrington, UK

USBR (2011). *Design Standard No. 13 Embankment Dams, Chapter 5:Protective Filters.* United States Bureau of Reclamation, Washington, USA

Dams – Benefits and Disbenefits; Assets or Liabilities?
ISBN 978-0-7277-6133-0

doi: 10.1680/dbdal.61330.215

Insights into the composition of Pennine type dams – experiences from two reservoir discontinuances

T SHERIDAN, Mott MacDonald, Sheffield, United Kingdom
V KING, Mott MacDonald, Sheffield, United Kingdom
D REBOLLO, Mott MacDonald, Leeds, United Kingdom
A JACKSON, Mott MacDonald, Leeds, United Kingdom

SYNOPSIS In recent years there has been an increased demand for a better understanding of the composition of our portfolio of 19th century embankment dams, with work being carried out to safeguard these structures and the communities who live around them. During works to discontinue two of these dams, investigations were undertaken to better understand the composition of the dams and gain an increased understanding of the potential risk of internal erosion, particularly associated with conduits through the foundation and other defects that may be present. A comparison has been made against the model of a typical 19th century Pennine type embankment dam.

INTRODUCTIONS AND OBJECTIVES

Beaver Dyke impounding reservoir is located approximately 7km west of Harrogate, West Yorkshire. Oakdale impounding reservoir is located approximately 2km southeast of Osmotherley within the North Yorkshire Moors. Due to the costs involved in upgrade works, the commercial decision was made to discontinue the reservoirs and return them to natural channels. Mott MacDonald was the designer on behalf of the undertaker Yorkshire Water. The works, completed by contractor JN Bentley, comprised excavating a v-notch through each embankment dam.

Both reservoirs are believed to have been constructed as typical 19th century Pennine type embankment dams. These are generally earthfill dams with a central puddle clay core, often with a cut off trench to a layer of less permeable strata beneath the dams.

Specifications for zoned fill construction became common practice from 1854 (Rigby *et al*, 2014).

Figure 1. Location of reservoirs (Basemap copyright © OS Opendata 2016)

Observations of the clay core and shoulders were made at the time of the excavation works and an investigation was planned to analyse samples of the embankment to gain insight into the material composition and properties of the dam construction materials.

This paper provides details of the investigation, the results of material testing and insight into the embankment dam construction. In both cases the results have confirmed that the dams are typical of the 19th century Pennine type embankment dam, although zoned fill construction of the embankment shoulders was not conclusively evident. At both dams the materials in the embankment shoulders were found to be generally fine grained and cohesive, however, there was evidence of coarser grained material layers or pockets that may have been susceptible to internal erosion.

BACKGROUND

<u>Beaver Dyke</u>

Construction

The available records suggest that the reservoir was constructed around 1890 as a typical 19th century Pennine type embankment dam. The embankment itself rose 16m above natural ground level with a minimum crest level around 170.5mAOD. On the upstream side the embankment had a gradient of approximately 1v in 3h. The downstream face was grassed with a gradient of approximately 1v in 2h. There was a single berm, 6m wide, located approximately 10m below the crest level. Below the berm, the embankment has a gradient of approximately 1v in 3h. The clay core was believed to be central to the embankment crest with a 12v in 1h gradient either side.

Records of modifications and remedial works indicated that the embankment crest and clay core were raised to restore freeboard in 1973. Since 1990 there have been a number of repairs, mostly associated with the spillway.

Geology

BGS 1:50 000 geological mapping showed no superficial deposits within the footprint of the reservoir; glacial till deposits were shown to the south of the site. The mapping indicated the bedrock geology is a sequence of sandstones, siltstones and mudstones of the Millstone Grit Series, Carboniferous in age. A fault was marked at the eastern end of the reservoir, running north to south.

Historical cross sections of the reservoir geology were provided to the designer by an adjacent landowner (original source unknown) and reproduced in Figure 2 below.

Figure 2. Historical geological cross section across Beaver Dyke Valley

The cross section indicated that the reservoir and slopes comprise predominantly sands, gravel and clay deposits, (marked 1 – 6), with 'blue stony clay' (no. 4) shown beneath the reservoir in the bottom cross section. The superficial deposits and weathered rock were underlain by blue shale (no. 7), brown shale (no. 8) and "hard rock" (no. 9), likely to be sandstone. Thin bands of coal (black) were marked on one of the sections. The cross sections also show a number of small faults.

Oakdale

Construction

Fewer records are available for this dam. It is believed that the reservoir was constructed around 1914, following a typical 19th century Pennine type embankment dam design. The embankment itself rose 18m above natural ground level with a minimum crest level around 184.7mAOD. On the upstream side, the embankment had a gradient of approximately 1v in 2.7h. The downstream face was grassed with a gradient of approximately 1v in 2h, although this steepened at the crest. No record of the presence or position of the clay core was available. It is believed that ongoing settlement following construction led to major works to restore freeboard however few records of these remedial works remain and the date of the works is unknown.

Geology

BGS 1:50 000 digital mapping shows no superficial deposits at the reservoir. The mapping shows the bedrock geology is a sequence of sandstones, siltstones and mudstones of the Saltwick Formation, Jurassic in age. A fault is marked at the western end of the reservoir, running approximately north to south.

SITE OBSERVATIONS

Beaver Dyke

During the excavation works the clay core was exposed in the embankment. The boundary between the core and the embankment fill was clearly observed due to the distinctive change in colour and material type, as can be seen in Figure 3 below. The core was observed as linear and there was no evidence of lateral deformation caused by movement or settlement of the embankment.

In contrast, the material exposed in the embankment shoulders was variable in composition. It predominantly comprised cohesive materials, clay and silt, with varying quantities of gravel and sand. The clays were blue grey and yellow brown in colour suggesting two sources, although the gravel was predominantly sandstone.

Figure 3. V-notch excavation through Beaver Dyke showing clay core

Oakdale

The material exposed in the excavations of the dam at Oakdale generally comprised a sandy, gravelly clay and silt within the embankment shoulders with a central puddle clay core. The margin of the clay core was well defined (see Figure 4) and lateral deformations were not evident. The material of the shoulders appeared relatively homogenous while distinct zoning of materials was not apparent.

Figure 4. V-notch excavation through Oakdale showing clay core

INVESTIGATION AND TESTING

Methodology

Bulk samples of the embankment fill material were taken during excavation of the V-notch, across the clay core and upstream and downstream shoulders of both embankments. The locations of the samples are shown for Beaver Dyke and Oakdale in Figure 5 and Figure 6 respectively.

Beaver Dyke

At Beaver Dyke the sampling locations were determined in order to investigate the nature of materials upstream and downstream of the clay core and also at different elevations in the embankment as illustrated in Figure 5 below. These included five samples at regular intervals along the clay core; a grid across the southern half of the embankment, both upstream and downstream of the clay core at 3m, 6m and 10m below top water level (TWL). Additionally, two samples were taken from materials surrounding the scour pipe and supply pipe at approximately 156mAOD, immediately downstream of the clay core.

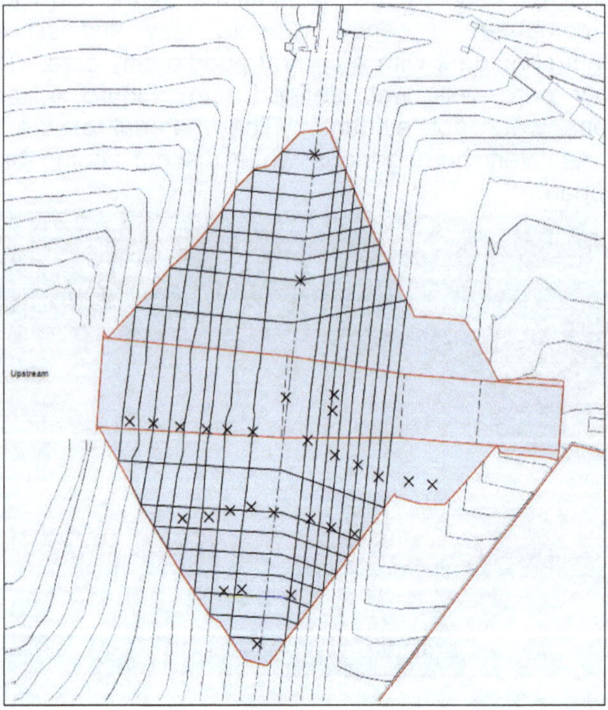

Figure 5. Sampling locations across embankment at Beaver Dyke

Oakdale

The selection of sampling locations at Oakdale was more limited due to difficulty safely accessing the embankment. Six samples were collected on the left hand side of the V-notch excavation including three samples from the downstream shoulder; one sample from the core; and two samples from the upstream shoulder.

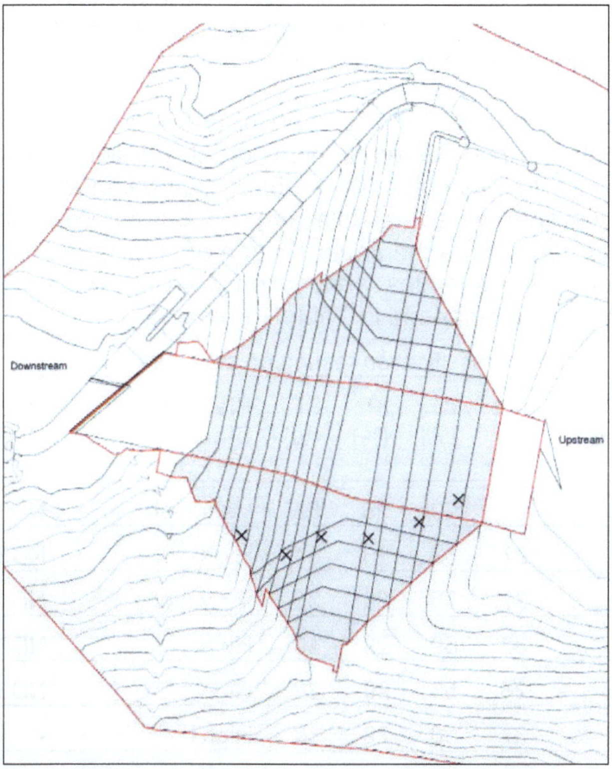

Figure 6. Sampling locations across embankment at Oakdale

Testing

Geotechnical testing was undertaken on the samples to determine the material properties and gain an understanding of the construction of the embankments. The testing comprised:

- grading analysis by sieving and hydrometer;
- plasticity testing (Atterberg Limits);
- determination of dispersibility (Pinhole test); and
- permeability.

RESULTS

Clay core

Grading

Grading tests were undertaken on three samples of the clay core at Beaver Dyke and one sample at Oakdale. The results show a similar range of particle sizes with 83% to 93% of the material sampled having a particle size less than 0.063mm (clay).

Plasticity and determination of dispersibility (Pinhole test)

Atterberg limits and determination of dispersibility testing was undertaken on five samples (01 – 05) of clay taken from the core at Beaver Dyke and one at Oakdale (06). At Beaver Dyke samples 01 and 05 were taken from the upper section of clay core, raised in 1973 and the remaining samples were taken from the original clay core. The results are summarised in Table 1 below.

Table 1. Results of classification testing on clay core

No.	Depth below TWL	mc (%)	IL (%)	IP (%)	PI (%)	Plasticity Classification	Dispersibility Classification
05	-0.30	35	79	26	53	Very high	ND3
01	0.06	27	53	20	34	High	ND3
04	5.94	37	75	33	33	Very high	ND1
02	6.01	33	69	29	40	High	ND2
03	11.76	26	53	25	28	High	ND3
06	Oakdale	37	58	22	36	High	-

The results indicate that the core materials comprise high to very high plasticity clay. Of the samples tested from Beaver Dyke, the properties of the raised section of clay core and the original clay core all fall within a similar range in terms of the plasticity index, liquid limit, plastic limit and moisture content, although there is not a distinct relationship between the plasticity properties with depth.

Pinhole dispersibility tests were also undertaken on samples from Beaver Dyke and as per the classifications in BS1377-5:1990 Table 2 (reproduced in Figure 7 below) indicate that the dispersibility of the original clay core varies between Class ND1 and Class ND3 (ND = non-dispersive), decreasing with depth. The samples from the raised clay core are both Class ND3. For a new dam, a clay material attaining to ND1 would be specified.

Figure 7. Dispersive classification of soils

Dispersive classification	Head	Test time for given head	Final flow rate through specimen	Cloudiness of flow at end of test		Hole size after test
				from side	from top	
	mm	min	mL/s			mm
D1	50	5	1.0 to 1.4	dark	very dark	≥ 2.0
D2	50	10	1.0 to 1.4	moderately dark	dark	> 1.5
ND4	50	10	0.8 to 1.0	slightly dark	moderately dark	≤ 1.5
ND3	180	5	1.4 to 2.7	barely visible	slightly dark	≥ 1.5
	380	5	1.8 to 3.2			
ND2	1 020	5	> 3.0	clear	barely visible	< 1.5
ND1	1 020	5	≤ 3.0	perfectly clear	perfectly clear	1.0

Permeability

Two permeability tests were undertaken on samples of the clay core from Beaver Dyke in a triaxial cell. The samples were recompacted at the 'as dug' moisture content using a 2.5kg rammer. The results indicate a permeability of between 6.8 and 5.8 x 10^{-11} ms^{-1} which is considered very low. It should be noted that whilst the permeability of the clay will be greatly influenced by the sample preparation/ remoulding, due to puddle clay construction methods, the samples tested are likely to be similar in consistency to the clay core when constructed. For new material in a clay core, a permeability of 10^{-9} ms^{-1} would be specified.

Chemical testing

Chemical testing was undertaken for a suite of metals. The results indicate very similar low levels of metals in both samples tested.

Embankment shoulders

Grading

The grading results of the embankment shoulders are shown in Figure 8. The results show that the materials in the two embankments were predominantly composed of silt/clay with a varying quantity of sand and gravel sized particles. The materials on the upstream side of both embankments have a wider range of particle sizes and are coarser grained, with the downstream materials composed of a higher proportion of clay and silt sized materials, with the exception of those sampled within the discontinuance channel area.

The results show that the materials in the two embankments were predominantly composed of silt/clay with a varying quantity of sand and gravel sized particles. The materials on the upstream side of both embankments have a wider range of particle sizes and are coarser grained, with the downstream materials composed of a higher proportion of clay and silt sized materials, with the exception of those sampled within the discontinuance channel area.

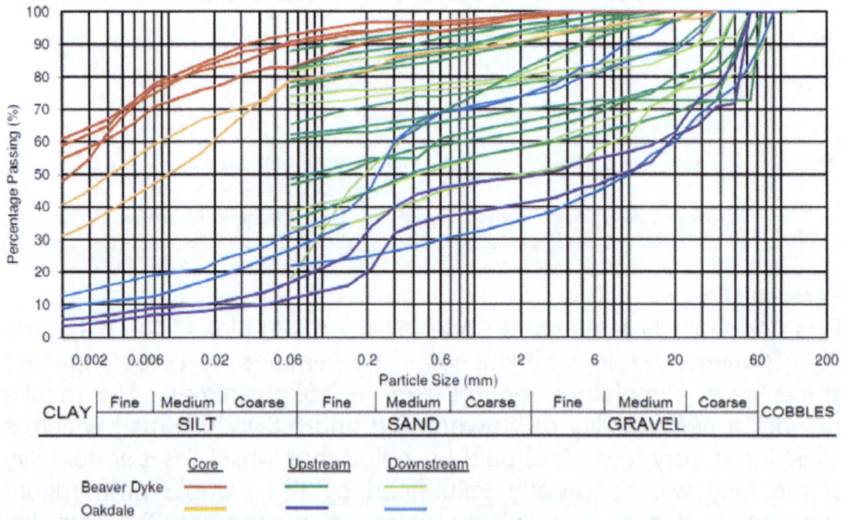

Figure 8. Grading of the embankment materials

There is no distinct pattern relating to the material and distance from the clay core, although the material appears to become coarser grained with depth. This may be a design feature but is more probably the order of materials dug from the borrow pit. At level 160mAOD the material is more variable in grading, but overall, the fill materials can be described as predominantly fine grained.

Scour pipe and supply pipe

Grading

At Beaver Dyke additional excavations below the discontinuance channel were undertaken to examine the condition of the scour and supply pipes and the condition of the surrounding material. The results indicate a well graded material comprising sandy gravelly silt with some clay. The materials are similar in grading to those of the embankment shoulders. There was a zone of firm to stiff clay (slightly gravelly) directly surrounding the pipes. The clay appeared to be puddled. There were no signs of degradation of the pipes themselves or of the surrounding materials and no evidence of internal erosion could be observed.

DISCUSSION

The two dams were built within a 30 year period and suffered similar problems with ongoing settlement in the years following construction.

At Beaver Dyke the embankment core was constructed of very low permeability clay that had a high to very high plasticity. This was raised in 1973 and the material used was similar in composition. It is likely that the clay was locally sourced as historical records indicate that clay deposits of yellow clay and strong brown clay were present in the area at the time of construction, likely to be the glacial till mapped on the slopes to the south of the site.

The embankment shoulders were constructed of clay and silt with varying quantities of sand and gravel. They included both blue-grey and yellow-brown clays. The material varied in grading, but the gravel and sand were predominantly composed of sandstone. It is likely that this material was sourced locally and included the blue gravelly clay, yellow clay and sands and gravel that are present in the area to the south of the reservoir. The inconsistency in the proportion of granular materials is due to local variations in the source material. Based on the material testing, it does not appear that the embankment fill was graded or zoned to have specific properties, although the upstream embankment was more granular in nature.

At Oakdale the embankment core was constructed of high plasticity clay. The downstream embankment shoulders would appear to have been composed of very clayey sand with coarser very clayey very sandy gravel in the upstream embankment.

The pinhole dispersion tests on samples of the clay core show variability between ND1 and ND3. The variation in dispersibility is likely to be due to the nature of the source material, as glacial till is naturally variable. Puddle clay specifications for new reservoir dams typically specify dispersion class ND1. At both dams the upstream shoulder materials were coarser than that in the downstream shoulder. This may have been due to the use of selected material, however a notable difference in the material was not observed at either site and it is possible that the finer materials have been washed out over time.

The results of the investigation confirm that Beaver Dyke and Oakdale reservoirs were both typical 19th century Pennine type embankment dams. Although they lacked the wide core, transitions and drains that could be expected in more recent examples of earthfill embankment dams (Novak *et al,* 2001) they did show clear evidence of the selection of low permeability materials to form the clay cores with coarser material used for the shoulders. Specifications for zoned fill construction became common practice in

1854 (Rigby *et al*, 2014), although this zoning was not evident at either Beaver Dyke or Oakdale.

Both dams had suffered from settlement after construction but neither showed signs of lateral deformation in the core or any significant evidence of internal erosion. Are these 19th century dams as high a risk as we assess? Or is it time to take a new approach when considering the risk of failure?

REFERENCES

BSI (1994). *BS1377-5:1990 Methods of test for Soil for civil engineering purposes – Part 5: Compressibility, permeability and durability tests,* Table 2. British Standards Institute, London.

Novak P, Moffat AIB, Nalluri C and Narayanan R (2001) *Hydraulic Structures.* London. Spon Press 2001.

Rigby PJ, Thompson AN, Jones DE (2014). 'The Nearly Perfect Nineteenth Century Embankment Dam' in *Maintaining the Safety of our Dams and Reservoirs. Proceedings of the 18th Biennial Conference of the British Dam Society.* pp88-100. ICE Publishing, London.

Dams – Benefits and Disbenefits; Assets or Liabilities?
ISBN 978-0-7277-6133-0

doi: 10.1680/dbdal.61330.227

Discontinuance of Small Reservoirs in Scotland

L J DUNNE, Atkins Limited
R MORRIN, Scottish Water

SYNOPSIS Much attention has been given to the discontinuance of small reservoirs in Scotland recently, for two main reasons:

- The phased implementation of the *Reservoirs (Scotland) Act 2011* (TSO, 2011a)
- The practical need for Reservoir Managers (referred to as "*Undertakers*" in the *Reservoirs Act 1975* (HMSO, 1975)) to reduce risk, and manage their long term financial and resource commitments associated with operating and maintaining small reservoirs.

This paper examines the processes involved in discontinuing small reservoirs and highlights some of the key challenges encountered on recent discontinuance projects.

Case studies are presented which relate to the discontinuance of five small reservoirs that are no longer used for their original purpose of water supply and, following assessment, each reservoir has been considered to represent a "liability" rather than an "asset". Discontinuance has been identified as providing the most appropriate long term solution for these reservoirs.

Many small reservoirs provide a degree of flood attenuation which protects downstream communities. Conversely, they can also represent a significant risk to downstream communities in the event of an uncontrolled release of water. Therefore, discontinuance of small reservoirs requires careful planning, consultation and investigation. The process can be lengthy and relatively expensive.

LEGISLATION CHANGES

Reservoir safety regulation is changing throughout the UK. In Scotland, the *Reservoirs (Scotland) Act 2011* will soon replace the *Reservoirs Act 1975*. When fully implemented, the *Reservoirs (Scotland) Act 2011* will introduce a risk-based regulatory regime based on the consequences of an uncontrolled release of water on downstream receptors.

This new legislation will bring about a number of key changes, one of the most significant of which includes the reduction in threshold for regulation of controlled reservoirs from 25,000m³ to 10,000m³ capacity. This means that smaller reservoirs in Scotland, which have previously been exempt from the provisions of the *Reservoirs Act 1975,* will be brought under regulation for the first time.

Many of these smaller, previously unregulated, reservoirs have fallen into poor states of repair due to lack of financial investment over the years. However, with these smaller reservoirs now coming under regulatory control, many Reservoir Managers are considering the long-term future of these reservoirs.

RISK DESIGNATIONS

Under the *Reservoirs (Scotland) Act 2011*, each reservoir will be classified according to whether it poses a threat to human health, economic activity, environment and cultural heritage, should it fail. The probability of failure will also be taken into account.

Risk designations of "high", "medium" or "low" will be assigned to each registered reservoir by SEPA based on the consequences of an uncontrolled release of water.

Depending on the risk level assigned, this may result in significant changes for Reservoir Managers as they maintain the safety of their small reservoirs. Reservoir Managers regulatory duties for each risk categorisation are summarised in Table 1.

Table 1. Reservoirs (Scotland) Act 2011 – Regulatory duties imposed upon Reservoir Managers

Risk	Description
High	Reservoir Manager must appoint a Supervising Engineer at all times. Reservoir Manager must appoint an Inspecting Engineer at least once every 10 years (or when recommended by the Supervising Engineer)
Medium	Reservoir Manager must appoint a Supervising Engineer at all times. Reservoir Manager only required to appoint and inspecting Engineer when recommended by the Supervising Engineer
Low	Reservoir Manager has no statutory requirements to appoint either a Supervising Engineer or Inspecting Engineer

PROACTIVE APPROACH

Provisional risk designations for smaller reservoirs within the $10,000m^3$ to $25,000m^3$ capacity range have not yet been designated by the Enforcing Authority, SEPA.

However, as a Reservoir Manager with one of the largest stocks of reservoirs in the UK (Table 2), Scottish Water has taken a proactive approach to identifying and managing its stock of "small" reservoirs in preparation for full implementation of the *Reservoirs (Scotland) Act 2011.* Atkins (the Designer) is assisting Scottish Water achieve its aim of reducing risk and long-term financial and resource commitments across its portfolio of dams and reservoirs, through:

- identifying and prioritising "redundant" reservoirs with capacity in the range $10,000m^3$ – $25,000m^3$
- undertaking studies and investigations to determine the extent of work required to bring them up to current standards
- appraising the relative merits of upgrading or discontinuing many reservoirs
- designing discontinuance works and / or upgrade works, as appropriate

Table 2. Scottish Water's Reservoir Portfolio

Reservoir Capacity	No. of Reservoirs
$10,000m^3$ – $25,000m^3$	31
$>25,000m^3$	264

The majority of the smaller reservoirs owned by Scottish Water are no longer used for water supply purposes and are considered to be "redundant". It is recognised that many of them have suffered from lack of investment over the years. Scottish Water is keen to embrace the risk-based approach to reservoir safety on these smaller reservoirs, whilst continuing to manage its full reservoir portfolio within available budget. Discontinuance can often represent the best long-term option for those which are redundant.

In many cases, various organisations are interested in purchasing these unused assets. Potential new owners include fishing clubs, private hydro scheme developers, local authorities and other outdoor amenity providers. As a responsible organisation, Scottish Water will not sell the reservoir until it has been made safe or discontinued, such that the reservoir is incapable of holding $10,000m^3$ of water above the natural level of any part of the adjacent land.

CASE STUDIES: DISCONTINUANCE - APPRAISAL, DESIGN AND CONSTRUCTION

Scottish Water has identified and prioritised the following reservoirs for discontinuation consideration through a rigorous redundant reservoirs programme (Judge, 2015). In each case, investigative studies have been undertaken or are ongoing and options appraisals carried out to determine whether discontinuance is appropriate.

Such initial studies and investigative works have included: flood studies; topographical and bathymetric surveys; silt surveys; ground investigation; environmental, and heritage studies. Early contractor and stakeholder involvement has been found to be particularly beneficial in identifying the key project risks, constraints and concerns.

Case Study 1: Bowling Reservoir – Discontinuance Design & Construction Challenges

Bowling Reservoir is located 4.5km east of Dumbarton, Scotland. It is triangular in plan, has a capacity of 11,000m³, and was originally constructed for water supply purposes. The impounding reservoir is retained by an earth embankment 7m high and 100m long. It is no longer used for water supply and is currently in very poor condition (Figures 1 and 2).

The original shaft spillway with bellmouth opening has previously been demolished and/or vandalised and the valve tower access bridge has fallen into a poor state of repair, such that it can no longer be used (Figures 3 and 4).

Flows continue to enter the reservoir from the small upstream catchment and the only means of discharging water from the reservoir is via a 200mm diameter scour pipe located in the base of the reservoir.

There is currently no formal spillway through the dam and water levels in the reservoir fluctuate on a regular basis. It has been calculated that flows exceeding the 1 in 10 year flood are likely to cause the reservoir to overtop.

There is also no means of recording water levels at present and the recent introduction of weekly observations by the Reservoir Manager has indicated that the water level approached crest level on several occasions during 2015. The ground along the toe of the dam is particularly soft and damp and it appears that some of the old redundant pipework through the embankment may be leaking.

Therefore the embankment dam is considered to be at risk of failure from overtopping and / or internal erosion.

Figure 1: reservoir partially full

Figure 2: reservoir empty

Figure 3: dam crest

Figure 4: drawdown pipework

A recent flood study determined there to be no significant downstream flood impact (due to loss of attenuation) should the reservoir be discontinued, and, following options appraisal, it was clear that the best long term solution for his redundant reservoir was discontinuance.

Detailed discontinuance design was undertaken which involved excavating a full height V-notch opening in the dam; reinstatement of the watercourse through the reservoir basin; silt treatment; and the reuse of the excavated material to infill and re-profile the reservoir.

Temporary works have now commenced, and full discontinuance work is programmed for spring / summer 2016. The Reservoir Manager and Designer are currently working with contractors and stakeholders to finalise construction methodologies and techniques such that environmental impacts are mitigated whilst ensuring works are carried out in a safe manner. Key challenges encountered to date are discussed below.

Challenge 1: Difficult Access

Bowling Reservoir is located near the top of a steep hill. Vehicular and plant access to the foot of this hill requires negotiation of narrow, private roads with tight bends, bridges/culverts with limited load capacity and a small watercourse with ford crossing, Then a steep ascent for approximately 1km through undulating hillside with numerous rocky knolls and rock outcrops is required, making plant access extremely difficult.

The design solution and construction methodology has been tailored to accommodate this, through maximising the re-use of available on-site material and limiting the amount of site deliveries and off-site disposal. For example, stone pitching retrieved from the V-notch excavations together with stone from dilapidated walls at the downstream toe of the dam and masonry/brick from redundant chambers are being used to line the reinstated watercourse and provide erosion protection as flows are conveyed through the V-notch opening in the dam.

Similarly, all excavated material from the embankment dam will be reused on site to infill the reservoir basin and re-profile the surrounding land. All silt will be treated in situ to avoid the need for off-site disposal.

Temporary works are required to create a safe access road to the reservoir. All-terrain vehicles will be used to transport personnel, plant, equipment and materials in limited quantities to reservoir.

Challenge 2: Limited Working Area
The triangular shaped reservoir can only be accessed by plant via the embankment itself as it is bounded on the northeast side by dense forestry up to the edge of the reservoir, and bounded on the southeast side by a near-vertical rock face.

The available working area for material stockpiles and plant manoeuvres is limited. Therefore, the Reservoir Manager, Designer, Contractor and stakeholders are currently working together to develop a phased construction approach which maximises available areas downstream of the dam for silt treatment and stockpiles, whilst ensuring the work can be carried out safely, achieving long term stability at the discontinued reservoir, and mitigating environmental impacts.

Challenge 3: Reservoir Draw-down
The reservoir has limited drawdown capacity and the only means of lowering the water level is via a single 200mm diameter cast iron scour pipe. In order to reduce the risk of dam overtopping, and facilitate the gradual lowering of reservoir water levels prior to construction, a temporary bypass pipe was installed to divert flows

from the small incoming watercourse around the reservoir to the receiving watercourse.

This temporary arrangement is monitored by Scottish Water operational staff a regular basis and emergency pumps can also be brought to site in the event of flood conditions, if required.

Challenge 4: Silt Management

The biggest challenge encountered in relation to discontinuance of Bowling Reservoir is the treatment of 2,500m³ silt which is currently retained in the base of the reservoir.

The silt is 0.45m to 1.0m deep and if left untreated following controlled reservoir breaching could pose a risk to public health and safety and could contribute to an environmental pollution incident.

Various silt treatment options and methods have been considered including:

- Removal of silt from reservoir basin to on-site treatment lagoon system with drying area
- In situ chemical silt treatment and consolidation
- Creation of partitioned cells and buried retaining structures in the reservoir basin to facilitate infilling and restoration

Site trials are underway at present to ensure the correct approach is adopted, which minimises any short and long-term environmental impacts. The outcome of these trials will inform the final design details and construction sequence/methodology and ensure compliance with licence requirements under the *Water Environment (Controlled Activities) (Scotland) Regulations 2011*. (TSO, 2011b)

Challenge 5: Land Issues

Scottish Water owns the footprint of the reservoir and has servitude access to the reservoir for operating, maintenance and emergency purposes. However, it does not own this access route or any of the surrounding land, which is currently used for livestock grazing.

There has been a significant and complex consultation process involving landowners, affected parties, interested buyers and stakeholders. Agreeing land access issues can be a lengthy process and early consultation with affected partied is advised.

Case Study 2: Greenlands Reservoirs Nos 1, 2 and 3

This case study describes an optioneering study and strategy development for the long term management of three small reservoirs, located in Dumbarton, Scotland. Key details of each of the dams are provided in Table 3 and illustrated in Figures 5-8.

Table 3. Key Details – Greenlands Reservoirs Nos. 1, 2 & 3

Reservoir	Dam Type	Height (m)	Length (m)	Capacity (m^3)
Greenlands No.1	Earth Embankment	4	50	15,500
Greenlands No.2	Earth Embankment	9	93	23,750
Greenlands No.3	Earth Embankment	10	70	19,000

The three reservoirs are interconnected with Greenlands No. 1 and Greenlands No.2 both discharging separately to Greenlands No.3. Greenlands No. 1 can also indirectly supply water to a separate reservoir system in the adjoining catchment, if required.

Greenlands Reservoir No.1, No. 2 and No.3 are all considered redundant assets, and as such, Scottish Water no longer maintains them for water supply. Access to each of these reservoirs is also relatively difficult, such that very little inspection, monitoring and maintenance is carried out at present.

Figure 5: Greenlands No.1 dam crest

Figure 6: Greenlands No. 2 spillway

Figure 7: Greenlands No.3

Figure 8: Greenlands No.3 crest

One interesting point regarding Greenlands Reservoir No.2 is that, like many other small reservoirs, it has previously been certified as discontinued under the *Reservoirs Act 1975*, when its capacity was decreased from 43,000m³ to 23,750m³ through spillway crest lowering. However, together with Greenlands Reservoir Nos. 1 and 3, this reservoir will shortly be subject to the requirements of the *Reservoirs (Scotland) Act 2011*, having a capacity of greater than 10,000m³.

The purpose of the study was to consider options to reduce Scottish Water's liability, risk, cost and maintenance/operational costs associated with these three redundant reservoirs and develop a strategy for their long term management. The study considered whether one or more reservoirs could be discontinued and identified the extent of works which would be required to restore the reservoirs to an acceptable standard.

A detailed flood study was undertaken at the outset to determine the flood impact to downstream communities should one or more of these reservoirs be discontinued such that their individual volume was less than 10,000m³. Eight different combinations of dam removal or spillway lowering were modelled and analysed and flood maps were prepared to compare pre and post-discontinuance.

The flood study concluded that the complete removal or discontinuance of Greenlands Reservoir No.2 was shown to have the greatest effect on downstream flood risk as it currently provides the greatest individual contribution of all three reservoirs to attenuating downstream flows. By comparison, Greenlands Reservoirs Nos. 1 and 3 provide relatively low levels of flood attenuation.

In order to provide the same level of flood protection to downstream receptors, approximately 15,000m³ flood storage capacity would need to be retained within the Greenlands reservoir system.

Four options were considered and appraised within this study:

Option 1: Retain & Upgrade Greenlands Nos. 1, 2 & 3 – All reservoirs made safe and further deterioration is prevented. However, Scottish Water has long term management, monitoring, supervision and inspection commitments when regulated under the *Reservoirs (Scotland) Act 2011*.

Option 2: Discontinue Greenlands Nos. 1, 2 & 3 – This option was ruled out at an early stage as the increase in flood risk to downstream receptors was not considered acceptable.

Option 3: Discontinue Greenlands Nos. 1 & 3 – Once discontinuance certificates are granted Scottish Water will be relieved of its duties under the Reservoirs Act. However, in order to mitigate any potential downstream flooding impacts, Greenlands No. 2 will require to be modified to accommodate additional flood storage of approximately 5,000m³, through reduction of the spillweir level and overflow channel.

Option 4: Discontinue Greenlands Nos. 1, 2 & 3 and create Flood Storage Reservoirs - all three reservoirs can be discontinued. However, to mitigate downstream flood risk two of the reservoirs will need to be converted to flood storage reservoirs, retaining a combined volume of 15,000m³. The disadvantage of this option is that these flood storage reservoirs will need to be maintained as such. The Local Planning/Flood Risk Authority may require agreements to be put in place to ensure this area cannot be developed for any other reason in future, thus potentially affecting its attractiveness to potential third party buyers.

The options appraisal considered a number of key issues for each reservoir, such as: reservoir safety; existing structural stability; the extent of restoration works required to bring it up to standard; downstream flood risk; environmental impact; heritage issues; visual impact; constructability; sustainability; health and safety; stakeholder requirements; current/future amenity value; site accessibility; long term maintenance and monitoring commitments; and cost (construction, operational, maintenance, supervision and inspection).

High level, indicative construction cost estimates were undertaken for all options. Costs ranged from £0.7M (upgrade works) to £2.5M (discontinuance), demonstrating that discontinuance is not always a low cost solution.

Scottish Water is currently engaging with stakeholders, landowners and other interested third parties with a view to discontinuing at least two of the reservoirs in the near future.

<u>CASE STUDY 3: Tighnabruaich Reservoir – Discontinuance Again!</u>
Tighnabruaich is an impounding reservoir (16,780m³ capacity), retained by an earth embankment, 120m long and 5m high. It is situated in a remote location and access to the reservoir is difficult.

This reservoir, like many other small reservoirs, was previously discontinued in 2009 under the *Reservoirs Act 1975*, such that it was rendered incapable of holding more than 25,000m³.

However, having a capacity of over 10,000m³, Tighnabruaich Reservoir will be subject to the requirements of the *Reservoirs (Scotland) Act 2011* and Scottish Water finds itself, once again, commissioning discontinuance works to lower reservoir capacity.

Figure 9: lowered spillway

Figure 10: potential access routes

Having discontinued the reservoir previously through lowering the spillway level (Figure 9), it was initially considered that further lowering of the spillway would be a relatively simple and quick process.

However, upon closer inspection, a couple of potential issues were identified, which are currently being investigated, and appear likely to delay the process. For example:

Site Access: The spillway was previously excavated and lowered using small plant delivered to site via helicopter due to the remote location of the site. However, in order to lower the spillway further to the depths required under this commission, it is likely that significant rock excavation will be involved, requiring heavier plant and equipment. Therefore identification of alternative temporary site access routes is currently being progressed together with consultation with local landowners and community groups.

Downstream development: In particular, a small hydropower scheme has recently been constructed downstream of the reservoir and discussions are currently being held with the operators regarding potential implications of reservoir discontinuance.

CONCLUSIONS

When fully implemented, the *Reservoirs (Scotland) Act 2011* will impose a number of changes in relation to how reservoirs are regulated in Scotland through implementation of a more risk-based approach to reservoir safety. One of the key changes associated with this new legislation is the reduction in capacity threshold for registered reservoirs from 25,000m³ to 10,000m³.

For Reservoir Managers such as Scottish Water, with a large number of small reservoirs in the range 10,000m³ to 25,000m³ capacity, this will mean increased financial and resource commitments if the reservoirs are classified as high risk.

Scottish Water has taken a proactive approach and has sought to identify and assess its stock of small reservoirs with a view to upgrading them or discontinuing them in order to reduce risk, financial and resource commitments.

The case studies presented in this paper demonstrate that reservoir discontinuance is not a low cost solution. It requires careful planning, consultation, and investigation. Gaining approval for discontinuing small reservoirs can often be a lengthy process.

Key challenges encountered when designing and constructing reservoir discontinuance works on small reservoirs include: silt management; access for plant and materials on remote sites; third party stakeholder engagement; downstream flood impact; temporary works design; construction methodology; high construction costs; and environmental impacts.

REFERENCES

HMSO (1975). *Reservoirs Act, 1975.* Her Majesty's Stationery Office, London, UK

TSO (2011a). *Reservoirs (Scotland) Act 2011.* The Stationery Office, London, UK

TSO (2011b). *Water Environment (Controlled Activities) (Scotland) Regulations 2011*. The Stationery Office, London, UK

Judge T (2015) *Managing Scottish Water's non-operational reservoirs*. Dams & Reservoirs Vol 25 Issue 2.

Section 5:

Experiences with drawdown exercises

Dams – Benefits and Disbenefits; Assets or Liabilities?
ISBN 978-0-7277-6133-0

doi: 10.1680/dbdal.61330.241

Recent Experiences in Design and Construction of Siphons to Supplement Reservoir Drawdown Capacity

N R KEMPTON, Mott MacDonald, United Kingdom
P BENNETT, Mott MacDonald Bentley, United Kingdom
J WILSON, Mott MacDonald Bentley, United Kingdom
A HOBSON, Yorkshire Water Services, United Kingdom
I SCHOLEFIELD, United Utilities Group PLC, United Kingdom

SYNOPSIS Recently there has been increasing focus on the ability of UK reservoirs to be adequately drawn-down in an emergency situation. This emergency planning has led to a growing demand for drawdown capacity assessments, consideration of the acceptable drawdown rate and, where existing facilities do not provide an acceptable drawdown, the design and construction of supplementary capacity.

It is not uncommon for the installed drawdown capacity at UK reservoirs to fail to meet the varying targets. With the reliability and adequacy of temporary solutions under question, permanent solutions are regularly preferred. As a result, the requirement for permanent improvement works such as the construction of siphon drawdown pipes is becoming prolific.

This paper shares some recent experiences in the design and construction of siphon pipes to supplement the drawdown capacity in order to achieve an acceptable drawdown rate. It presents four case studies of improvement works undertaken, including a variety of siphon options, a range of pipe diameters, pumped and suction priming systems and the use of various types of valves.

The experience presented is that of Mott MacDonald and the Mott MacDonald Bentley joint venture 'the Principal Designer' gained while delivering schemes for Yorkshire Water Services (YWS) and United Utilities Group PLC (UU) 'the Clients' assets under the AMP 5 and AMP 6 frameworks.

INTRODUCTION

The ability of a reservoir's water level to be lowered for precautionary or emergency purposes has been assessed as part of statutory Section 10 inspection reports for some time. With increasing attention on flood resilience, emergency planning and risk management in the engineering sector and society as a whole, the ability of reservoirs to be drawn-down to prevent or mitigate the consequences of dam failure is under widespread review.

Where drawdown capacity is not considered adequate by the Inspecting Engineer, measures to be taken in the interests of safety under section 10(6) of the Reservoirs Act 1975 (the Act) to supplement drawdown capacity are made as statutory recommendations.

The adequacy of temporary solutions in providing supplementary capacity is not always considered a suitable approach. This is often due to potential limitations with access and the availability of pumps during periods of widespread flooding. With permanent solutions regularly offering a more reliable solution, the requirement for improvement works such as the construction of siphon drawdown pipes are becoming prolific.

The optioneering and detailed design process can be lengthy and iterative with numerous alignment and arrangement configurations possible. The purpose of this paper is to try to share knowledge of options that are likely preferable or be considered.

DRAWDOWN ASSESSMENTS

The ability of a reservoir to be drawn-down from the Top Water Level (TWL) is dependent upon the capacity and operability of the installed drawdown facilities and the inflow under which the drawdown is required. To carry out this assessment information to complete the following three assessments is required:

- Determination of the outflow stage-discharge relationship;
- Assessment of the reservoir stage-storage relationship;
- Estimation of the inflows to the reservoir.

It is not uncommon for this information to be missing, incomplete, or outdated resulting in costly data gathering initiatives to enable a drawdown assessment to be completed.

The inflow under which drawdown is required is sometimes specified with the measure in the interests of safety in the Section 10 report. Varying targets of reservoir inflow have been recommended including zero inflow and the Q_{90} flow.

DRAWDOWN TARGET

Currently a guide on drawdown provision is being developed for the Environment Agency which intends to provide a risk-based approach to determining a suitable drawdown rate dependant on the potential downstream impact, the volume of water and the height (and type) of the dam. Until this Guide has been published, many undertakers and Inspecting Engineers have been using the 'Jonathan Hinks Formula', arbitrary values (such as 1m/day) or other 'rules of thumb'.

In the presence of forthcoming guidance, this paper does not further discuss the methods for determining the drawdown target. Neither does it discuss the inflow under which the drawdown is required.

DESIGN AND CONSTRUCTION OF DRAWDOWN SIPHONS - CASE STUDIES

Of the four case studies included in this paper, three have been successfully designed and constructed. At the time of writing, Warland reservoir siphon is about to commence the detailed design stage. All works have been or are being undertaken under the supervision of a Qualified Civil Engineer (QCE).

Table 1. Siphon Drawdown Case Studies Comparison

		Keighley Moor IR	**Wessenden Old IR**	**Eccup ESR**	**Warland IR**
Dam Height	m	15	17	24	17
Reservoir Capacity	10^3m^3	347	324	7,010	832
Existing Discharge @TWL	m^3/s	0.34	0.48	1.39 (supply only)	0.26
Inflow Condition		zero	zero	Q_{90}	zero
Target Drawdown	m	1m/day for 5m	1m/day for 5m	1m/day for 5m	1m/day for 5m
Pipe Length	m	166	65	112	60 (each)
Pipe Diameter	mm	500	400	1400	3 No. 600
Priming Method	-	Pumped	Pumped	Suction	Suction
Siphon Discharge @TWL	m^3/s	0.54	0.40	8.00	3.40 (total)

Case Study 1: Keighley Moor Impounding Reservoir (IR)

Figure 1: Overview of Keighley Moor reservoir and siphon.

Introduction

Keighley Moor Reservoir is located on Oakworth Moor in West Yorkshire. The dam is approximately 275m long, 15m at its highest and was completed in 1846. Historic ground investigations show the embankment to be a homogenous earthfill embankment with no clay core or cut off below the embankment.

The installed drawdown facilities at the reservoir are provided by a single cast iron pipe of 300mm diameter, HDPE lined in 1999, and of 64m length laid directly through the embankment.

Problem

Following a Section 10 inspection and drawdown assessment, it was considered that the installed facilities at the reservoir provided inadequate drawdown capacity. The following recommendations were made in the interests of safety:

- An emergency pumping plan be written;
- Additional drawdown facilities be installed.

Due to the remote site location and single lane access the Inspecting Engineer specified that a temporary solution was not adequate in providing the additional drawdown capacity at the reservoir. The Inspecting Engineer also specified the target drawdown as 1m per day for the top 5m, under zero inflow conditions.

Optioneering

Due to the requirement to provide supplementary drawdown capacity to a 5m depth, the option of gates in the overflow weir was quickly discounted. Through optioneering it was identified that a permanent siphon drawdown pipe of 400mm diameter over the centreline of the embankment was the preferred outline solution.

Detailed Design

During the detailed design optioneering phase, the identification and analysis of two further possible alignments were undertaken. Following a whole lifecycle cost the alignment was selected on a value based decision. The new alignment was also favoured as the remaining alignments included construction of the pipe on the steep embankment face. The siphon is buried for the entire length other than within the reservoir basin.

Pipe Details

As a result of the chosen re-alignment an upsizing of the pipe from 400mm to 500mm diameter was required to achieve the supplementary capacity. The associated cost increase was offset by reduced velocities enabling the use of a gate valve rather than a more expensive discharge control valve at the downstream end of the pipe.

Priming Arrangement

Due to the location and lack of a suitable power supply, a pumped priming system was selected. To allow for this pumped priming method, an upstream valve was required. A small mobile pump (kept off site) connects a permanent supply pipe and the filling hydrant to prime the siphon pipe, as shown in Figure 2. A second hydrant is included to allow air to escape from the siphon pipe during priming.

Figure 2: View of intake NRV (left), and the siphon crest priming arrangement (right).

Valve Types

For the inlet of the siphon a recoil-check non-return valve (NRV), as shown in Figure 2, was selected for its minimal maintenance requirements. At the siphon crest, 2 No. hydrant valves were included in chambers as previously described. At the downstream end of the siphon pipe a gate valve has been provided to operate the siphon.

Construction
The works to construct Keighley Moor Siphon commenced in May 2014 and were completed in early August 2014.

For the works, the reservoir was drawn-down to approximately 7m below the TWL. A contingency plan was put in place, which required monitoring of the weather forecast and reservoir level including 'trigger levels' upon which contingency actions were required to ensure the safety of the dam, staff and works.

Commissioning, Maintenance and Operation
Keighley Moor IRE siphon was satisfactorily commissioned on 27 November 2014 with the reservoir level at 1.73m below TWL. During the commissioning, the siphon was operated twice; firstly, terminating flow by closing the downstream valve, and secondly terminated by opening the crest hydrant to 'break' the siphon flow by introducing air.

<u>Case Study 2: Wessenden Old Impounding Reservoir (IR)</u>

Introduction
Wessenden Old is situated roughly 3km south of Marsden in West Yorkshire. The reservoir is retained by a 220m long earth embankment approximately 17m high. The reservoir was constructed in 1839 and contains a puddle clay core that was raised in 1934.

Problem
According to the 2011 Section 10 Inspection report, drawdown analysis of the reservoir showed additional drawdown capacity was required. The report contained the following measures to be taken 'In the Interests of Safety' (ITIOS):

- A plan be written and facilities provided to be able to achieve a drawdown of 1 metre in one day without inflow.

Temporarily the recommendation was met by the provision of a pumping pad adjacent to the supplementary spillway and emergency drawdown plan. However, a permanent solution was desired.

Optioneering
Following on from the successful completion of Keighley Moor emergency drawdown siphon, both the Designer (MMB) and the

Client (Yorkshire Water Services) were keen to capitalise on the efficiencies of standardised design. Therefore, optioneering work focussed on alignment options for the pipework and implemented learning from the Keighley Moor scheme, described previously.

Detailed Design

During detailed design optioneering it was identified by the client (MMB) that the provision of an artificial inlet sump could significantly decrease the length of pipework required. This reduction in length lowered the frictional losses sufficiently to enable the use of a smaller diameter pipe whilst still achieving the required discharge capacity and operational range. The siphon is buried for the entire length.

Figure 3: Wessenden Old IR Siphon inlet arrangement during construction

Priming Method

As per Keighley Moor siphon (described previously).

Valve Types

As per Keighley Moor siphon (described previously).

Construction

Wessenden Old Siphon construction commenced early March 2015 and was completed in June 2015.

Upon excavation to install the siphon-crest pipework, the clay core was not located where anticipated. The previous ground investigation had identified small raising works rather than the main clay core. A design revision was undertaken to enable the steel plate clay core cut-off detail to be relocated centrally to the main clay core.

Commissioning, Maintenance and Operation

Wessenden Old IR siphon was satisfactorily commissioned on 2 December 2015 at the TWL following delays in refilling of the reservoir due to nesting birds. A further test is to be undertaken at the lowest priming level (TWL-3m) when reservoir levels allow.

<u>Case Study 3: Eccup En-Route Storage Reservoir (ESR)</u>

Introduction

Eccup Reservoir is an en-route storage reservoir situated 8km north of the centre of Leeds on Eccup Beck, upstream of the Harewood Estate. The main inflow to the reservoir is via pumped mains and provides raw water supply to Headingly treatment works. The earthfill embankment dam with central puddle clay core is approximately 195m long, rises to 24m at its highest and was constructed in 1885.

During its first filling, leakage and settlement occurred resulting in the extension and deepening of the puddle clay cut-off trench and reconstruction of the clay core. To undertake these remedial works, brick buttresses were constructed at 30m centres along the length of the embankment. The reservoir was finally commissioned in 1897.

Problem

Following an inspection of Eccup Reservoir in July 2010, under Section 10 of the Reservoirs Act 1975 (HMSO, 1975), the Inspecting Engineer made the recommendations in the “Interests of Safety”, which included the following:

- The bottom outlet facilities shall be replaced and upgraded such that the reservoir can be drawn-down and fully drained, if necessary, at a rate of no less than 1m per day.

The inflow/outflow scenario was agreed with the client (YWS) and the QCE at the start of the project. The drawdown rate to be achieved was agreed as 1m per day for the top 5m from the top water level (TWL). The rate was to be achieved with the Q_{90} low flow inflow condition, no inflow from the pumped mains and includes the draw-off at the Eastern end of the reservoir (Headingley feed). The Client also specified that the siphon was to be self-priming at TWL.

Optioneering

The notional solution to provide the required drawdown developed during the investigation contract was for twin siphon pipes (900mm in diameter) over the embankment, discharging into the existing spillway channel.

Following an extensive detailed design optioneering stage, and whole life cost analysis, the siphon alignment was varied to a single 1400mm diameter pipe. This option was selected on a value based approach and required significant hydraulic design to ensure the required capacity and drawdown was feasible.

Detailed Design
The alignment was located so that the concrete channel containing the siphon crest was positioned on one of the existing brick buttresses to limit settlement of the pipe crest. A submerged discharge valve was selected on a value based decision and to control discharge flows to prevent downstream flooding and damage of structures. The siphon is buried through the crest only.

Priming Method
The siphon is self-priming at the reservoir TWL. For lower levels the siphon is primed through the use of a permanent suction priming system installed at the embankment crest. The downstream valve remains closed whilst the vacuum priming pump is used to draw water into the siphon from the reservoir until the siphon is full.

Valve Types
At the Client's request, a gate valve was included at the crest on the upstream side to enable isolation of the system. An air valve to allow the introduction of air into the siphon was also included at the crest to allow 'breaking' of siphon flow if the downstream valve could not be operated. The downstream valve is a submerged discharge valve (Figure 4) reducing from 1400m to 1200mm diameter.

Figure 4: View of Eccup ESR siphon gate valve (left) and terminal submerged discharge valve (right) during construction.

Construction
Construction of Eccup ESR siphon commenced in May 2014 and was completed in February 2015. During construction, the reservoir

could not be drawn-down to below the siphon inlet level due to ongoing supply demands. Therefore, to construct the siphon intake the upstream section of pipe was floated onto the reservoir, manoeuvred into position and then sunk onto pipe saddles constructed by divers. This is shown in Figure 5 with a view of the works cutting through the existing brick buttress for the crest works.

Figure 5: View of the intake section of siphon pipework being floated across Eccup ESR (left) and cutting through the brick pillar for the siphon pipe (right).

Commissioning, Maintenance and Operation
The siphon was successfully commissioned in February 2015 and followed a commissioning plan to identify the flow which could be released downstream without causing damage to existing structures.

Case Study 4: Warland Impounding Reservoir (IR)

Introduction
Warland IR is situated on the western slope of Blake Moor, Lancashire. Ground investigation data indicates the 1500m long (approx.) embankment to be of homogenous earth fill with no determinable clay core as briefly discussed by Rigby *et al* (2014). It was constructed in 1857 and subsequently reconstructed in 1923.

Problem
Warland IR has an existing joint scour/compensation main that runs through the embankment as well as an overflow weir at its right (most northerly) end. The drawdown rate achievable via the scour pipework is less than 100mm/day. As such, two recommendations have been made 'In the Interests of Safety':

- Within one year of the date of the report a written contingency plan shall be in place detailing how the reservoir may be drawn-down at an initial rate of at least 400mm/day with the aid of temporary pumps or siphons.

- Within six years of the date of the report steps should be taken to increase the existing initial drawdown rate to at least 400mm/day. At least half of this capacity should be provided by a permanently installed facility; the rest may be provided by temporary plant.

A contingency plan has been implemented by the Client (UU) which has the ability to draw the reservoir down by 480mm/day under no inflow. Due to the remote site location and access concerns during an emergency, the second recommendation has been redefined by the client to 'future-proof' the permanent drawdown facilities at Warland IR as not less than 1m/day for 5m with zero inflow.

Optioneering

The notional solution of one large (approximately 1200mm diameter) siphon pipe located at the highest, steepest part of the embankment has been altered significantly through optioneering. The preferred solution now comprises 3 No. smaller (approximately 600mm diameter) siphon pipes terminating in a series of gravity drains to be located along the embankment toe.

This results in a significant reduction in the scale of temporary enabling works to install the inlet pipework whilst also reducing the length of pipe to be buried on the steep slope. This arrangement has the added benefit of removing the need for a costly downstream submerged discharge valve. The arrangement enables the Client to individually 'exercise' and test the siphon pipes in a controlled manner, whilst complying with their routine valve testing regime with a reduced risk of downstream erosion due to excessive flows.

Detailed Design

At the time of writing detailed design has not yet been undertaken.

Priming Method

Suction priming is proposed to fully prime the system from empty within two hours. This option was selected due to a large lift of over 5m owing to historical works that lowered the reservoir TWL and the siphon crest currently being set just above the 100 year return period flood rise. This necessitates the need for an upstream valve. The priming system will be powered by on-site facilities with provision of a backup connection.

Valve Types

The outline design siphon system will be controlled by 3 No. eccentric plug valves located at the toe of the embankment, discharging into gravity drains. No upstream control valves are to be provided within the reservoir basin. This will negate the need for significant future drawdowns to maintain the siphon inlet.

A series of in-line 'tees' will be required at the crest for connection to the suction priming system and a valve to allow air to be introduced into the siphon crest 'breaking' the siphon during discharge.

Construction

Construction of Warland IR emergency drawdown siphon is proposed to start towards the end of summer 2016.

Commissioning, Maintenance and Operation

The siphon pipe will require commissioning, following design and construction.

CONCLUSIONS

The use of permanent siphon drawdown pipes to supplement reservoir drawdown capacity is often the preferred solution where the existing facilities do not provide an adequate drawdown of the reservoir water level.

The conditions and targets under which the drawdown is required are currently recommended on a case by case basis by the Inspecting Engineers as part of the Section 10 Inspection process. The publishing of standardised guidance on drawdown rates and inflow conditions will give confidence in future-proofing schemes and allow reservoir owners to plan and budget for future works.

Detailed optioneering and cost analysis is required for each individual site to identify the favourable solution due to the numerous combinations and alignments possible.

Mott Macdonald Bentley has successfully designed, constructed and commissioned three drawdown siphons. A fourth scheme is soon to start detailed design and further schemes are anticipated. The experience gained in delivering these schemes will enable efficient and considered design and construction in the future.

ACKNOWLEDGEMENTS

The authors acknowledge the guidance of the appropriate QCEs in specifying parameters for the works and the support provided by Yorkshire Water Services and United Utilities Group PLC, the

Clients, in delivering these successful schemes and for allowing these experiences to be shared.

REFERENCES

HMSO (1975). *Reservoirs Act, 1975.* Her Majesty's Stationery Office, London, UK

Rigby PJ, Thompson AN and Jones DE (2014). 'The nearly perfect nineteenth century embankment dam' in *Maintaining the Safety of our Dams and Reservoirs, Proceedings of the 18th Biennial Conference of the British Dam Society, pp 88-100.* ICE Publishing, London, UK

Dams – Benefits and Disbenefits; Assets or Liabilities?
ISBN 978-0-7277-6133-0

doi: 10.1680/dbdal.61330.255

Delivery of Drawdown Improvements at Anglian Water Reservoirs

R PETHER, Black & Veatch Ltd
I KIRKPATRICK, Anglian Water

SYNOPSIS In 2005 Anglian Water commissioned a study into emergency draw-down rates at all of its larger reservoirs. As a result of the findings, between 2010 and 2014 Anglian Water successfully constructed and commissioned schemes to increase drawdown rates at four reservoirs assessed as having insufficient capacity. A further significant improvement scheme is planned for a fifth reservoir.

In this paper the process through which these reservoirs were selected for improvement is described. We will go on to describe the options that were considered for increasing the drawdown capacity at each of the reservoirs and the preferred option in each case. Each scheme had its own challenges which will be discussed, some of which were unique whilst others are common to many reservoirs. The paper draws conclusions and describes lessons learnt for consideration on future schemes, either by Anglian Water or other undertakers considering drawdown improvements, perhaps as a result of the forthcoming guidance document setting out recommended drawdown rates for large reservoirs.

INTRODUCTION

In 2005 Black & Veatch was commissioned by Anglian Water to calculate reservoir emergency drawdown rates for sixteen of Anglian Water's larger reservoirs. The study found that five of the reservoirs had insufficient emergency drawdown capacity. Subsequently schemes have been completed at four of these reservoirs to improve the rate of emergency drawdown. The designer and contractor for these works was the client's @One Alliance, made up of a number of companies, including Black & Veatch.

A significant scheme is also required at a fifth reservoir, and feasibility studies for this work have now been carried out. Delivery of this scheme will be carried out by the current @One Alliance.

APPRAISAL OF EXISTING DRAWDOWN FACILITIES

The sixteen reservoirs included in the study are listed in Table 1 below.

Table 1. List of Reservoirs and characteristics

Reservoir	Height (m)	Volume (Ml)
Alton Water	22	9,463
Cadney Carrs	6	910
Caldecotte Lake	7	2,014
Covenham	15	10,670
Crookfoot Reservoir	22	1,068
Foxcote Reservoir	10	574
Grafham Water	26	55,494
Hart Lower Reservoir*	12	25
Hart Upper Reservoir*	9	75
Heigham Water	7	58
Hollowell Reservoir	12	1,926
Hurworth Burn Reservoir*	9	727
Pitsford Reservoir	22	15,743
Ravensthorpe Reservoir	11	1,634
Rutland Water	37	124,000
Willen Lake	6	2,043

* These reservoirs are no longer owned by Anglian Water

Geotechnical assessment

A geotechnical assessment of drawdown stability at the dams was made using published stability charts (Bishop and Morgenstern, 1960) and infinite slope analysis. It was assumed that each reservoir was at top water level prior to drawdown commencing and that no dissipation of pore water pressure occurs in the upstream shoulder during the drawdown period. For some of the older dams, where the shoulder fill was not recorded, the type of fill was inferred.

Because drawdown involves lowering the reservoir level and failure of the upstream face is only likely to occur after the water level has been lowered significantly, the threat of instability leading to catastrophic failure of the dam is remote. The recommended maximum rate of drawdown will be a balance between the degree of damage that may be caused to the embankment versus the risk of failure of the dam due to the situation requiring emergency drawdown. This decision would be made by the All Reservoirs Panel Engineer (ARPE) attending the incident.

Assessment of drawdown rates

Outflow rating curves (flow versus head) were developed for each of the scenarios along with curves illustrating the drawdown rate with time for those reservoirs with available capacity curves. For those reservoirs which did not have an available capacity curve, the drawdown rate with time was estimated using a derived capacity curve. This was done by using the data at normal and top water levels and assuming the following are true for the surface area (SA) and volume (V):

$$SA = ky^x \qquad \text{and} \qquad V = \frac{ky^{x+1}}{x+1}$$

where y is the water depth in the reservoir and V is obtained by integration of SA.

There was no agreed guidance within the reservoir safety profession on the desirable rate of emergency drawdown at the time of the 2005 assessment. However, based on various standards that were being adopted by reservoir owners at the time, it was proposed that the minimum capacity of the drawdown facilities should be such that 50% of the volume of a reservoir could be released in 10 days for impounding reservoirs, and in 20 days for non-impounding reservoirs. The difference was to take account of inflows into impounding reservoirs that could affect the rate of drawdown.

Five reservoirs were found to have drawdown rates that did not meet this criterion; these were Alton Water, Foxcote Reservoir, Grafham Reservoir, Pitsford Reservoir and Rutland Water.

For four of these five reservoirs a supplementary report was prepared to include the effect of inflows, using the average winter inflow. This negated the need for the conservative approach of requiring a reduction of 50% volume in ten days for impounding reservoirs and allowed the twenty day criterion to be used for all reservoirs. Foxcote was not included in this supplementary study as it has a very small catchment.

Aside from Foxcote, the other four reservoirs receive a large part of their inflow from pumping and hence the inclusion of the direct inflows and the relaxation of the drawdown requirement to twenty days led to an overall improvement in the assessment of the current availability of emergency drawdown capacity.

The studies were carried out in order to assist the Inspecting Engineers at the subsequent Section 10 Inspections to make a decision as to whether works were required to improve the rate of drawdown at each of the reservoirs. They were also used to inform the emergency drawdown plans that have been prepared for each of the reservoirs in terms of which valve arrangement would give the highest rate and what precautions are necessary to prevent failure of the upstream face during drawdown.

In order to provide further information for the Inspecting Engineers, an additional study was also carried out for these five reservoirs, identifying viable options to improve the drawdown facilities.

Options that were generally considered to increase the drawdown capacity at each of the reservoirs were as follows:

- Modify the existing scour and/or draw off facilities
- Use backflow through inlet pipework
- Use of mobile pumps
- New siphon(s)
- Penstock through spillweir
- Pipeline through low point in rim

DELIVERY OF DRAWDOWN IMPROVEMENT SCHEMES

As it was anticipated that drawdown improvement schemes would be required as Recommendations in the Interests Of Safety (RIOS) at Section 10 Inspections, it was crucial for all parties to note that the date set in each RIOS would then be a final end date for completion of the works at that reservoir. The schemes were to be delivered in accordance with Anglian Water's AMP5 capital delivery process by the @One Alliance.

The Inspecting Engineer required that each of the new draw-down facilities be demonstrated prior to issuing a certificate under section 10(6) and for this reason temporary discharge consents were required for commissioning at each of the reservoirs.

After each scheme had been delivered each relevant section of the reservoir drawdown rates study was revised and the draw-down plan revised accordingly for that reservoir.

The new drawdown arrangements at Foxcote Reservoir and Alton Water were covered in some detail in a previous paper (Tam and Humphrey, 2012) and therefore these schemes will not be discussed further. Details of the completed schemes at Grafham and Pitsford

Reservoirs and progress to date on the Rutland Water scheme are given below.

Grafham Reservoir

Grafham Reservoir was inspected in September 2010 leading to a RIOS to increase the drawdown capacity in order that the volume could be reduced by 50% in 20 days and a drawdown rate of 0.3m/day could be achieved. This work was required to be completed by April 2014.

An appraisal of the options concluded that the preferred option was to facilitate backflow through the inlet pipework. Three new 900mm pipes have subsequently been installed off of the existing 1500mm intake pipe to allow the inlets to discharge to the tailbay. The testing of this new installation causes issues as in order to demonstrate that the water discharged originated from the reservoir the pumping of water into the reservoir needs to be suspended.

There is an on ongoing issue at Grafham Water with killer shrimp, *Dikerogammarus villosus* (DV), within the reservoir. This is a non-native species of crustacean which kills many other animals, threatening the existence of those species. For this reason when testing either the three new 900mm drawdown pipes or the existing scour pipe, all of the water discharged has to be pumped or tankered back into the reservoir. This is by no means an easy task.

A 1.0m high steel weir plate has also been installed at the end of the outlet channel. The purpose of this is to stop any water being released to the receiving watercourse due to the presence of DV. If left in place this weir would have an impact on any emergency drawdown scenarios. However the intention is that it would be removed before any emergency release of water.

As a result of the improvement works the time to reduce the volume by 50% has been reduced from 27 days to 17 days and the calculated initial rate of drawdown is now 0.3m/day. The works were commissioned at the end of April 2014, just in time to comply with the deadline specified in the RIOS.

A simplified schematic of the drawdown facilities at Grafham reservoir is shown in Figure 1 below.

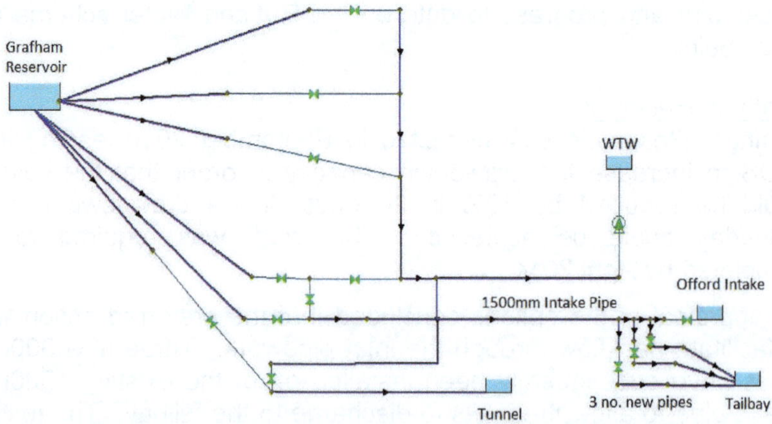

Figure 1: Schematic of drawdown facilities at Grafham reservoir

Pitsford Reservoir

Pitsford reservoir was inspected in November 2011 leading to a RIOS to increase the drawdown capacity in order that 50% of the volume of the reservoir can be released in less than 20 days with inflow equal to the winter average flow. This work was required to be completed by September 2014.

The preferred option identified for this reservoir was to install a new 750mm diameter scour pipe in the tunnel, to discharge downstream of the end wall of the chamber at the downstream end of the tunnel. It is also possible to use the supply pipe as a scour, but this involves demolition of the blockwork end wall of the valve chamber. The preferred option would therefore have included extending the supply pipe through the end wall of the chamber.

In order to achieve this it would have been necessary to core through the wall of the valve tower wet well, meaning that the tower would have to be drained and the outlet to supply interrupted. There were also health & safety concerns with working in the tunnel.

However, detailed design and construction of the works was carried out by the alliance and as part of this process an alternative compliant option was identified. This was to install three 600mm internal diameter siphons, each with a capacity of approximately $2m^3/s$. The advantages of the siphon option are as follows:

- Least cost option
- No need to drain the tower in order to drill through the wall of the wet well
- No interruption to supply

- Each of the siphons can be tested individually, meaning that a temporary discharge consent is only required for $2m^3/s$ rather than the combined $6m^3/s$.

However, there are also disadvantages to this option as follows:

- Mobilisation of temporary pumps to site required in order to prime siphons, increasing lead in time for drawdown. It is estimated that it would take approximately eight hours to commence drawing down the reservoir.
- The siphons can only work at reservoir levels at or above the 50% of volume level
- It had been intended to repair the middle draw-off main and guard valves whilst the tower was drawn down and this has not now been possible

The siphon scheme was commissioned in September 2014 (Figure 4) and has increased the calculated initial rate of drawdown to 0.2m/day whilst the time to reduce the volume in the reservoir to 50% has been decreased from 29 days to 15 days with the average winter inflow. A simplified schematic of the drawdown facilities at Pitsford reservoir is shown in Figure 2 below.

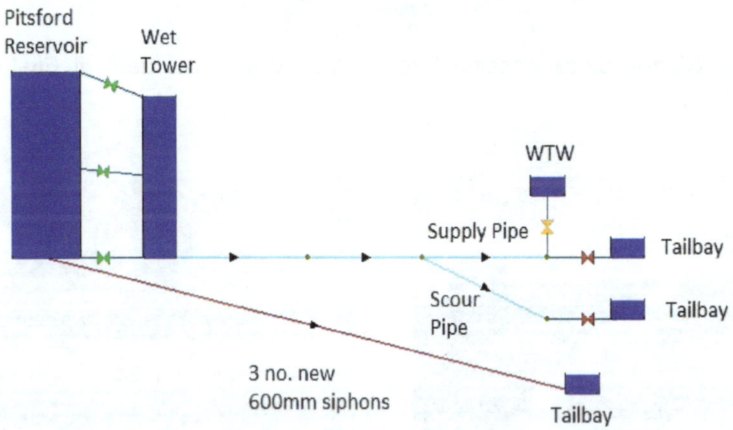

Figure 2: Schematic of drawdown facilities at Pitsford reservoir

In order to demonstrate the operation of the siphons to enable a 10(6) certificate to be issued, Anglian Water operations staff had to be fully trained in the operation of the siphons. A temporary discharge consent was required and in order to obtain this environmental mitigation in the form of gabions to prevent discharge of silt to the environment was required. These works were inspected by the Environment Agency.

One important element of the design of the siphons was the detail for passing the pipes through the narrow core of the dam. The details of this had to be agreed with the All Reservoirs Panel Engineer beforehand and a delegated inspection of the core was necessary at the time of construction. The construction detail is shown in Figures 3 and 5. Design of siphon systems such as this takes specialist hydraulic knowledge to ensure that the siphons can be operated successfully at the capacities required without cavitation occurring.

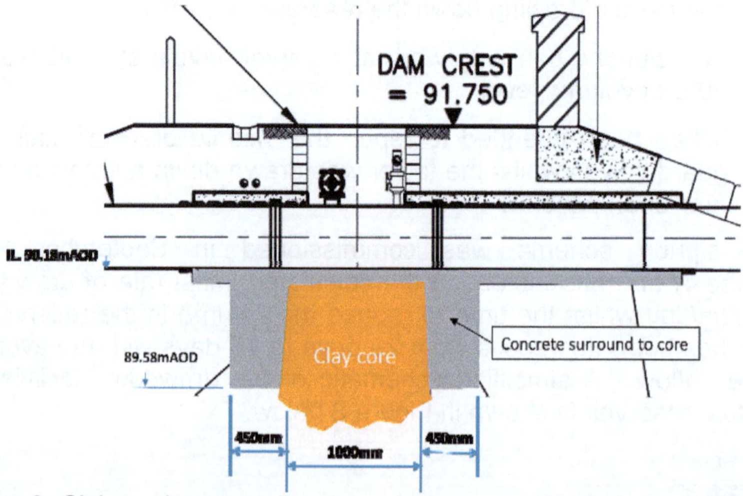

Figure 3: Siphon pipes passing through the core of the dam at Pitsford reservoir

Figure 4: Valve test during commissioning at Pitsford

Figure 5: Concrete protection to core during construction of siphons at Pitsford

Rutland Water

Rutland Water has a surface area of 12.6km² and the volume of the reservoir is 124Mm³. This makes achieving the guideline rate of drawdown very difficult. Putting it into context, the outflow from the spillway in a PMF is 13.5m³/s and the existing drawdown capacity is 22.5m³/s.

Rutland Water was inspected in September 2010. At that time no RIOS was made regarding the rate of drawdown at the reservoir in order to avoid setting a standard that would be impossible to achieve due to the volume of the reservoir. Instead a recommendation was made not requiring supervision by a qualified civil engineer within the meaning of the Act to 'undertake a study of means to increase the drawdown rate in an emergency'. This was an acknowledgement of the fact that the drawdown improvements would be major works and may take some time to plan, design and fund and also reflected the good condition of the dam, its inherent safety and the overall level of surveillance.

A simplified schematic of the drawdown facilities at Rutland Water is included in Figure 6 below.

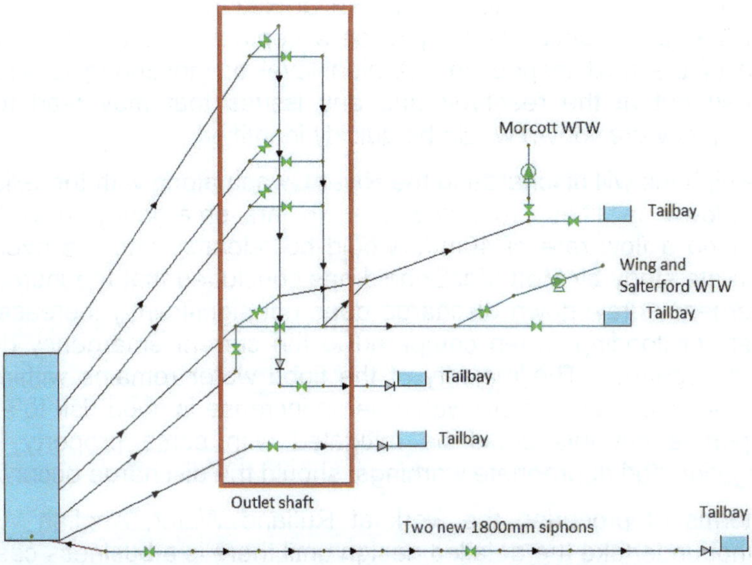

Figure 6: Schematic of drawdown facilities at Rutland Water

This includes a new 800mm diameter steel supply main that was connected to the existing outlet shaft in 2009. This pipe runs along the upper level of the 715m long existing outlet tunnel. Under normal operational conditions the flow from the new supply main feeds the new Morcott WTW. However there is also a tee in the new pipe and a further 800mm steel pipe isolated with a sluice valve which can be used to direct flow to the tailbay to supplement the emergency drawdown.

Further studies at Rutland Water have identified a preferred option to increase the rate of drawdown which is the construction of a new siphon arrangement. This work has yet to be designed in detail but it is anticipated that two siphons with a combined capacity of 20m³/s will be required. An initial study has concluded that the siphons will be approximately 290m long to suit the geometry of the site, and 1800mm in diameter. This will increase the drawdown rate from 20m³/s to 40m³/s and will allow the reservoir to be drawn down to 50% volume in 26 days including the average winter inflow.

The siphons would not achieve the original suggested criteria of 50% volume in 20 days. At Alton Water the preferred scheme did not quite meet the initial RIOS of the Inspecting Engineer but it was agreed that on the basis of the principles of ALARP (As Low As Reasonably Practicable) this option was acceptable when combined with increased monitoring at the reservoir to ensure that any issues that may lead to an emergency drawdown were identified promptly. Using the same principles of ALARP at Rutland Water it is thought that the siphon option is likely to be acceptable to the ARPE at the time of the next inspection. A high level of monitoring is already carried out at the reservoir and any issues that may lead to an emergency drawdown would be quickly identified.

The siphons will discharge to the River Gwash along with the existing drawdown facilities, meaning that in an emergency drawdown scenario a flow rate of 40m³/s would be released into the river. A separate study by Mott MacDonald has concluded that the increased emergency draw-down discharge does not significantly increase the extent of flooding, when compared to the current emergency draw-down scenario. The majority of the flood water remains within the natural flood plain. There would be an increase in flood risk to a few properties but this could be mitigated with some property level mitigation and appropriate warnings, should the discharge occur.

In terms of procuring the work at Rutland Water, Anglian Water cannot undertake the detailed design until there is a business case to justify the works. As there is no RIOS requiring the work at present it will not be possible to put forward a business case until the next Section 10 Inspection has taken placc, despite the fact that the previous Inspecting Engineer has indicated that a RIOS is likely at the time of the next inspection. This inspection is currently planned for 2020 and although an early inspection could be carried out there are no plans to do so at this stage. In the meantime a full Environmental Impact Assessment will be required as the reservoir is a Site of Special Scientific Interest, is designated as a European Special Protection Area and is internationally recognised as a globally important wetland RAMSAR site.

Whilst the siphon option is currently the preferred option at Rutland Water, it is possible that, as was the case at Pitsford, an alternative will be put forward by the alliance during detailed design.

CONCLUSIONS AND LESSONS LEARNT

Over the past ten years Anglian Water has delivered four successful schemes to improve drawdown rates at their reservoirs. This process started with a study of all of their larger reservoirs which informed the subsequent Section 10 Inspections allowing the Inspecting Engineers to make RIOS to improve the standard of drawdown. The inspecting Engineer's requirements varied at each of the reservoirs, in part because of the different circumstances in each case, but also because at present there are no standard industry guidelines for the drawdown requirements at a large reservoir. This situation may change in the future with the publication of the new drawdown guidelines, although it is likely that much will still be left to the individual judgment of the Inspecting Engineer.

A number of common options were considered at all of the reservoirs, but the preferred option has been different in each case. This is partly because of the varied existing facilities at each of the reservoirs which make some options more favourable than others. It is also dependent on the magnitude of the increase in drawdown that is required, which is mainly influenced by the volume of the reservoir and hence the water that must be released within a specified time period.

At some of the reservoirs there have been more than one viable option, and in these cases the final preferred option has been selected in accordance with Anglian Water's asset creation process, advised by their design and construction alliance. The business case for carrying out the works at each of the sites is reliant on a RIOS being made by the Inspecting Engineer at the Section 10 Inspection in order to provide a legal justification for the expenditure.

Environmental issues have been a concern at all of the sites, particularly during the testing and commissioning stages when discharge consents are required. Options that allow staged testing of the drawdown facilities provide a significant advantage at this stage. For example at Pitsford a discharge consent is only required for

$2m^3/s$ as the three 600mm siphons can each be tested independently without releasing the combined $6m^3/s$. This will be particularly significant for the forthcoming Rutland Water scheme, where it is intended that the required increase in drawdown capacity of $20m^3/s$ will be provided by two siphons, each with a capacity of $10m^3/s$.

An Environmental Impact Assessment for the Rutland Water scheme will be carried out in the next few years and it is anticipated that the work will be designed, constructed and commissioned within three years of the next Section 10 Inspection, which is due in 2020.

REFERENCES

Bishop A and Morgenstern N (1960). Stability Coefficients for Earth Slopes. In *Géotechnique 10:4, pp129-153.* Institution of Civil Engineers, London

Tam M and Humphrey M (2012). Improving Anglian Water's emergency response for reservoir safety. In *Dams: Engineering in a Social and Environmental Context; Proceedings of the 17th Biennial Conference of the British Dam Society, pp 474-485.* ICE Publishing, London, UK

Dams – Benefits and Disbenefits; Assets or Liabilities?
ISBN 978-0-7277-6133-0

doi: 10.1680/dbdal.61330.267

Scour Releases for UK Reservoirs – A Case study

A PEPPER, Mott MacDonald Ltd, Cambridge, UK

SYNOPSIS As part of the inspections carried out under Section 10 of the Reservoirs Act 1975 (HMSO, 1975), the Inspecting Engineer is required to assess the efficiency of the reservoir drawdown capacity and proper function (SI 2013-1677, Schedule 5, viii (HMSO, 2013)). However, within the UK there are a number of large raised reservoirs which do not have their scour (drawdown) valves regularly tested under full head conditions. One reason that many undertakers in England do not regularly open these valves is due to concerns that the Environment Agency (EA) may have regarding the environmental impact resulting from the release of water from a reservoir.

This paper outlines a case study in 2015 where the EA granted permission for a single release of water from a scour valve on a statutory reservoir, and the lessons learnt. During the test the scour valve did not close as anticipated, highlighting the need to regularly exercise reservoir scour valves under full head conditions.

INTRODUCTION

The reservoir in this case study is an impounding large raised reservoir which had not had the drawdown valves tested under full reservoir head for more than a decade, although they were tested annually under balanced head without problems. By operating the valves under these conditions, and not full head, it is not possible to confirm that that the reservoir could be drawn down using the valves.

In 2012 an attempt to open the bottom outlet at this reservoir was aborted for a number of reasons, including an uncontrolled escape of highly turbid water due to the valve not being able to be closed when required.

In order to mitigate these concerns, Mott MacDonald Ltd. was appointed by the Undertaker to evaluate the risks and recommend a procedure which would be acceptable to the EA and enable the valves to be opened.

This paper first outlines the environmental assessment and identified mitigation measures to minimise the environmental impact of the scour valve operation, allowing the Undertaker to apply for a discharge consent under Section 166A of the Water Industry Act 1991 (HMSO, 1991).

This paper then summarises the scour test, which did not go as expected, resulting in a discharge over and above that which was anticipated.

Finally, this paper discusses the lessons learnt and suggests steps which should be taken for subsequent scour tests at this reservoir.

SITE DESCRIPTION

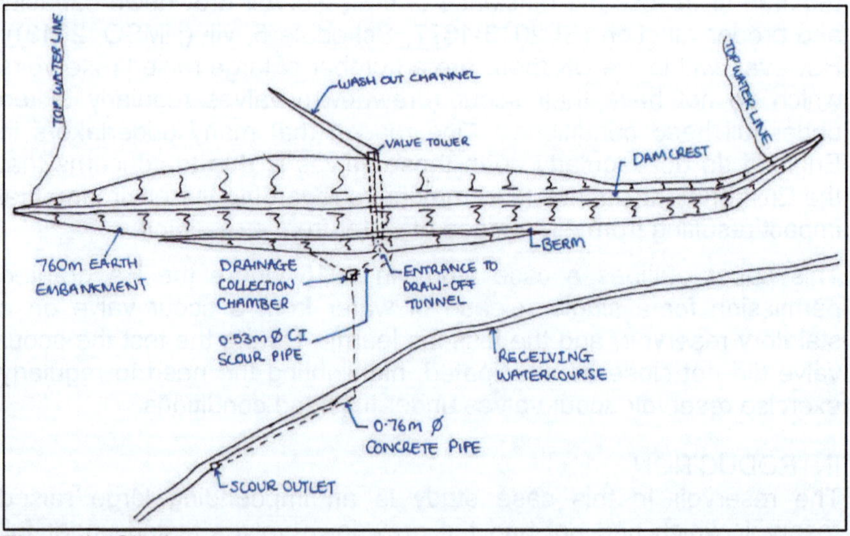

Figure 1. Site Layout

The reservoir presented in this case study has the following characteristics as shown in Figure 1:

- maximum capacity of 2,758,000m³ at Top Water Level (TWL);
- 15.8m high, 760m long embankment dam constructed in 1956;
- cast iron 0.53m diameter scour pipe with a sharp entrance at the base of a dry-well draw-off tower at the upstream toe of the dam;
- two inline scour valves separated by 0.6m. The downstream duty valve is operated electrically whilst the upstream guard

valve is manually operated via a handwheel located at the base of the tower. Note that under normal operation the upstream guard valve is fully open and the flow is controlled by the downstream automated duty valve

- downstream of the valves the pipe extends along the length of the draw-off tunnel for almost 120m, to a drainage collection chamber located beyond the toe of the dam;
- from the collection chamber, a 0.76m diameter concrete pipe conveys flow for 250m to an outlet on the right bank of the receiving watercourse.

Based on the above information and taking account of all of the relevant friction losses, the maximum discharge from the scour pipe with both valves fully open and the reservoir at TWL was calculated to be 1.2m³/s.

The receiving watercourse is a natural channel which runs alongside an agricultural field for 200m before flowing through a short culvert under a field access track (Figure 2). Over this stretch the channel consists of mud and grasses with a substrate of silt and vegetation. Beyond the field culvert, the brook flows for 4km to its confluence with a much larger river system.

Figure 2. View facing upstream from the field access culvert

ENVIRONMENTAL ASSESSMENT

Ecology

A desk study was carried out as part of the Preliminary Ecological Appraisal (PEA) to establish the environmental baseline conditions for the reservoir and the surrounding area within a 2km radius of the scour outlet.

The PEA indicated that downstream of the scour outlet there were no designated sites, sensitive habitats, protected or notable plant and animal species likely to be adversely impacted by the scour test.

In addition to the desk study, macroinvertebrate sampling was carried out in the receiving watercourse downstream of the scour outlet. These samples indicated that the receiving watercourse supported a low number of invertebrates with a minimal conservation value.

Flooding

A full Flood Risk Assessment (FRA) was undertaken in 2012. This concluded that for the initial 1km downstream of the reservoir there were no properties or infrastructure at risk from fluvial flooding of 1% annual exceedance probability (1 in 100 year return period).

As the scour test was planned to be a controlled release of water contained within the channel, it was considered that there was no excess risk of flooding to these areas.

Oxygen Content

Due to the depth of the reservoir thermal stratification of the water can occur, particularly in the warmer summer months. Releasing water with low Dissolved Oxygen (DO) content can result in the deterioration of the water quality in the receiving watercourse, potentially killing fish and other fauna.

Sediment Release

Due to the length of time since the scour valves had been fully tested it was possible that a significant quantity of sediment behind the scour valves could be released into the downstream watercourse. This could have potentially resulted in damage to aquatic stream life as well as allowing sediment-laden water into the river system downstream.

MITIGATION MEASURES

Ecology

To minimise any impact on nesting birds, the works were planned to take place in winter. However, as this is within the hibernation period for hedgehogs, checks for hibernating animals were undertaken prior to the test.

Flooding

To prevent local flooding along the brook, the capacity of the watercourse was assessed, taking into account the base flow and the additional inflow from the scour test. At a point in the channel 130m downstream of the scour outlet, the depth of water in the brook was to be measured immediately in advance of the valve test. If the water depth was greater than 0.2m the test would have been postponed to prevent out-of-bank flow.

Sediment Release

To avoid an uncontrolled release of sediment into the receiving watercourse the field access culvert across the brook 200m downstream of the scour outlet was selected as a location to temporarily impound the discharge. This would allow the initial sediment release to be intercepted and removed.

Temporary Dam

A temporary dam was created by placing straw bales across the inlet to the culvert. The straw allows water to slowly drain through it, limiting the sediment flowing downstream. The culvert was selected as the temporary dam location as it provided a safe location for installation of the bales, as well as providing lateral resistance to the force of water acting on the bales.

A second set of straw bales was placed 10m downstream of the culvert, providing a secondary impoundment and enabled a visual inspection of the water quality immediately downstream of the culvert (Figure 3). Both sets of straw bales were securely restrained in position using metal stakes.

If at any point during the test the water flowing through the second wall of straw bales was observed to be sediment laden the scour valves were to be closed immediately.

Figure 3. Straw bales in place prior to the test

Receiving Watercourse

Based on the topographic survey undertaken as part of the FRA the capacity of the watercourse was estimated to be 2.8m³/s. As this capacity was more than double the estimated maximum scour flow (1.2m³/s) the test could be undertaken with both valves fully open whilst keeping flows in-bank.

To prevent local flooding due to the temporary impoundment, the volume of the receiving watercourse and the timings to open and close the valves were considered important.

The volume of water which could be stored in the impounded watercourse was estimated to be 320m³. If both scour valves were fully opened it would take 4½ minutes to fill this volume. This did not include for the time to open and close the valves nor any allowance for the normal stream flow.

From discussions with the Undertaker, it was understood that it would take 1½ minutes to open and close the automatic valve. By holding the valves fully open for 1 minute, some of the sediment build-up behind the valves would hopefully be removed and the water may then run clear.

Limiting the scour valve test to a total of 2½ minutes would result in roughly 140m³ of spare capacity within the brook. This provided for an allowance for slightly elevated flow within the channel as well as any short delays in closing the automated valve.

The water behind the bales would then be given time to drain through the straw, and any remaining sediment removed. Due to the anticipated quantity of sediment behind the scour inlet, this process was assumed to be required multiple times during the day until the scour water ran clear.

Channel Erosion

Due to the high velocity exiting the scour pipe (>5m/s), there was a risk that erosion of the channel and banks immediately downstream of the scour outlet could occur. As well as potentially undermining and damaging the outlet structure, erosion of the channel would increase the sediment load transported in the watercourse.

To limit this erosion the Undertaker installed a Reno mattress at the scour outlet prior to the scour valve test.

Oxygen Content

It was considered that the flow of water exiting the outlet would be highly turbulent and this would introduce additional oxygen to the water. Additionally, the DO level would increase as the water travels along the channel to the straw bales. It was therefore judged that there should not be a low DO content in the water flowing downstream of the culvert.

As a precautionary measure it was recommended that the scour valves be tested in the colder months when the water in the reservoir should not be stratified.

Release of Fish

When scour valves are opened, there is a possibility that some fish may be washed from the reservoir with the discharged water. To limit the number of fish released, an electric fish scaring device was suspended from a boat in the reservoir, as close to the scour pipe inlet as safely possible. By undertaking fish scaring shortly before the scour valves were opened the number of fish surrounding the scour inlet was minimised.

As it was expected that a small number of fish would still pass through the scour pipe and into the impounded section of the receiving watercourse, a specialist contractor was employed to capture the released fish and return them to the reservoir.

Invasive Species

Under the Wildlife and Countryside Act 1981 (as amended) (HMSO, 1981), it is an offence to 'plant or otherwise cause to grow in the wild' any non-native, invasive plant species listed under Schedule 9.

The straw bales would allow the water to slowly drain through the material, slowing the progress of the released water to the remainder of the watercourse. The flow downstream of the impoundment would then be unlikely to disturb the plants.

No invasive macrophyte surveys of the reservoir were carried out, leaving a risk that invasive species present could be transported into the receiving watercourse. However, as the overflow from the reservoir already transfers water from the reservoir to the receiving watercourse, any invasive species in the reservoir would already be found in the downstream watercourse. Therefore, opening the scour valves would not alter the risk of invasive species entering the brook.

SCOUR TEST DESIGN

Based on the issues and associated mitigation measures considered above, Table 1 below provides the recommended scour valve testing regime developed for this reservoir.

Table 1. Recommended Scour Valve Testing Regime

Step	Description
1	Ensure the EA Watercourse Inspector is aware of the programme at least a week in advance of the planned scour test.
2	Measure water level in receiving watercourse 130m downstream of the scour outlet; the test can go ahead if the level is less than 0.2m above the bed. The test will need to be postponed if heavy rain is forecast for the test day.
3	Test guard and duty valves independently under balanced head conditions.
4	Securely install and restrain straw bales to block the culvert and a second wall of bales across the channel downstream of the culvert.
5	Fully open the manual (upstream) valve.
6	Open automatic (downstream) valve to 100% and then close, for a maximum total time of 2½ minutes. Visually monitor turbidity.
7	Wait for water to drain through the straw bales then remove sediment deposits. Visually monitor turbidity.
8	Repeat steps 5 – 7 as necessary until scour water runs clear..

During discussions with the EA a third line of straw bales was requested to be installed 10m downstream of the second set as an additional measure.

SCOUR VALVE TEST

The scour test took place in the winter of 2015. Prior to the test, consent had been granted by the EA for a release of water from the reservoir under Section 166A of the Water Industry Act 1991. This enabled the valves to be exercised up to a maximum of eight times in a single day, or until water exiting the scour pipe ran clear. Due to the length of intervening time since the previous full head scour test, it was unknown how many tests would be required for the scour water to run clear.

Valve Opening

Prior to commencing the scour test, both the upstream manual guard and downstream automated duty valves were tested independently under balanced head. The timing of the actuated valve was different to that which was previously understood, taking an additional two minutes to open and close the valve. To prevent localised flooding, the automated valve was not to be held open, but opened and immediately closed.

The water depth in the receiving watercourse was measured as per the recommendations in Table 1 and found to be 0.09m. As this was less than the 0.2m restriction, the test proceeded as planned.

At the onset of the test a number of people were positioned at the valve tower and control centre, at the scour outlet, and at the straw bales. After the balanced head test, the automated valve was left in the closed position and the manual valve reopened. The test began with the actuated valve opening as expected. Once the valve reached 100% open, the actuated valve started to close as anticipated.

Due to the length of pipework between the reservoir and the scour outlet, it was two minutes before water appeared in the receiving watercourse. The automated valve had begun the closing procedure and was 20% closed when water emerged at the scour outlet.

As expected with a reservoir scour release, the initial flow of water was seen to be sediment laden and carrying a small number of fish (see Figure 4).

Figure 4. Initial sediment laden flow

The water ran clear after 30 seconds. At this stage the total volume released was contained within the channel, impounded behind the straw bales. Had the test continued as planned, water would have been fully retained within the receiving watercourse.

However, the actuated valve could not be closed to more than 75%. The valve was reopened and a number of attempts made to fully close the duty valve before the process to lower the guard valve was started.

Due to the pressure head acting on the guard valve, it took a few men several minutes to close the manual valve. Once the guard valve was partially closed the pressure on the duty valve was reduced, allowing the latter to fully close.

The actuated valve was closed 17 minutes after opening, leaving the scour valves partially open for nearly 15 minutes longer than expected. As a result of this excess flow, water spilled out of the brook in a number of locations (Figure 5).

Figure 5. Water spilling out of the watercourse

Ten minutes after the test began the access track behind the culvert was overtopped (Figure 6). One minute later the second and third lines of straw bales failed (Figure 7). The first line of straw bales (those blinding the culvert) was not destroyed, although the upper layer is thought to have lifted.

Figure 6. Overtopping of field access culvert

Figure 7. Collapse of second and third line of straw bales

After the water had drained through the bales, the sediment deposits were removed from the channel. A walkover downstream revealed the marks indicating raised water levels within the watercourse of less than 0.2m. There was also no evidence along the brook of either lateral migration or sediment deposition or erosion, nor any ecological concerns.

CONCLUSIONS AND LESSONS LEARNT

Conclusions

Despite the delay in closing the valves, the scour test was successful, with clear water appearing after 30 seconds, confirming that under full reservoir head the valves could be fully opened should the reservoir need to be drawn down.

The procedure outlined in this case-study has hopefully set a precedent within the EA to allow similar tests at this and other reservoirs in the future, allowing essential safety tests to be completed without causing excessive damage to the environment.

Lessons Learnt

Had the actuated valve closed as anticipated the volume of water entering the receiving watercourse would have been much reduced.

Due to the difficulty in closing the manual valve against full reservoir head, more time was required to lower the valve, compared to balanced head.

There are a number of lessons which have been learnt and corrective actions taken as the valve did not close as expected:

- Refurbish and repair the downstream actuated valve.
- The process to close the manual valve should start at the same time as the actuator valve to relieve pressure on the actuated valve. Several confined space trained personnel should be on standby to relieve those turning the manual valve.
- Each row of straw bales should be staked to the river bed, rather than relying on the stakes through from the upper level.
- Prior to the next scour test, samples of the river bed should be taken to allow a pre- and post-test comparison of the stream bed sediment. This is likely to be a one-off sampling to demonstrate minimal impact on the watercourse.

The key problem during this scour test was the inability of the valves to close under full reservoir head. As the valves had not been tested under these conditions for several years it is possible that rather than an inability to close, the valves may not have been able to open.

This would be a more significant concern from a reservoir safety perspective, preventing or reducing the drawdown capacity of the reservoir in an emergency. This case study has therefore highlighted the importance of regularly testing reservoir scour valves under full reservoir head.

REFERENCES

HMSO (1975). *The Reservoirs Act 1975*. Her Majesty's Stationery Office, London, UK

HMSO (1981). *The Wildlife and Countryside Act 1981.* Her Majesty's Stationery Office, London, UK

HMSO (1991). *The Water Industry Act 1991*. Her Majesty's Stationery Office, London, UK

HMSO (2013). *Statutory Instrument 2013 No.1677 – The Reservoirs Act 1975 (Capacity, Registration, Prescribed Forms, etc) (England) Regulations 2013*. The Stationery Office, London, UK

Dams – Benefits and Disbenefits; Assets or Liabilities?
ISBN 978-0-7277-6133-0

doi: 10.1680/dbdal.61330.281

Managing the Environmental Risk from Reservoir Draw Down

D ARMOUR, Mott MacDonald Ltd.
M HEWITT, Mott MacDonald Ltd.
J MALIA, Scottish Water
R MURRAY, Mott MacDonald Ltd.

SYNOPSIS Pollution incidents caused by uncontrolled discharges of sediment rich water from reservoirs can impact on an owners' reputation and performance measures. Such incidents can also affect programme and have significant cost implications.

Scottish Water is developing a Reservoir Operation and Maintenance Strategy (ROMS) comprising work procedures and instructions, similar to those that it has implemented to control activities on its water distribution network. This paper describes the development of processes to identify the risks and manage the controlled drawdown of reservoirs for the purposes of maintenance and capital works, ranging from nominal water level lowering over a period of a few weeks for minor repairs to complete reservoir emptying for the purposes of dam breaching. It includes lessons learned from previous incidents, an outline of the approach to identifying the threats within the reservoir and the receptors in the downstream environment and hence the level of risk that the drawdown may pose. It describes the suggested minimum mitigation and monitoring measures to be incorporated in the drawdown plan dependant on that level of risk.

INTRODUCTION

Scottish Water engaged Mott MacDonald Ltd to assist with the development of a strategy for the management of planned reservoir drawdown for the purposes of investigations and works. Scottish Water has recognised that the planned drawdown of a reservoir can cause numerous problems; the most notable being the release of sediment-laden water via the bottom outlet from the depths of the reservoir, causing a pollution incident in the downstream channel (Figures 1 and 2).

Figure 1. Sediment in downstream watercourse from reservoir drawdown.

Such incidents have occurred in recent years on three reservoir projects (scour valve repairs and dam breaching). The effects of a pollution incident include fish/invertebrate kill, habitat damage and general impairment to the appearance of the natural downstream river system. Environmental Pollution Incidents (EPIs) such as these are reportable to the Scottish Environment Protection Agency (SEPA) and carry with them the risk of prosecution and adverse reputational matters. The Reservoir Operation and Maintenance Strategy (ROMS) will enable better management of activities at Scottish Water's 300 plus reservoirs. A consistent risk based approach to planned drawdowns and other activities at reservoirs such as routine valve exercising, surveillance and monitoring will result in improved legislative compliance and maintain reputation.

Figure 2. Suspended sediment in a drawn down reservoir from rainfall/runoff over exposed bed.

LESSONS LEARNED

Recent EPI incidents have occurred due to a number of issues; lack of a bypass/bywash channel (inflow diversion and clean water dilution), monitoring frequency and records, impracticable water quality targets, inadequate means of silt capture/treatment (Figure 3), limited knowledge of the hazards (levels, nature and mobility of silt), and reliance on >100 year old infrastructure to manage flows. It is recognised that a greater understanding of the sites and the downstream receptors is essential for the successful management of a drawdown project. Starting a process to manage drawdown early in the life of a project will flag up the issues, enable appropriate mitigation and monitoring and reduce the risk of an EPI.

Figure 3. Inadequate means of treatment in place

DEVELOPMENT OF THE DRAWDOWN PROCEDURE

The first step in developing a consistent and practicable reservoir drawdown process was to establish procedural limits and key roles and responsibilities, understand the likely hazards and receptors, and finally prepare a Work Procedure (WP). The WP is the umbrella document that describes the tasks required through a series of Work Instruction (WI) documents and standardised forms and monitoring logs to plan and manage a drawdown with a clear understanding of the risks present. The tasks required by the WP are summarised in Figure 4. The key aims of the WP were to ensure that it is recognised as a mandatory process that must be followed for all drawdown projects, that it is not prescriptive and that the level of input is proportionate to the size and risk profile of the project.

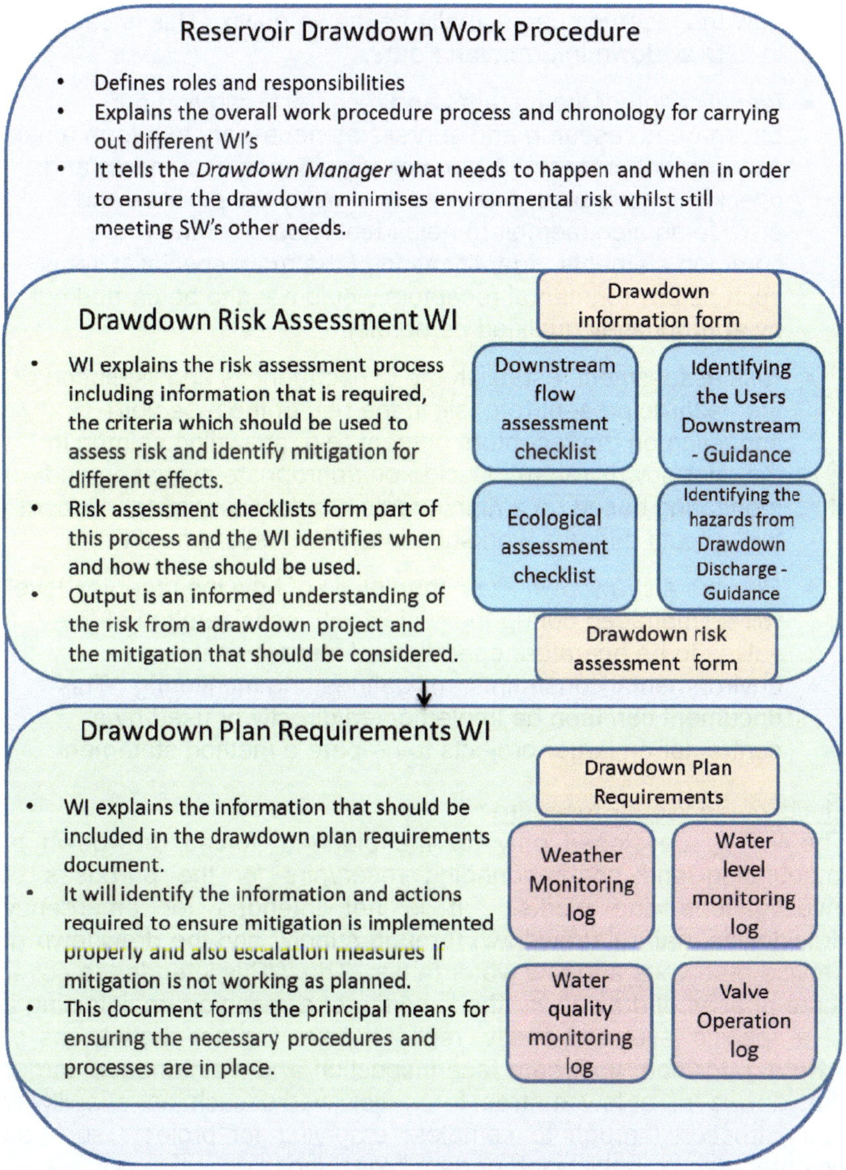

Figure 4. Overview of the ROMS reservoir drawdown work procedure

The WP can be broken down into a number of discrete exercises as follows;

- *Data gathering*; assembly of detailed information to understand how the reservoir can actually be drawn down. This is captured in a 'Drawdown Information Form'.
- *Identification of the hazards and receptors*; carrying out background research and surveys as necessary to inform a risk assessment process. A number of guidance documents and checklists have been prepared to simplify the process and provide an aide memoir to help identify some of the more common elements. Identification of the more specialist items such as environmental receptors would need to be carried out by appropriately qualified personnel.
- *Risk assessment*; establish the consequences and likelihood of the hazards present (e.g. silt in the bed of the reservoir) impacting on the receptors present (e.g. spawning salmon in the river downstream). Decide on appropriate mitigation and monitoring based on a high, medium or low risk rating. Record the outputs during a workshop in a 'Risk Assessment Form'
- *Prepare a Drawdown Plan*; the details of how the reservoir level will be managed during the project with consideration of the valves to be operated, operational / reservoir safety / environmental constraints, mitigations and monitoring. This document can then be implemented directly or used by a contractor on larger projects to prepare a method statement.

Limits of the Work Procedure

The WP is developed only for the planned, forced drawdown of impounding and non-impounding reservoirs for the purposes of investigations and works. It is not intended for emergency drawdowns, natural drawdown through supply, and the drawdown of service reservoirs and raw water tanks. The WP is intended to cover three phases of drawdown; the lowering of a reservoir, maintaining it at a specific level and finally, refill. It covers minor drawdowns of <1m e.g. for dam upstream face inspection and maintenance, larger drawdowns of a few metres for major works such as a spillway reconstruction through to complete emptying for projects such as upstream face rehabilitation or dam breaching.

It is necessary, before starting the WP, to confirm that some general aspects of the drawdown have already been established such as an outline target drawdown level and programme. These outline targets need to be agreed in advance by all project stakeholders such as Operations, Water Resources, Reservoir Safety, and the Environment/Planning Teams. Further to this an initial environmental screening exercise will need to have been carried out to provide fundamental ecological information. For optioneering exercises the cost of implementing mitigation measures arising from the subsequent ROMS drawdown process need to be allowed for when comparing options.

KEY ROLES

It was important to establish specific roles for every drawdown project. The two roles created are that of the Drawdown Manager and the Drawdown Supervisor.

The Drawdown Manager is responsible for ensuring that the WP is implemented including;

- Initial data gathering, risk assessment and preparation of a Drawdown Plan
- Acceptance of Contractor's method statements subsequently based on the Drawdown Plan

The Drawdown Manager role is always fulfilled by the Reservoir Engineer within the Reservoir & Supply Demand Team who has been assigned responsibility for that reservoir.

The Drawdown Supervisor is responsible for ensuring, by checking and auditing, that the outputs of the Drawdown Plan and method statement are implemented on site and records are being maintained.

The role of the Drawdown Supervisor might be fulfilled by the in-house delivery project manager, operations team leader or by the Drawdown Manager (in the case of small drawdowns for the purposes of inspection, investigation or minor repairs).

HAZARDS, RECEPTORS, LIKELIHOOD AND CONSEQUENCE

Common hazards associated with a drawdown project are as follows;

- Silt/sediment in the bed of the reservoir/forebay of the bottom outlet that has the potential for being mobilised and drawn through the pipework and released downstream.
- Poor oxygen content and low temperatures of water abstracted from lower levels and discharged downstream.
- Poor quality chemistry of stored water and sediments e.g. through historical land use, natural geology, or temperature (e.g. algal blooms).

- Changes to the natural flow regime, e.g. releasing large flows of water that exceed the normal flow regime of the downstream channel, potentially causing flooding and habitat destruction.

These can all be assessed by a combination of desktop studies, background research and site survey. The level and extent of assessment should be proportional to the size and risk profile of the drawdown project. Guidance has been prepared in the identification of such hazards and the criteria required for a 'Low' likelihood of their presence. For example, if the calculated maximum discharge from the dam pipework is less than the annual flood for the reservoir catchment, then the likelihood of flooding property downstream will be 'Low' given the downstream area would experience this magnitude of flow on a regular basis. Where the presence of a hazard is not known the default position is to assign a 'High' likelihood of its presence.

A receptor on a drawdown project is anything that could be affected by the release of the hazards present in the discharging water with consequences such as fish kill, habitat damage or economic loss. Receptors include;

- Ecological receptors; statutory/non-statutory designations, aquatic habitat quality, migratory and resident fish, otters, freshwater pearl mussels, water voles, water dependant nesting birds.
- Downstream users; abstractors (hydropower, distilleries, water bottling, industrial cooling), recreational users (angling clubs, canoe clubs), general users of riverside areas (dog walkers, joggers).

Again these can be similarly researched. Guidance has been prepared in the identification of common receptors and the criteria required for a 'Low' consequence of their exposure to the hazards present. For example, if an initial ecology impact assessment confirms key species (e.g. resident fish and freshwater pearl mussels) are located far enough downstream that no impacts are likely then the consequence of a release of sediment is 'Low'.

RISK ASSESSMENT

At this stage it is necessary to pull together the information from the previous stages (i.e. assessment of likelihood of hazards being present and the consequence to the receptors if the hazards are realised) to assess the risk of adverse impact or effect.

It is considered that the risk assessment will generate four potential outcomes for each combination of likelihood and consequence:

- Low likelihood and low consequence
- Low likelihood and high consequence
- High likelihood and low consequence
- High likelihood and high consequence

The risk assessment process is to be carried out as a workshop exercise with the key stakeholders in attendance. The output of the workshop would be a completed risk assessment document with a high, medium or low risk rating for each scenario that will assign specific monitoring requirements. Mitigation should also be determined from the workshop and documented in the risk assessment.

A work instruction explaining how to complete the risk assessment is part of the WP.

Table 1 shows the risk categorisation to be applied for reservoir drawdown projects. Two examples follow.

Table 1. Risk categorisation

		Likelihood	
		Low	High
Consequence	Low	**Low Risk**	***High Risk***
	High	**Medium Risk**	***High Risk***

Example 1. Low likelihood + high consequence = Medium Risk

Reservoir drawdown is carried out using a scour valve which has been regularly flushed (to limit build-up of silt in the forebay), a bywash is present (provides dilution and bypassing of reservoir inflows) and the drawdown depth is limited to the top 1/3 of the reservoir (sediment less likely to slip), and probably within the normal operating regime of the reservoir.

Receptors, e.g. freshwater pearl mussels in the river immediately downstream, are present that would be impacted by sediment release.

This medium risk scenario could be encountered when carrying out pitching repairs, investigation works or minor spillway repairs at a reservoir where there is only a low level draw-off or scour available.

Example 2 - High Likelihood / low consequence = High Risk

Reservoir drawdown can only be carried out using a scour valve which has not been regularly operated (built up sediments could be drawn through the pipework); there is no by-wash to dilute discharged flows; and the drawdown required is half of the total impoundment depth (higher likelihood for an underwater sediment slip).

Receptors are considered far enough downstream to not be affected by sediment release and the 'low' criteria set out in the Ecology Assessment checklist and Downstream Flow Assessment are met. Reputation may still be affected by an unintentional release of silt that would still be reportable regardless of the limited impact on the environment.

This high risk scenario could be encountered when carrying out major spillway reconstruction, wave wall installation at a small reservoir in a large catchment where there is only a bottom draw-off or scour available.

MONITORING AND MITIGATION

Appropriate monitoring is essential for a successful drawdown project. Each level of risk will have a different degree of monitoring associated with it. The minimum expected requirements for monitoring are set out below. The specific water quality attributes to be monitored will be dependent on the hazards and receptors identified.

- Low Risk: Daily water quality monitoring, daily water level monitoring and daily weather monitoring.
- Medium Risk: Twice daily water quality monitoring, twice daily water level monitoring and daily weather monitoring.
- High Risk: Continuous water quality monitoring, twice daily water level monitoring and daily weather monitoring.

Based on the level of risk for each hazard and receptor it is necessary to identify appropriate mitigation measures that should be applied to the drawdown operation. The mitigation will be informed by the specific characteristics of the reservoir and the downstream watercourse.

The approach taken to mitigation needs to consider options to eliminate the hazard in the first place. Should elimination not be practicable, mitigation options to reduce or manage the hazard shall be applied.

Table 2 provides a selection of typical mitigation methods that may be appropriate to apply.

Table 2. Example mitigation methods

Risk	Mitigation Measure	Description
Sediment Release	Physical Removal	Removal of sediment from the scour forebay by vacuuming using pumps once water level has been reduced to a lower level using higher draw-offs. Pumping silt over land/ into bunded lagoons to dry.
Sediment Release	First flush management	Initial release undertaken in a 'pulsed' fashion, with measures immediately downstream to capture and filter sediment laden water.
Water quality and sediment release	Chemical dosing	Can be used to alter pH and in some cases can be used to bind and settle contaminants/nutrients. Flocculants can be used to coagulate sediment where feasible (depending on scale of dosing required).
Water quality and sediment release	Downstream holding pond / lagoon	If space permits, provide a storage volume (see below) that may be used to capture a release of silty water before it affects the downstream environment. 
Rainfall runoff on exposed bed	Silt nets	Netting installed along the length of the reservoir to capture silt laden runoff that would otherwise pass downstream.
Water quality, sediment release and flooding	Managing release rate	Where water quality impacts are likely, reducing the drawdown rate will increase dilution in the receiving waterbody. In extreme situations, stopping the drawdown completely may be required.

Risk	Mitigation Measure	Description
Water quality, sediment release and flooding	Managing inflow rate	Drawdown upstream reservoir to help control inflows (e.g. throttle upper reservoir on the upper draw off and use this as an overflow), Install a temporary by-wash (e.g. piped / channelled; see below) to capture inflows and divert to the downstream channel thus avoiding the reservoir bed and need to use the scour pipework at full capacity. 

DRAWDOWN PLAN

Preparation of the drawdown plan is the final stage of the WP, it is where the fine details of the management of the reservoir level will be kept, along with a robust communications plan an escalation procedure, and standard forms for recording key variables such as valve status, water level, water quality and weather forecasts. A work instruction explaining how to complete the drawdown plan is part of the WP.

TESTING AND ROLL OUT

Between January and March 2016 the WP was successfully tested on a number of live drawdown projects; the breaching of Caaf and Bowling reservoirs and the drawdown of Waulkmillglen and Ryat Linn reservoirs for the purposes of CCTV survey of the outlet pipework. For these tests consultants were engaged by the Drawdown Managers to complete a number of the tasks such as the drawdown information form and the downstream flow assessment, risk assessment and preparation of a draft drawdown plan. Other tasks were kept in-house such as the determination of the downstream users and ecological receptors.

Constructive feedback sessions were then arranged to discuss further modifications and improvements to the WP. A key improvement was to make the process flexible such that it did not become too prescriptive such that small, low risk drawdown projects would require completion of unnecessarily detailed information.

The updated WP, as of March 2016, is being rolled out to the key team members (in house delivery project managers, reservoir engineers, resources and operations) and suppliers (framework consultants and contractors).

Following the roll out the WP and related Work Instruction documents will be branded and embedded within Scottish Water's management system.

Institution of Civil Engineers

publishing

SECTION 6:

ISSUES WITH MECHANICAL COMPONENTS OF DAMS

Dams – Benefits and Disbenefits; Assets or Liabilities?
ISBN 978-0-7277-6133-0

doi: 10.1680/dbdal.61330.297

Spillway and Dam Gate Reliability – Harmonising the Approach to Mechanical and Electrical Systems?

K GRUBB, KGAL Consulting Engineers, Poole, UK
R DIGBY, KGAL Consulting Engineers, Poole, UK
P JONES, KGAL Consulting Engineers, Wakefield, UK
B IMISSON, KGAL Consulting Engineers, Wakefield, UK

SYNOPSIS There is much innovative work going on around the world in respect of risk-based approaches to dam safety. Whilst these techniques are being used widely in the civil engineering aspects of dam design and inspection, there is no single, accepted, best-practice approach for the mechanical and electrical aspects of protection gates.

In 2014 the main ICOLD commission set up a hydromechanical sub-committee to specifically consider these issues. The lead author sits on this committee as the UK representative and is heavily involved in the writing of a "Best Practice Approach to Protection Gate Reliability" which is expected to be launched in 2016.

The purpose of this paper is to set out the various approaches being used elsewhere and to show the direction of travel being adopted by ICOLD and which will likely become accepted practice in the near future. In so doing the paper addresses the following questions:

- What is an acceptable risk to individuals and society?
- Does this vary if:
    - Equipment is old or new?
    - Equipment is in the developed or third world?
- How does hydromechanical equipment reliability relate to dam reliability overall?
- How will this be translated into contract specifications?
- What are the implications for Inspecting and Supervising engineers?

INTRODUCTION

Over some time, the civil engineering world has harmonised much of its approach to dam safety from a civil engineering viewpoint. It has done this through the enthusiasm and drive of its professional membership, through such forums as ICE, BDS and ICOLD.

It has also been "pushed" from time to time by the consequences of well-documented accidents and incidents. As a consequence legislation is in place in the UK and elsewhere around the world covering the inspection and supervision of existing dams.

ICOLD also provides bulletins and guidance on the acceptable "risk-based" design for new dams.

Issues relating to civil engineering are not covered in this paper, which seeks instead to consider the mechanical and electrical implications of spillway and low level outlet, dam-protection gates. Such gates have been the subject of many recent papers in respect of reliability, but have not been dealt with in a harmonised way internationally. This situation could now be changing and this paper seeks to outline the position and direction of travel.

Although the lead author sits on the ICOLD hydromechanical sub-committee, which is currently addressing the issues of protection gate reliability, it is important to state that this paper does not seek to state ICOLD's future policies, which are subject to due process. Instead the paper discusses the issues and the possible ways in which they should be addressed in the future.

There has been much debate in the industry in respect of such issues as:

- How can the same "rules" apply in developed and third world countries? This includes whether it is reasonable to expect an adequate level of future maintenance from all geographic zones.
- Should the same rules apply to new and existing dams?
- Is it acceptable to rely on human intervention to enable secondary systems?

WHERE ARE WE NOW? - SOME HISTORY

The Traditional Approach

Individual engineers and organisations have looked extensively at the question of gates and their related systems and sought to ensure that they were safe enough. Traditionally this was dealt with through a combination of specifying redundancy and ensuring that the contractors chosen were of "proven" experience.

With respect to redundancy, there has been a "rule of thumb" that after calculating the number of spillway gates required, one additional gate is specified. Clearly this is not a "scientific" approach.

Most dam protection gates have traditionally been purchased through some form of EPC "design and build" contract. Here specifications often had a "shopping list" of features that were deemed to achieve reliability.

Some purchasers have used "size rules" for pre-qualifying contractors so that there was an assumption that things would be satisfactory if the contractor had previously "performed".

<u>International Standards</u>

There are many international standards that relate to the design of gates and their systems. These are generally non-specific on functionality and can be applied to the complete range of gates including navigational, flood, river control and hydropower sectors. Hence the risks from their failure would have a wide ranging set of consequences from a danger to life through to pure economic loss.

Whilst the need for reliability is often dealt with within these standards, there are no risk based rules.

There have been a number of standards introduced that take a risk-based look at machinery and/or their control systems. These include Harmonised European Standards and also ISO standards. Examples include:

- IEC 61508: Functional Safety of Electrical / Electronic / Programmable Electronic Safety-related Systems
- BS EN IS0 13849: Safety of Machinery – Safety Related parts of control systems
- BS EN 62061: Safety of Machinery: Functional safety of electrical, electronic and programmable electronic control systems.

The standards quoted above take a risk-based approach to their subject and have been harmonised in respect of the Machinery Directives enacted into UK law via the Supply of Machinery (Safety) Regulations.

Note that some of the above standards use the concept of a Safety Integrity Level (SIL). There are four such levels with SIL 1 being the lowest (probability of failure of 10^{-1}) and SIL 4 being the highest (probability of failure on demand of 10^{-4}). BS EN ISO 13849 adopts a Performance Level (PL), though it is possible to cross refer these performance levels.

In the case of machinery used to protect a dam, reliability will usually be defined as "the Probability of Failure upon Demand", rather than a rate of failure per 1,000 hours (say).

The above codes tend to focus on the safety of machinery in respect of the user. A machine that fails to perform its job regularly but shuts down safely will often be considered to be acceptable. This approach does not work for dam protection.

Recent Focus on Probabilistic Analysis

Over the last 15 years, some organisations have looked at reliability from a probabilistic failure rate point of view. In this respect much good work has been undertaken by Scottish and Southern Energy in Scotland, BC Hydro in Canada and the US Corps of Engineers in the USA (apologies to others not mentioned here). This work has considered the risk to lives in the event of various types of failure, the required level of reliability for those potential risks and the actual reliability levels delivered.

There have been a number of papers published which have detailed the probabilistic approach taken on specific schemes.

How does probabilistic analysis work?

First some definitions:

Hazard: - That which has the potential to do harm

Consequence: - The likely outcome of the hazard

Risk: - For our use in reliability this is the arithmetic combination of the probability of an event and its consequences.

Safety Related Machinery: - Machinery which would have safety related implications should it fail to perform as required. Clearly this can be applied to the whole of a spillway protection gate.

There is no such thing as a machine with zero risk. This is because we are incapable of designing a machine that cannot fail or train a human that never makes a mistake. We are all used to dealing with risk at an everyday level without necessarily quantifying what that risk is. In the UK (for which the authors have some published statistics), the risk of dying in any particular year:

- From a road traffic accident is 6×10^{-5} per annum
- From an accident in the home is 4×10^{-4} per annum

- From a natural disaster (per individual) is 2 x 10^{-6} per annum

Therefore the concept of defining what an acceptable risk is has become the norm. The acceptable level of any risk will depend on a number of factors including the extent to which we choose to take it (dangerous sports) and the extent to which we are unknowingly subjected to it (the potential for a dam failure).

No universal rule for the acceptable risk associated with dam safety can be stated here, as different nationalities have inherently different natural background risks.

International codes to date generally differentiate between the risks posed by a dam to the individual and that faced by society at large. Individual risk is generally related to the totality of risk relating to the dam. Societal risk is generally related to each specific event that might occur.

For the individual, current practice is to consider a level of risk (of death) that he or she is exposed to normally and ensure that the risks from the dam do not exceed that. Figures for acceptable risk often used in developed societies are 10^{-3} where there is some control over the risk (wild water rafting); 10^{-4} for a minimum tolerable background; 10^{-5} as a general rule (sometimes used as the general risk to individuals and 10^{-6} as an aspiration to be achieved "as low as reasonably practicable (ALARP) (sometimes related to the individual most at risk).

Societal risk is usually covered by assessing the loss of life (LOL) that would result from a dam failure due to an event and adopting an acceptable value for the annualised lives at risk (ALR). The ALR is then found by dividing the LOL by the population at risk (PAR). This reflects society's aversion to disasters involving multiple fatalities. The values adopted are generally in the range 10^{-2} and 10^{-3}.

Dam failures where there were no potential fatalities or damage to the environment would generally be assessed on purely financial grounds.

Thus the engineer responsible for designing the gated spillway and its associated operating systems first needs to analyse the consequences of failure and then determine how reliable the specified gate systems need to be.

Note that current ANCOLD advice recognises the higher costs of addressing existing dams versus designing new ones. They thus accept the principle that acceptable risks for an existing dam can be up to ten times higher than for a new one. Clearly this will apply for schemes in their region and it is for others to consider their position elsewhere.

Manual intervention needs to be considered also in risk terms. If the flood event to be considered is, say, the one in ten thousand year event, then how many people could be considered to have turned up for work on that fateful day? Again the answer to this question could reasonably vary according to the circumstances of the dam.

International Commission on Large Dams (ICOLD)

In 2014, the International Commission on Large Dams formed a technical committee for hydromechanical equipment. The terms of reference given to the committee were:

- To conduct an analysis on current practices in the design, implementation and maintenance of hydromechanical equipment.
- To develop guidelines for the design, management and maintenance of hydromechanical equipment for each type of hydraulic structure.

The committee is currently developing a bulletin entitled "Best Practices for Achieving Reliability of Flood Discharge Gates". A draft of this document currently largely exists, though no deadline for its publication has been set.

INCIDENTS AND ACCIDENTS

It is worth dwelling on the reasons why reliability is an issue. Dams have the ability to cause massive loss of life if they fail and there are a number of dam failures which have been recorded. In a recent presentation by Leyland (2014) on the safety aspects of dams, the author quotes the following statistics:

Table 1. Relative Safety of Hydropower as a Technology

Technology	**Deaths/TWh Generated**
Nuclear	0.04
Hydro	1.4
Coal	60

Leyland also postulates that hydropower does not have the same safety culture as nuclear, including the need to share experience of failure. In this respect he may have a point.

In fact published statistics are hard to find in respect of deaths due to gate failures. For instance, there is a commonly quoted statistic that 30% of dam failures are due to a failure of the spillway to pass the flow, however ICOLD has not been able to establish the provenance of this statistic.

It has been established that there have been dam failures due to spillway inadequacies, which include a failure to open during the flood event.

It has also been well documented that deaths by drowning have occurred due to the unexpected/unwanted opening of discharge gates.

During 1987 there were extreme floods in South-Eastern Norway. There were many difficulties experienced and a subsequent paper noted the following statistics in respect to the problems:

Table 2. Problems Experienced during Floods

Problem	Percentage Occurrence
Power Failure	50%
Communication Problems	23%
Spillways not Opened	19%
Damaged Access Roads	17%
Clogging of Spillways	10%

It should be remembered that many large dams cannot be abandoned and store huge amounts of energy with significant populations in the downstream vicinity. Risk is thus passed on to future generations.

Many of the studies undertaken on existing gate systems have revealed common-cause failures, so that the appearance of redundancy is illusory.

In conclusion, there are warning signs with respect to the reliability of our current portfolio of dam assets, if we want to see them.

HEALTH AND SAFETY – AN ENGINEERS RESPONSIBILITY

Internationally, Health and Safety legislation requires that an engineer has a duty to consider the risks associated with his design in respect of safety. They are thus required to consider the risks associated with the reasonable use and misuse of the equipment under design, and:

- Avoid those risks that can be avoided
- Reduce those that cannot be avoided so that they are as low as reasonably practicable (ALARP)

- Protect from and manage those remaining risks
- Record the results of their risk assessment so that all parties are aware of the risks that they own.

Engineers should be aware that contractors will almost certainly not provide more functionality than they have specified, hence there can be no "hiding" behind an Engineering, Procurement and Construction (EPC) contract from the point of view of the contract specification. Some requirement to validate that an appropriate reliability level has been achieved is important.

Potentially, fatalities arising from a failure to consider reliability needs, and ensuring that they were met, could leave an engineer open to a prosecution involving corporate manslaughter.

THE WAY FORWARD

General

From this point on it needs to be emphasised that the authors are speculating on the "natural" outcome likely to derive from the current position and the direction of travel with respect to safety and reliability.

How to Procure Dam Protection Gates

Given the current state of the art and the direction of travel in relation to ICOLD, the process for specifying gated spillways should be as follows:

1. Determine the various hazards arising from one or more gate failures, remembering that unexpected opening can be fatal in some circumstances (and cause economic damage) as well as a failure to open.
2. Determine the consequences of such gate failure modes in terms of risk – how many people, likelihood of failure leading to fatality, etc. This will probably require different Safety Integrity Levels (SIL) for the different failure modes.
3. Thus determine the reliability levels required of the gates and their associated systems.
4. Design the gates and establish by analysis that the actual reliability levels of the gates and systems meet the requirements.
5. Build, test, maintain, etc.

6. After a suitable period re-examine the recorded hazard and risk analysis as things change with time (the number of people living downstream, for instance).

Where the contract specification includes a working design from the specifier, then the full results of their hazard analysis and safety review should be included with the systems specification.

Where the contractor is to design the system to meet a reliability standard, then the contract specification should clearly set out the standard to be attained, together with a requirement for the system designer(s) to provide a numerical reliability study which demonstrates that the standard will be attained.

Reliability as a Journey not a Destination

There is a need to document the hazard analysis process. This fulfils two essential needs:

1. It provides clear evidence that a risk analysis has been undertaken and that the engineers responsible have fulfilled their health and safety related duties;
2. It enables others to re-visit the assessment at a later date and update it in the light of possible changes, viz. new hazards that have appeared, changes in the assessment such as increased population in the dam vicinity.

The documentation should be seen as live throughout the life of the dam and be updated every 5-10 years.

Implications for Inspecting and Supervising Engineers

There is no reason why the current UK system of Inspecting and Supervising Engineers will need to change. However the routine operation of a gated spillway is not proof of its inherent reliability, only that it worked on the day it was tested.

However, the skill set needed to undertake a probabilistic reliability study is not one that a civil engineer usually has. More likely they will need specialist advice from a reliability expert(s).

The need to undertake studies for existing installations as well as update those studies on a periodic basis is likely to become a requirement. This will require an expansion of the existing expertise in this area, which is currently small.

CONCLUSIONS

This paper has provided an overview on the current position with respect to the reliability of mechanical and electrical equipment used on dam protection gates.

The need for improved emphasis on reliability is demonstrated by the current "statistics" in respect of dam and gate failures and the known consequences of such events.

The civil engineering community has been taking a risk-based approach to acceptable levels of failure from some time. Arguably the current European Machinery Directives also require this approach. The predicted future requirements will logically build on these existing trends.

ICOLD is currently developing guidance for mechanical and electrical machinery systems that is likely to embrace the probabilistic approach that is already being adopted as best practice in a number of organisations.

The risk-based approach helps to overcome the need for absolute rules by applying principles which can apply to all countries, since the risks associated with future maintenance and dependence on manual intervention can be evaluated. This also applies to the existing background risks to people who live in particular countries and their relative acceptability.

By adopting ongoing evaluation, reliability will be seen to be a journey rather than a destination. Inspecting and Supervising Engineers will become part of this process.

Increasingly there will be encouragement to share lessons learned from failures.

REFERENCES

Leyland B (2024). *Gated Spillways are they Safe Enough?* PowerPoint Presentation to British Dam Society, March 2014.

Dams – Benefits and Disbenefits; Assets or Liabilities?
ISBN 978-0-7277-6133-0

doi: 10.1680/dbdal.61330.307

Incident management and repair of a ruptured scour main at Talybont dam caused by a pressure wave

T A WILLIAMSON, Dwr Cymru Welsh Water, Newport, UK
A L WARREN, Mott MacDonald, Bristol, UK

SYNOPSIS Talybont reservoir is impounded by a 30m high earthfill embankment dam. A 760mm diameter water supply main and 635mm diameter scour main are located in a tunnel that passes beneath the dam. A branch off the scour main supplies a hydropower turbine.

When the power to the turbine pump failed one evening, a pressure shock wave in the scour main caused a section of the pipe to rupture catastrophically. The incident was detected when the reservoir level dropped by an unusual amount overnight. The repair was undertaken in difficult conditions due to the confined space environment of the tunnel and manual handling issues of heavy pipe sections; the inability to isolate the main due to the guard valve being jammed in a partially open; the supply-critical main having some corrosion and weakened pipe joints near the section of scour main that was to be repaired; and the unknown condition of the remainder of the scour pipework and valve, which had been subjected to the pressure wave.

This paper describes how the incident was managed; it presents the options considered to undertake the emergency repair and the post-incident analysis and lessons learnt.

INTRODUCTION

Talybont reservoir is situated in South Wales in the UK. It is impounded by a 30m high earthfill embankment dam with a puddle clay core, built between 1932 and 1938. The surface area of the reservoir is recorded as 127.2 hectares and the storage capacity as 11,654,000m^3. The dam is owned and operated by Dwr Cymru Welsh Water primarily for the purposes of supplying water. Original drawings show a small hydropower turbine building was constructed at the toe of the dam. Records show this was not in operation for many years; then in recent years the plant was brought back into service with modern generating equipment to supply power to local dwellings. The turbine comprises a 36 kW crossflow turbine.

The supply and scour pipework are located in a concrete-lined tunnel that passes substantially below existing ground level beneath the dam. A valve tower is located over the tunnel upstream of the dam crest and the tunnel is plugged immediately upstream of the tower shaft. The 635mm diameter cast iron scour main is on the left side of the tunnel (looking upstream) and the 760mm diameter steel supply main is on the right as shown in Figure 1. The scour main section can be isolated by a single gate valve located immediately downstream of the tunnel's plug.

Figure 1. Talybont tunnel looking upstream, with the scour main on the left

Near the tunnel tailbay (at the toe of the dam), a 300mm branch off the scour main on the right side supplies a micro-hydropower turbine; and a 200mm branch off the left side of the scour serves as a bypass as shown in Figure 2. The valve on the bypass pipe is engineered to open automatically when the turbine stops generating to relieve pressures in the scour main. It also provides compensation flow to the downstream watercourse.

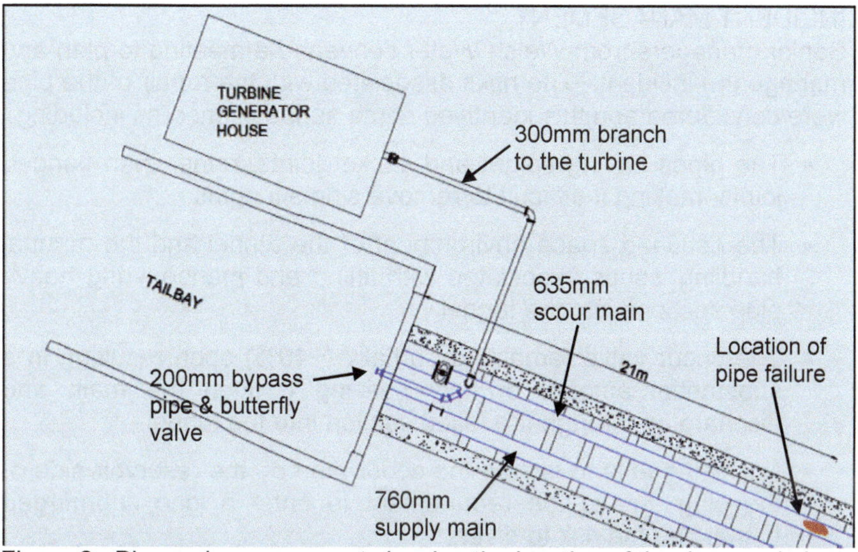

Figure 2. Pipework arrangement showing the location of the damaged pipe

THE INCIDENT

At around 21.00 on 25 June 2015 the turbine inlet valve was manually operated to reduce flows. Shortly after, the gauged river compensation flow measured a short distance downstream from the dam showed a significant increase in flow, from 25Ml/d to 150Ml/d.

The following morning a Welsh Water operative received a call that the reservoir level had fallen by an unusual amount. On visiting the site the operative found that a significant amount of water was discharging from the tunnel. The first action was to close the scour valve and although it was not possible to close the valve completely, the flow from the tunnel significantly reduced. A detailed inspection found that a large section of the scour pipe had failed (burst out from the underside of the pipe) 21m from the tunnel portal. Part of the failed pipe section is shown in Figure 3.

Figure 3. Section of the pipe that burst from the underside of the pipe

INCIDENT MANAGEMENT

Senior managers from Welsh Water convened a meeting to plan and manage the incident. The risks associated with the repair of the pipe were considered and this identified some serious concerns including:

- The pipes having spigot and socket joints, rather than flanged joints, making it difficult to remove single lengths.
- The confined space environment of the tunnel and the manual handling issues associated with lifting and manoeuvring heavy pipe sections into the tunnel.
- The scour valve remaining partially (~10%) open resulting in a substantial amount of flow passing through the main and discharging through the failed section into the tunnel.
- Not being able to isolate the scour pipe on the reservoir side of the plug due to the requirement to enter a long submerged tunnel and the risk to divers.
- Concern regarding the structural condition of the rest of the scour main and valve, which had been subjected to the pressure wave.
- The supply-critical steel main having some corrosion and weakened pipe joints running parallel to the section of scour main that was to be repaired.

Since the Talybont incident was not considered to be an emergency that would affect public safety, Welsh Water managed the event internally, i.e. the emergency services were not involved. However, as it was considered best practice, the Civil Contingencies hierarchical framework for controlling incidents was set up, using the recommended command structures.

Due to the high-risk nature of this work to all personnel involved and the risk to water supply, a "Gold Incident" (the highest incident category level in Welsh Water) was declared and teams were set up in two locations known as "Gold Command" and "Silver Command".

Gold, Silver and Bronze Command structures are used by emergency services in the UK to establish a hierarchical framework for the control of major incidents. The concept and explanations of this have been reinforced since the introduction of the Civil Contingencies Act 2004 (HMSO, 2004).

A Gold Commander is in overall control at Gold Command, which is located in a distant control room, where the strategy for dealing with the incident can be formulated away from the pressures of the incident. Silver Commander manages implementation of the strategy following direction given by Gold and formulates action plans, which are completed by Bronze. Silver is normally located near the incident. Bronze Commander controls the resources at the incident and will be located at the site.

For the Talybont incident, Gold Command comprised senior members of Welsh Water staff, with their Director of Operations taking the role of Gold Commander. This 'post' was set up in one of Welsh Water's main offices, whilst Silver Command was located near the incident site within the Water Treatment Works just 100m from the tunnel.

The Silver Command team included Silver Command (Welsh Water's Head of Water Assets); Welsh Water's Dam Safety Manager; an All Reservoirs Panel Engineer from Mott MacDonald; and designers from Welsh Water's Capital team. Bronze Command comprised the contractors (Lewis and UTS), operatives from the Dam Safety team and Production team and Mines Rescue, who provided a confined spaces rescue team.

EMERGENCY REPAIR OF THE RUPTURED SCOUR MAIN

Repair options

Various solutions were considered for repairing the scour main. Some of these were considered to be too technically challenging and would have required works to the dam structure or working within the valve tower or the reservoir itself.

One of the options considered was to replace the pipeline up to the scour valve at the base of the valve tower; however, the thrust on the valve would have been carried only by the connection to the pipe section through the concrete plug. If this had failed, full reservoir head would flow into the tunnel, endangering anyone working there. Additionally, this might have damaged the supply pipe such that there would be free flow through both the scour and supply pipe and a loss of water supply to thousands of customers.

Other options that were considered unsuitable included:

- Placement of a limpet dam structure over the tunnel intake. There was a concern that depressurising the upstream tunnel section under full reservoir pressure may cause structural damage to the tunnel and destabilisation of the upstream shoulder of the dam.

- Freezing the scour main upstream of the breached section. This required full bore flow in the pipe and cast iron was not considered to be suitable for freezing.
- Inserting an inflatable flow-stop within the pipe. The flow-stop sock would be inflated to reduce flows and water diverted through bypass pipework. These are considered workable on mains up to 2 bar. However, the pressure in the scour was nearly 3 bar.

<u>Planning the preferred repair solution</u>
After considering all the options, the preferred solution was to undertake the repair under the existing flow conditions, i.e. without attempting to plug or divert the water flowing in the scour main.

The Silver and Bronze teams worked together to develop the solution, reporting back several times a day to Gold Command. Initially, it was considered that a plastic pipe would be used for the repair due to its light weight and easier manual handling capabilities; however, the available diameters of these pipes meant that the existing concrete pipe supports would have to be re-profiled. It became evident during the operation that the contractors would not be able to safely re-profile the supports, once the failed pipe section had been cut out and removed, due to the force of water that would be directed straight at this working area. Therefore, steel pipe sections were manufactured in the contractor's workshop to suit the existing pipe diameter so modifications to the supports were not required.

It took over a week to prepare for the repair, including developing detailed risk assessments and method statements, together with contingency plans. Additional staff and materials (above the identified resourcing requirements) were brought on to site as reserves.

Within Silver Command, drawings were pinned on the walls, action plans were written up on white boards with step by step instructions, and briefings were held until all team members fully understood their role and responsibilities. With everything in pace, Gold Command gave authorisation for the repair to progress.

<u>Implementing the repair</u>
On the day of the repair, the first step was to undertake a final briefing in Silver Command to ensure everyone was clear about their role and responsibilities and the process to be followed.

Materials and safety equipment were checked, then contractors entered the tunnel to undertake the repair. After weeks of planning and preparation, the whole repair took less than six hours.

The new steel pipe was rolled into the tunnel on timber boards in two sections to reduce the handling weight and then joined together within the tunnel using a tensile coupling. Then the failed pipe section was cut out and removed. Figure 4 shows the flow discharging from the cut section, whilst Figure 5 shows the contractors lifting the steel pipe into place using a lifting system that utilised a number of lifting points anchored into the tunnel roof lining and lugs on the pipes.

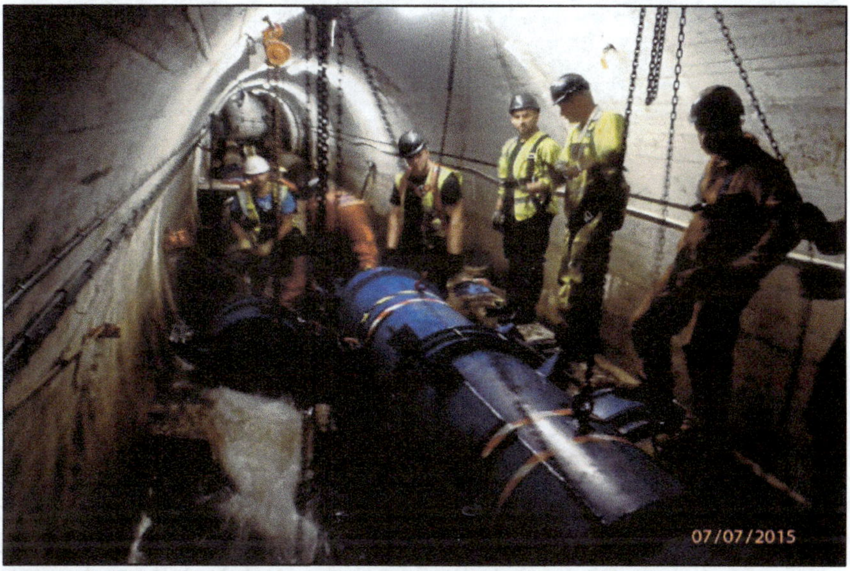

Figure 4. Flow discharging at the removed section of the scour main

Following the repair the main was re-commissioned by Welsh Water's Dam Safety team through a sequence of operations to slowly fill and pressurise the repaired main. This process took a further six hours, with each step in the process being authorised by Silver Command through 4-way radios with staff stationed at the tunnel entrance, base of the tower, top of the tower and in Silver Command.

Once the main was fully pressurised the scour valve was freed off and fully opened, closed and opened again under balanced head conditions. The compensation valve was then opened. The valves to the turbine were locked closed, pending further investigations.

Figure 5. Lifting in the new section of pipe to replace the failed section

POST-INCIDENT ANALYSIS

On the night of the incident when the turbine inlet valve was manually operated to reduce flows, the "Daily Power Report" showed that the power reduced from 16kW to 4kW in about one second. Then the next reading, two hours later, showed the plant was running at zero; it had stopped generating power.

It is normal for the turbine to automatically shut down when there is a distribution network power issue. The hydropower plant is engineered so the bypass butterfly valve opens to release flows to the river and then the turbine inlet valve closes. It is not known whether this process happened as designed when the generator stopped or indeed what happened during the two hours when flows reduced and stopped. It is speculated that either the bypass valve failed to open or it opened too slowly to prevent a significant pressure rise in the scour main.

Upon removal of the failed pipe section from the tunnel, a crack along the underside of the pipe was visible and corrosion was noted on the surface of the pipe section. A forensic exercise of this failed pipe section was undertaken in an attempt to gain a better understanding of the cause of the Talybont incident. A visual examination of the pipe showed evidence that the scour main had lifted off the pipe supports when the incident occurred causing minor damage to the supports. Tensile strength and hardness tests and metallographic examinations were also undertaken and these showed:

- The material strength was a typical Grade 150 iron to BS 1452;
- Some minor defects in the pipe were noted; and
- The reduction in wall thickness due to corrosion was mainly on the inside and extended up to 4mm. The external surface only showed a minor degree of corrosion.

The forensic examination report concluded that the condition of the pipe had been affected by corrosion damage, but this was not excessive and had not contributed significantly to the pipe failure. There was no evidence to indicate the location of the fracture initiation point, but it was clear that the fractures had been present for some time due to the corrosion seen on the fracture faces, particularly on one section where the corrosion indicated cracking had initiated from the inner surface. Due to the presence of water on the tunnel floor and the restricted space between this and the pipe invert, it is highly unlikely that any historic leakage could have been detected. Seepage water enters the tunnel from the valve tower and tunnel joints making it difficult to detect leakage from fine pipe cracks. Clearly fractures in the pipe had been present for some time prior to the incident.

The extent to which previous turbine operations and pressure changes had contributed to the pipe deterioration is unknown.

In conclusion, the pipe failure at Talybont was believed to have occurred due to load rejection of the hydropower turbine where a pressure wave (water hammer) was initiated, which further fractured the pipe section to the point where it finally ruptured.

LESSONS LEARNT

Preventing pressure wave damage caused by installation of micro-turbines on existing dams

Design for a number of possible transient and operational scenarios
Detailed analyses and modelling of all possible hydraulic and pressure transient scenarios within a system should be undertaken before installing new equipment to a conduit. For example, where turbines flows are manually controlled, a range of flows should be considered and how this may impact the system.

Do not over-simplify the design of a system
In some systems, analysis may be complex owing to the nature of both the pipework and equipment within the system. All pipe connections and valves should be analysed to consider the performance under adverse conditions including failure of system protection equipment.

Specialist expertise should be employed to design the system, utilising current and appropriate computational methods and software, which analyse the stresses and strains throughout the system. The designer needs to analyse the whole system for a range of scenarios to ensure that the predicted performance is valid.

Check materials for loadings under adverse conditions
Analysis should include checking the current quality, age and condition of all materials in the system to ensure they withstand loads induced by pressure transients. One of the main criteria affecting pressure rise (depending on the turbine type) is the turbine guide vanes closing time. It is necessary to consider through transient analysis the minimum closing time so maximum pressure rises remain within permissible limits (Lliev *et al,* 2012).

Provide fail-safe protective measures
There are a number of protection measures which can be designed and installed within hydropower systems to reduce the damaging effects of pressure waves. Options include:

- Installation of pressure-relief valves;
- Installation of safety membranes designed to rupture under excessive pressure (Çalamak & Bozkus, 2012);
- Modification of pipe thickness, stiffness or diameter to accommodate maximum pressure rises;
- Increase the turbine guide vane closing time; and
- Use of a surge shaft a short distance upstream of the turbine.

At Talybont, there was some concern that the bypass valve did not operate as it was designed to. If used as a pressure relief protection device, these need a high degree of engineering and maintenance to ensure they continue to work efficiently under adverse conditions.

Additional to the lessons learnt as a result of the root cause of the Talybont incident, there are some other valuable lessons that can be learnt. These may be of interest to dam owners and operational teams and as well as to incident managers.

Valve exercising
The incident demonstrated that some of the valves within the tunnel had not been tested, probably because they were not considered important for dam safety. Some of the valves on the schematic and exercising logs were inconsistently labelled. It was also discovered that one of the valve wheels showed that it was clockwise closing, when in fact it was anti-clockwise closing. These details, which can seem trivial during normal operations, can make a big difference during an incident. Some advice regarding valve exercising includes:

- Accurate and detailed records on valves should be maintained. Valve schematics should show all valves, including abandoned valves. They should be clearly labelled, together with their type, direction of operation and number of turns to open/close.
- Where valves are designed to function under non-balanced head conditions they should be tested under non-balanced head conditions frequently and records kept describing the conditions and ease of operation, as well as the status of the valve.
- It is advisable to exercise all valves (except abandoned valves) at least annually including wash-outs, by-pass and supply main valves to provide isolation and the ability to continue supply in the event of a pipe burst.
- It is useful that all valves, access doors and padlocks are locked with a common key that is available to all key staff.

Pipework inspections
During the incident, it was identified that there was no record of the corrosion on the supply main in the records. Due to restricted space below the pipes and the water backing up from a v-notch at the end of the tunnel with water levels just below the pipe inverts, it made detecting other pipe leaks and defects in the pipe difficult. Some advice regarding pipework inspections includes:

- Regular close visual inspection of pipework, including examination underneath, and a monitoring regime of non-destructive testing and surveying should be agreed.
- It is advisable to consider how a system could be isolated if there was a burst at various locations as this could affect the integrity of a tunnel or tower and/or cause a rapid drawdown failure. Contingency plans can be put in place and any valves identified as being needed in the plans can be exercised regularly.

- Critical valves as identified in the point above should be repaired as a priority if they are found to be faulty during exercising.
- The Talybont incident underlined the value for having a secondary means of upstream isolation in the form of a guard valve on pipelines.
- Working lights are essential in tunnels and towers to allow easier and more accurate examination and to assist with maintenance.
- Regular removal of silt from tunnels is good practice, including ensuring that leakage does not back up within a tunnel, which may prevent the detection of pipe leakage.

Incident management

At times during the Talybont incident there was some confusion regarding terminology such as what was meant by upstream and downstream and which was right and which was left. It should also be highlighted that dam incidents are generally complex events and can continue for much longer than expected. In the case of Talybont, many staff worked 12 days without a rest day. Some advice regarding incident management includes:

- At the start of the incident, clarify typical terminology to be used in briefings.
- Bronze, Silver and Gold Commanders should be identified. All site communications should go to Gold Command through Silver.
- It is important for incidents that continue over several weeks for rotas to be agreed and for key staff to take rest days.

CONCLUSIONS

A pressure wave may occur when a fluid in motion is forced to stop or change direction suddenly. This often occurs when a valve is closed suddenly at the end of a pipeline system, and a pressure wave propagates through the pipeline.

The pipe failure incident at Talybont was believed to have occurred due to load rejection of the hydropower turbine on a number of occasions, where pressure waves initiated and propagated a fracture in the cast iron pipe to the point where it finally ruptured.

The consequences of a pipeline failure within a dam can be catastrophic and might lead to loss of life, environmental damage, non-availability of equipment, loss of reputation, increased insurance premiums, emergency call out resources and increased maintenance costs. It is essential that all equipment within dams is protected from adverse operational conditions which may exceed the original design provisions. Where hydropower facilities make use of dam conduits, the maximum adverse conditions should be evaluated, making allowance for the possible non-availability of protection systems.

As well as describing measures to reduce the risks of pressure wave damage, this paper also provides some other important advice for dam owners to consider when managing, maintaining and inspecting dams. Namely, the importance of regular valve exercising, good record keeping and adherence to appropriate pipework monitoring regimes. It also includes some advice on managing incidents in a structured manner according to established hierarchical frameworks.

ACKNOWLEDGEMENTS

Acknowledgement is given to Dwr Cymru Welsh Water for their permission to share the learning from the Talybont incident. Also to Lewis Civil Engineering Limited, who managed the emergency repair work and provided some of the photographs; and to all parties for their contribution in ensuring a safe and successful outcome at Talybont.

REFERENCES

Çalamak M and Bozkuş Z (2012). *Protective Measures against Waterhammer in Run-of-River Hydropower Plants.* Teknik Dergi, Volume 23, Number 4, pp. 6187-6202.

HMSO (2004). *Civil Contingencies Act 2004*. The Stationery Office, London

Lliev V, Popovski P & Markov Z (2012). *Transient Phenomena Analysis in Hydroelectric Power Plants at Off-Design Operating Conditions.* International Journal of Engineering Research and Applications, Volume 2, Issue 6, pp.1480-1488.

Dams – Benefits and Disbenefits; Assets or Liabilities?
ISBN 978-0-7277-6133-0

doi: 10.1680/dbdal.61330.321

ice
Institution of Civil Engineers
publishing

Bosley Reservoir, outlet improvements

D H BROWN, Canal & River Trust

SYNOPSIS Works were carried out at Bosley Reservoir in Cheshire during 2015/16 to increase the dam freeboard, to understand better the risks associated with downstream control of the outlet and to improve the drawdown capacity.

These works, which included the design and construction of two siphons and the investigation of the original draw-off system, are discussed in this paper.

INTRODUCTION

In his report of a statutory inspection of Bosley Reservoir in 2006, Jonathan Hinks expressed concern that the reservoir could not be drawn down quickly enough using the draw-off valve and that the emergency plan placed too great a reliance on pumping. During the critical first 24 hours there would be little reduction in water level until pumping was established. He considered that there was insufficient guidance on the subject at the time and suggested the 'Hinks formula' (Hinks, 2009)) as a means of determining a suitable draw-off rate for a reservoir. He determined that the next inspection should be carried out within five years by which time he expected that the reservoir profession would have reached a conclusion about drawdown matters. He also raised concerns about the risks associated with downstream control of a pressurised draw-off pipe through the dam.

Martin Airey undertook the next inspection in 2011 and concluded that both items were indeed of concern and that measures in the interests of safety were needed. He was appointed as qualified civil engineer (QCE) to supervise the implementation of these measures in the four-year timescale given, although there were times when his absence abroad lead to the transfer of the role to Tim Hill.

The Canal & River Trust constructed new siphons and other works during the winter of 2015/16. Keir Construction was the contractor and the designer was Arcadis (formerly Hyder).

BOSLEY RESERVOIR

Bosley Reservoir is the main supply of water to the Macclesfield Canal. The canal was authorised by Act of Parliament in April 1826. Tenders to build Bosley and Sutton Reservoirs, two of the five originally planned, were invited in 1828 and construction followed soon afterwards. The canal opened in November 1831. Thomas Telford was responsible for the route of the canal but the engineer was William Crosley. The general layout of Bosley reservoir is shown in Figure 1.

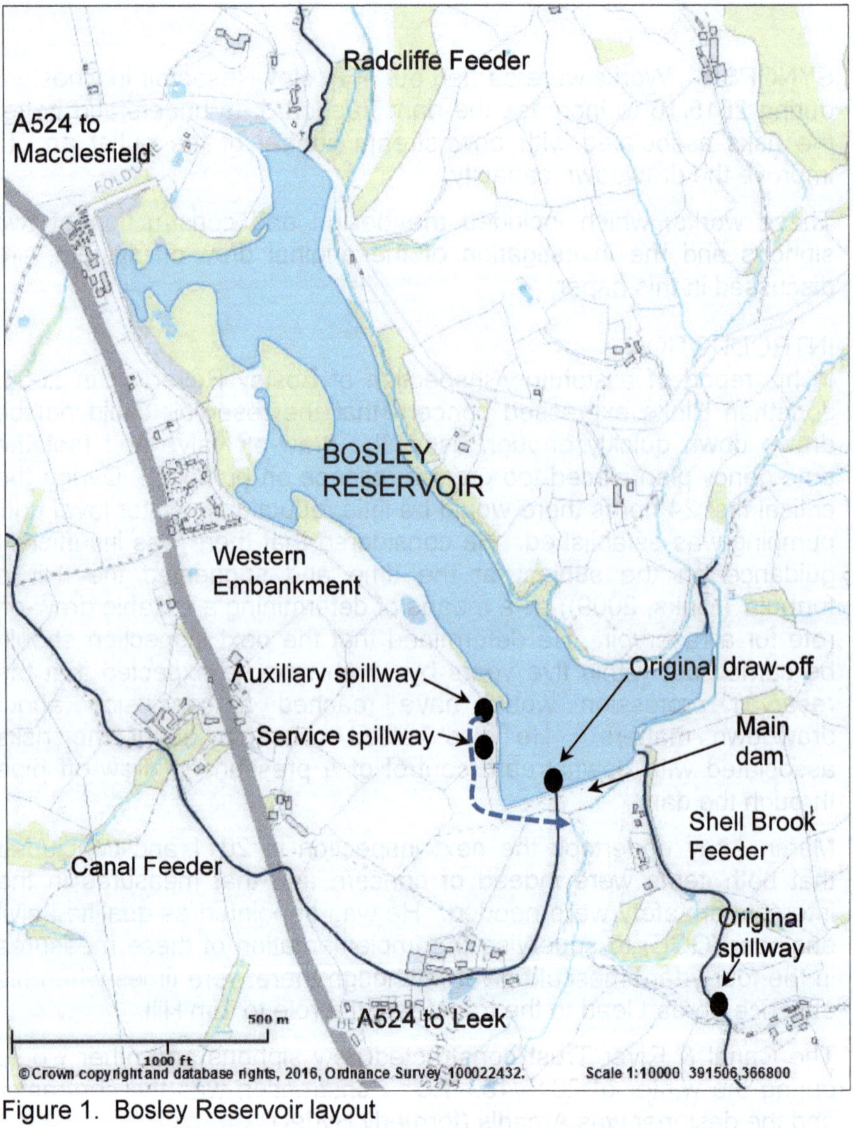

Figure 1. Bosley Reservoir layout

The main headbank, which is 18.5 m high and 270 m long, was built across the valley of the Bosley Brook but it was necessary to build a low flank embankment along the western side of the reservoir, where the land falls away, to retain the water.

Until 1985, when a new spillway was built 600 mm lower than the original weir, 1.8 million m^3 of water were stored. The holding is now 1.5 million m^3.

The area of the reservoir is 32 hectares. The direct catchment area is 5.2 km^2, comprising open moorland. There are also two indirect catchment feeders, Radcliffe and Shell Brook, giving an additional catchment area of 5.86 km^2.

The geology is shales, thin coals and grits of the Millstone Grit Series with a thin cover of boulder clay overlying the bedrock. The dam has a wide core of puddle clay.

CHALLENGES

In his report dated 23 March 2012, the Inspecting Engineer recommended the following measures in the interests of safety:

An updated flood study and flood routing analysis shall be carried out to check the adequacy of the freeboard so as to confirm (or otherwise) that overtopping will not arise during the passage of the PMF design flood event.

A permanent means of increasing the emergency drawdown capacity of the reservoir shall be provided so that the reliance on temporary pumping plant is reduced.

A CCTV survey of the cast iron draw-off pipe shall be carried out to assess the internal condition. Depending upon the findings of this survey it will be possible to determine the scope of any further remedial action that may be required, such as pipe lining or the provision of upstream control, in order to ensure the integrity of the pipe.

FLOOD STUDY

The 1986 statutory inspection had included a flood study of the reservoir. The overflow arrangements had been modified in 1985 by the construction of a new reinforced concrete service spillway at a lower level than the original spillway and a reinforced earth auxiliary spillway. The original spillway was remote from the dam on one of the indirect catchment feeders. When the study had been reviewed at later inspections it had been concluded that the freeboard provision was marginal.

A new study was carried out in-house by the Trust during 2013. The catchment areas and characteristics were reassessed using the FEH CD ROM. The inflow hydrograph for the PMF was derived using the FSR approach, but with the parameters determined using FEH.

A flood routing exercise using an ISIS model was undertaken to pass a PMF inflow of 73.8m³/s through the reservoir and over the main and auxiliary spillways. The original spillway was ignored for this flood routing because the approach to that weir is via a narrow overgrown cutting on one of the reservoir feeders. The stillwater flood surcharge level was 1.09m above the level of the service spillway and the wave surcharge was re-assessed at 0.92m.

It was concluded that the main dam had insufficient freeboard and needed to be raised by about 400mm.

HEADBANK RAISING

The main dam was raised during the works in 2015/16. The core was located by trial pits with a view to extending it upwards using puddle clay. The contractor proposed that it would be simpler to use piles to raise the watertight element. The core was exposed, a line of 1.4m long trench sheets was driven and then fill was placed around the piles to form the raised crest. An access track was built along the new crest using the 'BodPave' system to allow vehicles to reach the new siphons (Figure 2).

Figure 2. Crest raised

DRAW-OFF PIPE INVESTIGATION

When the reservoir was built it was normal practice to lay the draw-off pipe in a shallow trench under the dam and to control flows with a single valve at the downstream end. This is a hazardous arrangement introducing water under reservoir head into the body of the dam (Brown, 2009 and Hughes *et al,* 2015). A defect in the pipe would lead to rapid erosion of the embankment fill. It is particularly difficult to inspect pipes which cannot be easily drained. A submersible remote operated vehicle CCTV camera was introduced into the pipe from the downstream end through a gland fitted downstream from the valve. Visibility was minimal; the sonar was inconclusive due to tuberculation and the camera could not pass a point midway under the upstream shoulder 47m from the downstream end of the pipe.

Divers removed a section of grille from the upstream end of the pipe and introduced a remote operated vehicle (ROV). This also indicated that there appeared to be an obstruction under the upstream shoulder. The possibility of a collapsed pipe or a displaced joint was suspected. It was expected that there would be no remedy for this defect and that the pipe would need to be grouted up. The siphon design allowed for modifications to allow the new system to be used to feed the canal instead of the existing draw-off pipe.

During the siphon construction works the water level in the reservoir was reduced to 10% of the capacity. This gave the opportunity to place a plate over the upstream end of the pipe and carefully drain the pipe, giving a much reduced risk of problems. This enabled a CCTV survey to be carried out in the dry. It was discovered that there was compacted silt in the invert of the pipe where it had settled under the weight of the embankment. This was removed by water jetting. A further CCTV survey identified that there were no visible defects in the pipe, even at the point of the apparent obstruction. This gave confidence that the risks associated with downstream control of an unlined pipe could be tolerated for a few more years.

It is operationally preferable to continue to use the original draw-off rather than carry out the siphon modifications.

The next stage of the works will be to drain the reservoir again to 10% and install a cofferdam around the pipe inlet to hold back the silt and to protect the fishery. The pipe can then be lined using a cured in place liner and a guard valve fitted at the inlet. Guidance will be taken from the conduits guide (Hughes *et al,* 2015).

DRAWDOWN IMPROVEMENTS

The original draw-off comprised a single 380mm diameter cast iron pipe, 70m long. At the downstream end two in-line 305mm valves situated in a short masonry tunnel controlled the flow. The maximum outflow was gauged at 0.14m³/s, considerably less that that calculated using the hydraulics of the system. This was no doubt mainly due to the condition of the intake grille which, when removed by divers, proved to be a considerable restriction to the flow (Figure 3). It was replaced with something more suitable.

The reservoir inflow, calculated as the average non-separated flow (ANSF) was similar in magnitude to the draw-off pipe capacity. Therefore no reduction in water level would be possible until pumping could be installed. It has been shown (Brown *et al,* 2010 and Windsor, 2012) that it is practical to establish up to 1m³/s of temporary pumping within 24 hours. Twice that amount would have been needed to reduce the reservoir holding to 50% within 5 days, the Trust's adopted criterion (Brown, 2009). During that first critical 24 hour period, nothing could have been achieved. This was confirmed as unacceptable by the Inspecting Engineer in 2012.

Figure 3. Grille removed by divers

The Trust considered the options for addressing this problem using an in-house panel of experts. Tunnelling options were rejected on the grounds of difficulty, cost and the effect on the dam. Sluices in the spillway were rejected because only the top water would be removed and pumps would still be needed to further reduce water levels. A variety of siphon options were considered. The Trust has a policy of laying the upper part of the pipework just below top water level, so that the siphon is always available for use when the reservoir is full without the need for a permanent priming pump or bringing in temporary equipment (Figure 4). It also allows the water level to be drawn down further before the siphon breaks, there being the atmospheric pressure limit on siphoning depth to somewhat less than 10m. Multiple pipes were preferred to a single large pipe so that they could be tested separately without leading to problems downstream. Also if they were to be operated at reduced capacity there was less likelihood of losing the prime. Siphons of this design had been built at Birkenburn Reservoir in Scotland by British Waterways, the Trust's predecessor, in 2010 and earlier at Killington Reservoir in Cumbria and proved satisfactory.

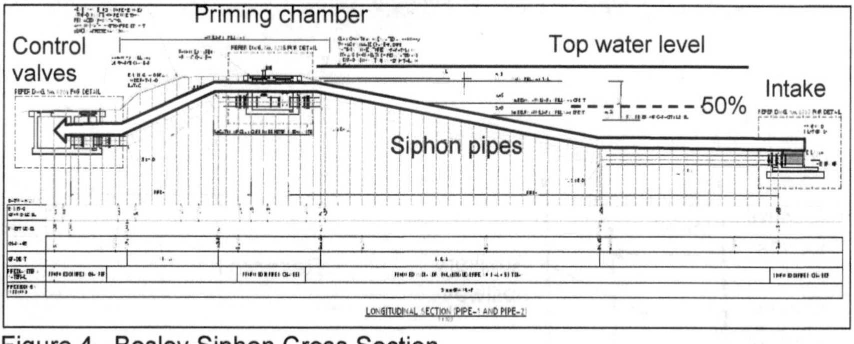

Figure 4. Bosley Siphon Cross Section

Siphons at the highest point of the main dam would have had greater differential head but would have been more damaging to the embankment so a location at the auxiliary spillway was chosen. This also allowed existing pipework to be employed to take the discharge away.

Twin 625mm diameter HDPE pipes were chosen to reduce the stored volume by 50%, drawing the reservoir down by 3.0m in five days. The intakes are 6.1m below top water level and are protected by grilles. A priming chamber with guard valves is situated under the auxiliary spillway crest. The control valves are situated in a dry chamber at the downstream toe of the auxiliary spillway and discharge into a stilling chamber which remains full of water to ensure that air cannot enter the system at this point (Figure 5). The siphon discharge enters the existing system below the main spillway (Figure 6).

Figure 5. Pipework being laid under the downstream shoulder

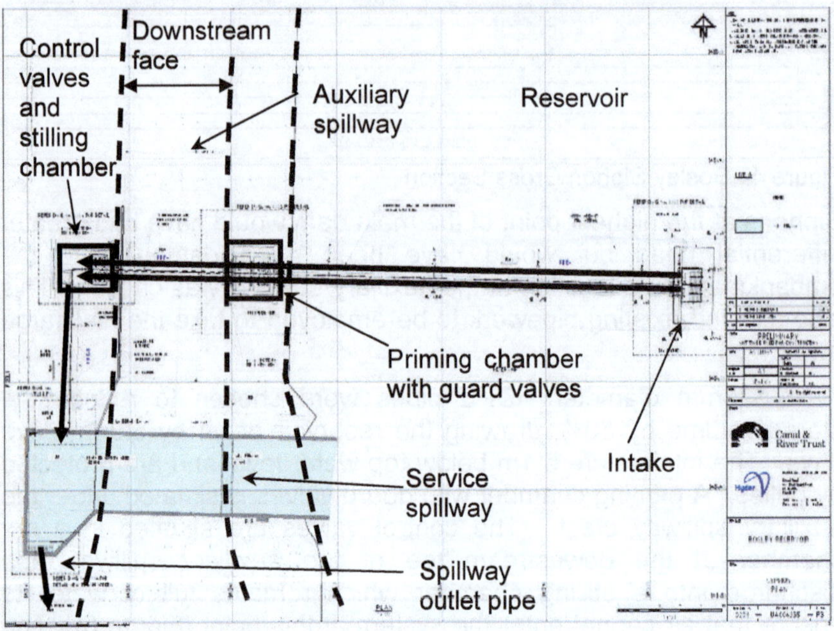

Figure 6. Siphon layout

In order to allow the siphons to be built the Armorloc revetment blockwork protecting the auxiliary spillway from erosion had to be temporarily dismantled. The water level was drawn down to 10% to minimise the possibility of the reservoir spilling whilst the scour protection was absent. Water levels were monitored and a contingency plan prepared for pumps to be brought in should the level have risen to an unacceptable degree. The priming chamber was situated in the area of the core of the dam. Puddle clay was reinstated around the chamber, the outside faces of which were built to a slight angle to the vertical to promote a good seal as the clay consolidated. Once the pipework was in place it was subject to pressure and vacuum testing before it was buried. Defects would have been much harder to rectify later.

Had it proved necessary to grout up the original draw-off, a facility would have been necessary to allow the siphons to be used to supply the canal. A smaller diameter control valve was therefore fitted in parallel with one of the 625mm diameter downstream valves. The upstream end of the same pipe was also designed to allow it to be capable of extension to a greater depth in the reservoir.

Priming of the siphons, should it be needed, is done using a portable air compressor. Air passing through a venturi creates a vacuum which draws the air from the head of the siphon (Figure 7). There is a 150mm connection to each siphon pipe in the priming chamber for this purpose.

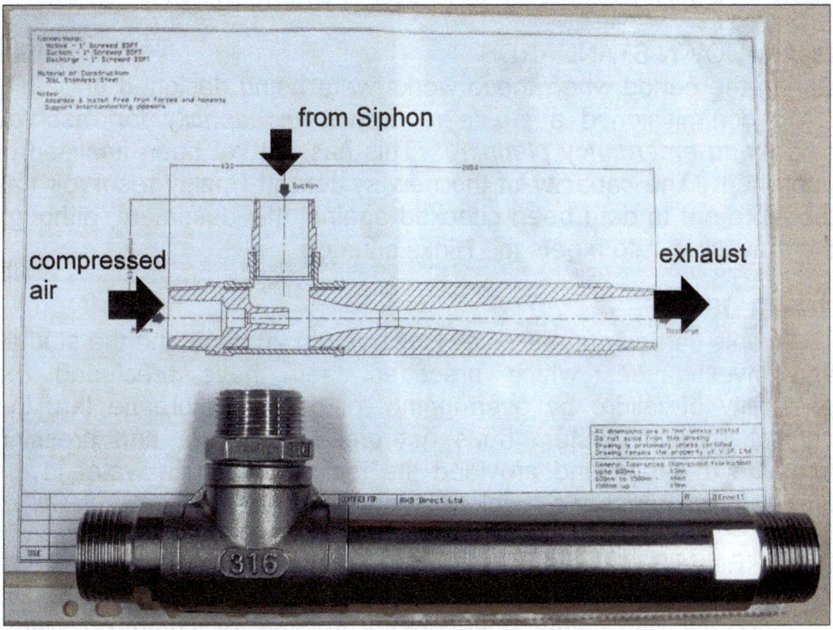

Figure 7. Priming system

The siphons were successfully tested on 19 February 2016 to the satisfaction of the qualified civil engineer (Figure 8).

Figure 8. Siphon test, discharge into service spillway system

DRAW-DOWN STANDARDS

During the period when these works were being designed and built, Defra commissioned a *Guide to drawdown capacity for reservoir safety and emergency planning*. This has not yet been finalised or published. The capacity of the new system at Bosley reservoir has therefore not to date been checked against this document, although the drawdown rate meets the Hinks criterion

CONCLUSIONS

The works carried out at Bosley Reservoir in 2015/16 and the studies and investigations which preceded them have precluded the possibility of failure by overtopping, given an assurance that the original draw-off system does not present a major and pressing threat to the dam and provided the facility to reduce water levels quickly in the event of a problem being identified.

REFERENCES

Brown DH (2009). Low Level Outlets 3: British Waterways' approach, in *Dams & Reservoirs Volume 19 Number 1 March 2009.* Thomas Telford, London

Brown D, Gardiner K and Williams N, (2010). *Exercising of Emergency Draw-down Plans* Long-term benefits and performance of dams. Thomas Telford, London, 2010

Hinks J, (2009). Low Level Outlets 1: formula for target capacity, *in Dams & Reservoirs Volume 19 Number 1 March 2009.* Thomas Telford, London

Hughes AK, Bruggemann DA, Gardiner KD, Blower T, Deane M and Nash D, (2015). *Dam and reservoir conduits. Inspection, monitoring, maintenance and repair.* CIRIA, London

Windsor DJ, (2012). Pebley Reservoir (Derbyshire) Emergency Drawdown Exercise in *Dams: Engineering in a Social & Environmental Context – Proceedings of the 17th Biennial Conference of the British Dam Society pp193-205.* ICE Publishing, London

Dams – Benefits and Disbenefits; Assets or Liabilities?
ISBN 978-0-7277-6133-0

doi: 10.1680/dbdal.61330.333

Slaithwaite Reservoir Improvement Works

M COOMBS, Arcadis
N POVEY, Canal & River Trust (seconded from Arcadis)

SYNOPSIS David Brown (CRT Principal Reservoir Engineer) identified the need for the drawoff pipe at Slaithwaite reservoir to be lined in 2009 (Brown, 2009). In addition the drawoff valve itself, located at the head of a tunnel approximately 40m long, was proving increasingly difficult to operate and was located within a confined space. In order to address these issues a project was undertaken to extend the existing drawoff pipework, replace the valve, provide a guard valve and investigate the upstream inlet arrangements.

The opportunity was undertaken to include further improvement works, identified in the 2011 Section 10 inspection and by the Supervising Engineer, during the project. This included replacement of a bellmouth penstock to allow safe operation remotely from the reservoir bank, repairs to the spillway, grouting of a downstream culvert and refurbishment of a pedestrian footbridge. This paper outlines the works undertaken, the difficulties encountered and the solutions subsequently developed.

INTRODUCTION

The Canal & River Trust (the Trust) is the undertaker for 72 'large raised reservoirs' in England. They have an average age circa 200 years, and represent some of the oldest reservoirs in the country.

Arcadis Consulting Ltd (previously Hyder Consulting) is the framework design consultant for the Trust and has undertaken design work on various Trust reservoirs, working in conjunction with key Trust personnel - in this instance Alex Holt, Principal Engineer on the project.

Slaithwaite reservoir is located immediately upstream of the town of Slaithwaite in the Colne Valley, 7km west of Huddersfield, West Yorkshire. The reservoir has open access to the public, but no vehicular access. The footbridge and crest provide a Public Right of Way footpath for the general public and is a route regularly used by school children. The reservoir also has a large fish population and an active angling group.

Slaithwaite Village is a conservation area and although the site has no statutory heritage designations, the Trust considers all its structures to be of heritage value. The site is in close proximity to residential properties, as shown in Figure 1.

Figure 1. Slaithwaite Reservoir

SLAITHWAITE RESERVOIR

The reservoir is a large raised reservoir (Category A) constructed circa 1797 to supply the Huddersfield Canal and impounds the waters of Merry Dale Clough, a tributary of the River Colne. The reservoir has a storage capacity of 310,000m³, a surface area of 46,000m² and a normal top water level of 167.04mAOD. The dam is 17m high, of earthfill construction and is reported to have a puddle clay core.

Water enters the reservoir from the west, from an inlet in Merry Dale Clough. It flows down the clough channel into the reservoir. Where the water enters the channel there is also a by-pass channel that runs from west to east along the north of the reservoir. This is used to allow water to enter the industrial mill buildings' ponds, just north of the footbridge.

The reservoir has a bellmouth overflow chamber which leads to a tunnel that discharges to the far east of the site. The spillway runs over this tunnel and the footbridge over the spillway. Downstream of the spillway is a series of drop shafts (an upper and lower) and channels. A plunge pool is situated at the bottom of the lower drop shaft with a 40m tunnel downstream, while an auxiliary tunnel runs over the top of this. The general layout is shown in Figure 2.

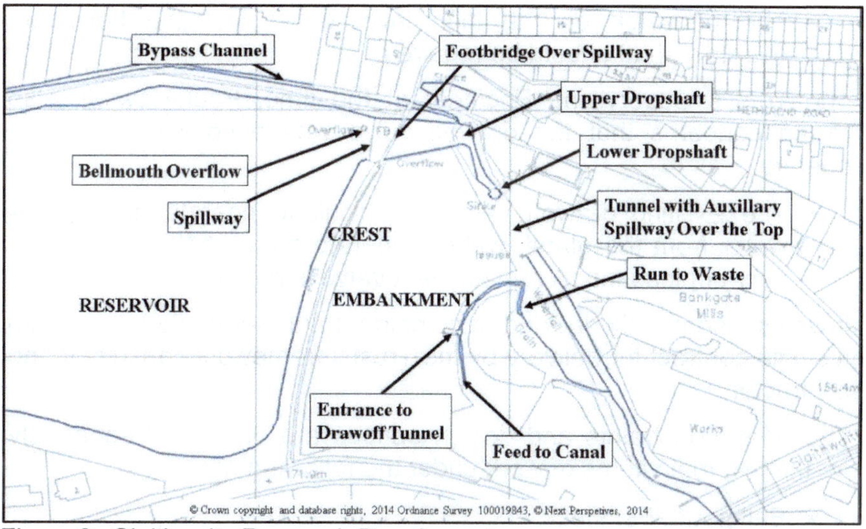

Figure 2. Slaithwaite Reservoir Dam Layout

HISTORY

The reservoir was constructed from 1795 to 1799 by the Huddersfield Canal Company. The construction engineer was Benjamin Outram.

Leakage was a problem during construction and in a 1797 report by Robert Whitworth stated "The leakage, at present, is rather too much, and as the Water rises it may expect to increase, but, if it do not leak more than the supply of the canal will require, it will do very well.' In June 1799 work was undertaken to make the bank watertight.

Over the course of its life the reservoir has supplied a number of mills and both the canal and railway. An iron pipe connected to the outlet is believed to have run as far as Huddersfield Station where it powered a hydraulic turntable mechanism.

A drawing dated 1968 shows details of modifications to the spill weir and channel, including extending the weir, widening the upstream end of the channel and lowering its invert level. Work was also undertaken in 1991 to raise and widening the crest, widen the spillway discharge channel and construction of a new reinforced concrete channel over the original spillway discharge tunnel.

SECTION 10 INSPECTION

The most recent Section 10 inspection was carried out on the 5 May 2011 by Andrew Rowland of Black and Veatch.

There were no recommendations in the interests of safety, however one of the main recommendations is given below:

- The penstock on the high level outlet into the bellmouth overflow shaft should be repaired.

The Trust's Supervising Engineer at the time, Paul Howlett, also raised concerns that the drawoff valve was becoming difficult to operate:

- The existing drawoff valve is difficult to operate and sprays water from the gland, wetting operators and making operation very unpleasant. A replacement valve is required.

It was subsequently decided to combine a number of work items into a single contract for delivery by the Trust's framework consultant and contractor.

The brief, constraints and design options are discussed separately for the two main items of work under the following headings:

- Bellmouth Penstock
- Drawoff Valve and Safe Operation

<u>Investigations</u>
Prior to any work being undertaken on the project a number of inspections were undertaken (or available) as follows:

- A hydrographic survey was conducted by British Waterways in March 2004.
- An Inspection for Assessment on the footbridge was conducted on the 15 November 2011
- A dive survey was conducted on the 21 December 2011.

A record drawing from 1978 is shown in Figure 3, which provides a section through the embankment and used to direct the dive survey.

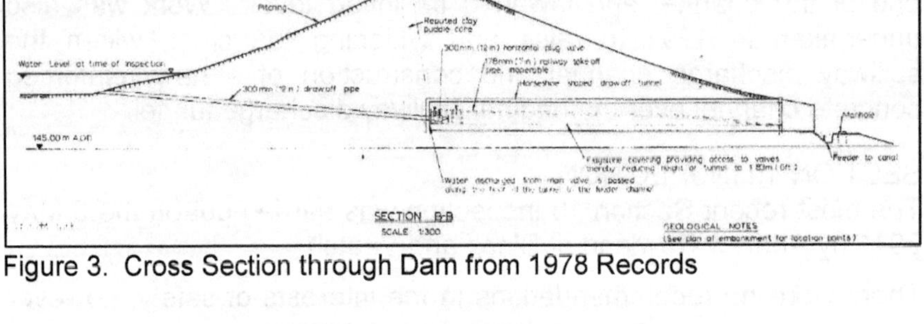

Figure 3. Cross Section through Dam from 1978 Records

BELLMOUTH PENSTOCK

The reservoir has a low level scour valve drawoff and a high level penstock outlet (into a bellmouth overflow structure). The penstock is located at the base of the bellmouth structure and is manually operated (Figure 4).

The bellmouth structure is directly in front of the spillway, with a footpath running alongside the bypass channel. Therefore unless water levels are low the penstock cannot be reached without wading out into the reservoir across the spillway. This is obviously undesirable with significant health and safety issues.

The penstock also forms part of the emergency drawdown plan, and may therefore have to be operated when the water level is at its highest. In this situation the spillway will be in full operation, and a boat would be required to reach the penstock.

In order to rectify this situation a number of operational issues needed to be resolved:

1. An improvement in the ease of operation of the penstock - due to the penstock condition it was decided that gearing needed to be included in the solution.
2. The ability to provide both manual and remote operation. Manual operation to cater for the failure of any remote operation system.

Various options were considered together with the location of any remote operational system. It was decided that:

- Gearing should be incorporated to increase the ease of operation,
- The penstock should be replaced,
- A platform should be provided to allow manual operation to be undertaken at the bellmouth if required.

A number of alternative methods for the remote operation were considered, drawing on experience from sewage treatment, marine and petrochemical industries, as follows:

<u>An Electric Operating System</u>
This was dismissed for a variety of reasons;

- There is no power at the site and it would have been expensive to bring a cable to the location to provide a continuous power supply,

- A portable generator would be an alternative but is less than ideal since there is no vehicle access to this location,
- If the power supply failed for some reason a battery backup would be required – this would take up space and have to be maintained,
- The system was intended for emergency use and therefore infrequent operation, and as such did not justify the expense.

A Hydraulic Operating System

Such a system could be used and improve operational ease in a similar way to an electric system, with the advantage of only requiring a small oil tank and a portable generator to operate. But this was also discounted due to:

- The hydraulic hoses could potentially leak,
- As above a portable generator would be required,
- Suitable space would be required for the oil tank and generator ideally out of the rain and sufficiently robust to avoid vandalism.

A Cable Wire System

This was the option selected (Figure 5). It does have some disadvantages; it is not as smooth or rapid an operating system as the other two, but it also a number of distinct advantages:

- The system is simple with no power supply required,
- It only requires a small operating mechanism which can be located within a manhole,
- It can be operated manually via a simple T key,
- It can be operated at a more rapid rate if required via a normal drill with an appropriate adapter,
- Operating costs are lower compared to the other two options.

All of the above options require a hose, cable or wire from the penstock structure to the preferred location on the footpath. To facilitate this a simple cable duct was installed from the headstock of the penstock to the bank and cables laid within it. In order to retain the ability to manually operate the penstock at the bellmouth (should the remote operating system fail) the headstock was designed such that it can be removed and a T key used to operate the penstock from above.

Figure 4. Existing Outlet Penstock.

Figure 5. Replacement Penstock

DRAWOFF VALVE AND SAFE OPERATION

Investigations via dive surveys were made of the upstream inlet arrangement. Historical drawings indicated a 12" pipe but the dive survey indicated an 800mm diameter vertical pipe with a 90° bend, located some 45m upstream from the centre of the dam crest, approximately 14m below top water level (Figure 6).

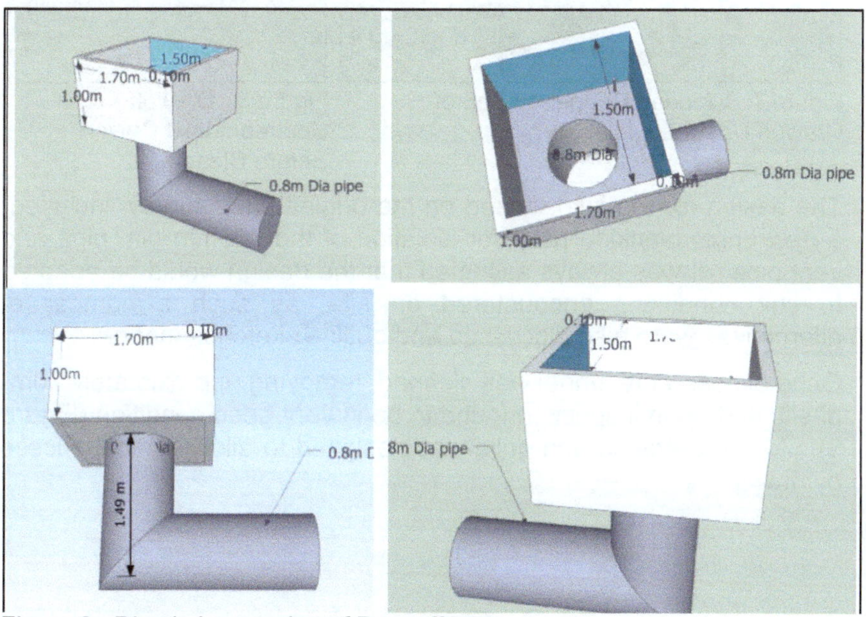

Figure 6. Diver's Impression of Drawoff Inlet

It was therefore decided to lower the reservoir and install a cofferdam to isolate the inlet area, whist still maintaining a level of between 5%

and 10% within the reservoir. This is sufficient to avoid the need for oxygenating the water or undertaking a fish rescue.

The installation of the cofferdam allowed the inlet to be exposed, but the resulting structure proved to be significantly different from the historical records. Underneath approximately 2.4m of silt there was a 1.5m high by 1.3m wide oval masonry tunnel, 40m long. There was a wooden hopper structure as per the diver sketch, but the 800mm vertical pipe identified by the divers was in fact the vertical flow path down through the silt with a subsequent horizontal route along the top of the tunnel (Figures 7 and 8).

Figure 7. Wooden Hopper on Top of Drawoff Upstream Inlet

Figure 8. Drawoff Upstream Inlet Culvert Before Cleaning

The design for the inlet, based on the original dive survey, included; a new grille, blanking plate for isolation of the 800mm dia. pipe and vent pipe. It was always assumed that the design would be adapted to suit conditions encountered on site, as such a number of alternatives were also discussed and outline sketches made.

Once exposed the tunnel was cleaned, removing approximately 70m^3 of silt, and upon inspection found to be in very good condition (Figure 9). A new entrance and grille was designed to allow vertical access by divers.

Figure 9. Drawoff Upstream Culvert After Cleaning

Downstream Outlet

The tunnel ended in a headway with a 30" cast pipe assumed to be through the core to a cast iron bulkhead at the headwall of the downstream tunnel, where the existing valve was located. In order to strengthen this pipe and reduce the risk of potential deterioration a new 500mm diameter PE liner pipe was installed and grouted in place with a flanged end so it could, if necessary, be plated off in the future.

The bulkhead arrangement at the downstream tunnel was very unusual. It had two drawoff pipes cast into it, one a 9" blanked off pipe and the other a 12" pipe connected to the drawoff valve. The 9" pipe is believed to have been the supply to the railway line to operate the hydraulic turntable at Huddersfield station (Figures 10 and 11).

The existing pipe work and bulkhead were removed. A new sleeve pipe from the upstream tunnel, blockwork wall and guard valve were installed (Figure 12). Pipework was also installed along the tunnel to the entrance where a scour valve was located (Figure 13). Immediately downstream of the valve, pipework directed water back into the outlet channel underneath the tunnel floor.

The new arrangement has dramatically improved the situation, with the guard valve able to be closed to allow the scour valve to be maintained and the scour valve located just inside the door of the tunnel entrance thus reducing the confined spaces risk and removing the need to traverse the tunnel (apart from when exercising the guard valve).

Figure 10. Existing Drawoff Valve in Downstream Tunnel

Figure 11. Bulkhead with Railway Pipe Blanked Off

Figure 12. New Guard Valve at Bulkhead

Figure 13. New Drawoff Valve At Downstream Tunnel Entrance

CONSTRUCTION

The works were delivered by the Trust's framework contractor, Kier May Gurney Ltd with an overall cost, including design and management costs, of the order of £1.6m. The work was undertaken in two phases;

- Phase 1 between July 2012 and December 2012 which comprised the penstock, footbridge and drop shaft leakage works.
- Phase 2 between July 2013 and May 2014 which consisted of the inlet and outlet works.

There was a requirement for the water level to be lowered significantly during Phase 2 construction to a level of approximately 10% and maintained at this level throughout the works. In order to facilitate this process a diversion of the inlet flow was undertaken by installation of a temporary weir on the main inflow. An additional temporary weir with a hydraulic hose duct was installed downstream of the main spillway to allow a flow of water to bypass the reservoir but still enter the canal.

As the works were undertaken in the summer there was a need to manage the water resource carefully, to ensure that the canal remained in water and that the normal supply from Slaithwaite was not wasted.

<u>Other Works</u>

Other works carried out under the same contract included:

- Pressure grouting of the base slab culvert downstream of the second dropshaft and plunge pool,
- Refurbishment of the footbridge over the spillway; as part of this refurbishment a community initiative to develop interpretation panels to be hung on the bridge was undertaken involving the local community and primary school. These panels tell the story of the reservoir and canal history and its importance to the local community (Figure 14).
- Installation of a land drain on the eastern embankment, discharging via a V-notch into the scour outlet channel,
- Minor repairs to the spillway joint sealant and wave erosion protection.

Figure 14. Refurbished Footbridge.

LESSONS LEARNT

As with all projects there are lessons that can be drawn from the work undertaken for the benefit of similar works undertaken elsewhere. The key points being:

- Diver surveys will only give an indication of what is present.
- Make sure you have a plan B i.e. sketches of alternative options.
- Consider the gearing on any form of remote operation.

CONCLUSION

The works have significantly improved the safe operation of the scour valve and bellmouth penstock, as well as providing a guard valve. In addition there have been enhancements to monitoring (V Notch), leakage reduction and the general amenities offered by the reservoir.

The refurbished bridge was officially opened on 29 November 2012 and there were representatives from the local school and community, the Trust, Arcadis and Kier in attendance.

The works undertaken will allow the continuing successful operation of one of the oldest reservoirs in the country well into the future.

REFERENCES

Brown DH (2009). *Unprotected Pipelines in Reservoir Embankments.* Unpublished Internal Report, British Waterways, UK.

SECTION 7:

NEW AND UPGRADED RESERVOIRS

Dams – Benefits and Disbenefits; Assets or Liabilities?
ISBN 978-0-7277-6133-0

doi: 10.1680/dbdal.61330.347

Butterley Spillway Improvement Works

R WOODS, Mott MacDonald Bentley
C MORRISON, Mott MacDonald

SYNOPSIS Following on from a partial spillway failure in 2002 during a flow event significantly below Probable Maximum Flood (PMF), and a lengthy planning process, works have now started on site for the much-needed spillway improvement works at Butterley Impounding Reservoir. This paper will describe the project journey, the challenges faced by the project team and the design details.

The existing spillway is Grade II listed and an iconic structure within the local landscape. The solution that has currently been awarded planning permission has been reached through compromise between the Client, the All Reservoirs Panel Engineer (the Inspecting Engineer), The Project Team and Local Planning Authority.

The upper two-thirds of the spillway will be refurbished including replacement of the existing masonry invert and raising of the right hand side spillway wall. The lower third of the spillway will be completely rebuilt to provide a more consistent gradient. A number of existing features are to be retained and incorporated into the new spillway including masonry piers, copings and curved wing walls to help preserve the character of the original spillway.

INTRODUCTION

Butterley Impounding Reservoir (IRE) is located in Marsden, near Huddersfield in West Yorkshire and is the lowest of four water supply reservoirs in the Wessenden Valley, below Wessenden Head, Wessenden Old and Blakeley Reservoirs. These four reservoirs and their associated catchwater areas are all owned and maintained by Yorkshire Water Services (YWS) Limited.

Butterley IRE provides a raw water supply to Longwood Water Treatment Works (WTW) and Blackmoorfoot WTW at a rate of up to 20 Ml/d and has a capacity of 1,773,000m³. As a result of its size, it is categorised as a "large raised reservoir" by The Reservoirs Act 1975 ("the Act") (HMSO, 1975) and therefore it must be operated, inspected and managed in accordance with the requirements of the Act.

The reservoir is designated Category A in accordance with "Floods and Reservoir Safety" (ICE, 1996) which recommends the spillway be designed to safely pass the PMF.

RESERVOIR HISTORY

T&C Hawksley was commissioned by Huddersfield Corporation Waterworks to construct Butterley Reservoir, with works commencing in August 1891. In 1901 the reservoir was completed; however as it was being filled with water for the first time a leakage occurred when only 36ft of the intended 94ft depth was achieved. Following an investigation into the reasons for the leakage T&C Hawksley was dismissed by Huddersfield Corporation Waterworks and G H Hill was appointed to carry out remedial works. The remedial works involved the construction of wing trenches on each side of the dam and were completed in June 1906.

CONSTRUCTION

The reservoir is retained by an earth-fill embankment approximately 34 metres high and 229 metres long with a clay core (Figure 1).

Figure 1 – Upper spillway/lower spillway location

A masonry and concrete overflow weir and tumblebay structure is located on the right hand (eastern) side of the embankment which is connected to a curved masonry spillway channel. The upper section of the spillway (approximately two-thirds of its total length) is made up of a masonry block invert in a stepped formation with masonry keystones approximately every four metres. The lower section of the spillway contains two steeper cascades made up of larger masonry blocks.

The spillway walls are also of masonry construction with stepped chamfered copings placed on top. Three intermediate masonry pillars are situated throughout the length of the wall with a curved wall and terminal pier located at the bottom of the spillway.

The overflow was first registered on 11 July 1985 as a Grade II listed structure, and now falls under the Planning (Listed Building and Conservation Areas) Act 1990 (HMSO, 1990) for its special architectural and historic interest.

PROBLEM DEFINITION

In July 2002 a flood event caused a number of masonry blocks in the spillway to be dislodged from the concrete backing between the two cascade sections in the channel (Figure 2).

This flood event was determined to be somewhere between approximately a 1 in 12 year and a 1 in 100 year event, which is significantly less than the PMF event (approximately 241m^3/s) which the spillway should be designed for.

Figure 2 - Damage caused to spillway invert in 2002

Following a statutory inspection pursuant to Section 10 of the Act the Inspecting Engineer recommended that a study of flow depths and velocities was undertaken in relation to the existing overflow channel. Physical modelling work was commissioned by Yorkshire Water in order to determine the hydraulic capacity of the overflow. The model identified the following deficiencies:

- The sweeping curve of the spillway channel concentrated flows around the outer wall, creating a depth well in excess of the spillway wall height and thus allowing significant overtopping.
- The pillars in the spillway were built such that they protrude around 200mm into the channel. High velocity flow travelling around the outer wall of the spillway hits the protruding pillars and plumes, sending water onto the adjacent ground surface even at very low flows. At the PMF, some of the plumes would be expected to be nearly 20m high.
- At the cascades of steps it was found that flows separated from the base and leapt the whole of each flight of steps causing the water to severely impact onto the channel downstream. This could lead to impact damage of the spillway structure and negative pressures causing masonry blocks to be plucked out of the spillway, as per the damage caused in 2002.
- As the channel straightens, flow sweeps off the bend forming a large cross wave. This cross wave impacts onto the dam side turret at the start of the second cascade and sends water out-of-channel. This then flows overland across the embankment to the scour channel.

These defects identified by the physical model had to be resolved by instruction of the Inspecting Engineer, who was subsequently appointed as the Qualified Civil Engineer (QCE) to oversee the design and construction of the works, and in order for Yorkshire Water to fulfil their statutory obligations under the Act.

MASONRY SPILLWAY INCIDENTS

There have been some relatively recent incidents, at Boltby Reservoir, North Yorkshire in 2005 and at Ulley Reservoir, South Yorkshire in 2007 which have highlighted the potential risk of damage to stepped masonry spillways during high flows. As a result the Environment Agency produced the "Guidance for the Design and Maintenance of Stepped Masonry Spillways" (Environment Agency, 2010). This highlighted that the presence of masonry invert steps can result in the production of negative pressures which may be sufficient to dislodge or remove the blocks if insufficient restraint is present.

In addition, the guidance also indicates that once one block has been dislodged or removed the remaining sections of the masonry are then at increased risk, and rapid deterioration of the structure can occur. It was therefore clear that the masonry invert blocks would not be satisfactory to safely pass a PMF event.

FEASIBILITY

Twelve options were developed and analysed in order to determine the best method to address the issues highlighted from the physical model and to satisfy the local residents and planners. Some of these options included: constructing a new culvert channel down the west side of the embankment; creating a new spillway over the centre of the embankment; mining out a drop shaft and tunnel; and even the discontinuance of the reservoir was considered.

Through evaluating each proposed solution it was concluded that the preferred option to take forward was the reconstruction of the existing spillway, including a new stepped concrete invert and ramp to bridge the cascade, raising of the spillway walls and setting back the faces of the piers.

The main benefit of this option was that it utilised the majority of the existing spillway fabric, with a few modifications which would then enable the spillway to meet the safety requirements set out by the QCE in terms of capacity and performance under high flows.

The other solutions involving shafts, culverts and spillways over the top of the dam could meet the requirements set out by the QCE, however they would all increase the risk to the structure by creating additional potential seepage paths through and past the core, and in the case of the tunnel options require construction in areas which are known to being geologically 'difficult'. These solutions would thus be unacceptable in terms of reservoir safety when there was a solution which did not increase the risk to the structure. It was felt that when dealing with such a high consequence of failure, i.e. major loss of life, the risks should be eliminated where possible. These risks could be eliminated by construction of an option that does not involve a new excavation through the water tight element of the dam.

The preferred option did, however, compromise the existing structure by the loss of some of the historic fabric, but design changes were made to retain the essential form of the historic structure, and its relationship with its landscape setting, whilst also retaining as much as possible of architectural design features and the historic materials of the original.

PLANNING PERMISSION AND LISTED BUILDING CONSENT

Due to the Grade II listing of the spillway at Butterley IRE, development of the structure to allow it to meet current and future needs turned out to be challenging. Public consultation began in early 2012 and unfortunately the improvements proposed at that time did not meet with a positive response.

YWS and Mott MacDonald Bentley (MMB) spent 2012 and early 2013 developing the design in consultation with the local planning authority, English Heritage and Natural England to name but a few. In June 2013 both planning and listed building applications were submitted for improvement works to the spillway and clay core raising. The improvement works to the spillway entailed building a new reinforced concrete spillway within the existing one and smoothing out the steep cascades with a gradient similar to that of the existing upper spillway. Elements were incorporated into the design to mitigate the visual impact on the existing structure, through reusing the coping stones and lining the external face of the channel in masonry.

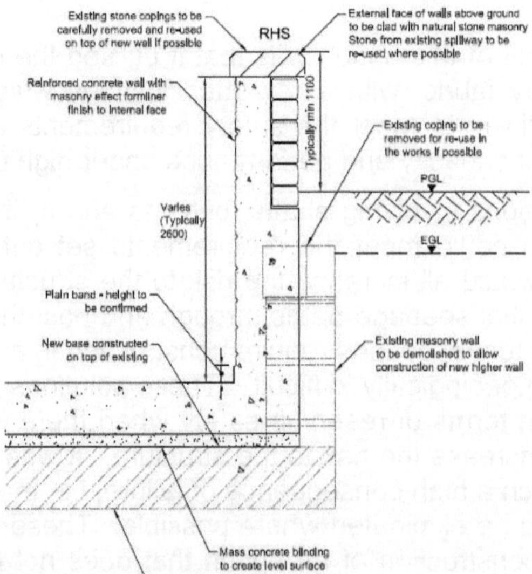

Figure 3 – Extract from drawing PR02-07-230 (P1) Proposed new spillway wall section (Original Planning Submission)

While the local planning authority (LPA) was reviewing the planning and listed building applications, YWS and MMB continued to liaise with third parties to further refine the design and retain as much of the existing structure as much as possible. The design was updated to retain the existing masonry spillway walls and raise them with masonry clad concrete. The revised design was then submitted to the (LPA). However, despite these improvements the LPA was not convinced that the substantial harm proposed to the listed structure was completely necessary. Planning permission was rejected in early 2014, meaning that the only option was to go back and refine the solution and retain the existing features wherever possible.

By this point the Section 10(6) compliance date for the improvement works had lapsed and it appeared that Planning (Listed Building and Conservation Areas) Act, 1990 was overruling the Reservoirs Act 1975. A 1 in 1000 year drawdown of Wessenden valley was implemented to provide safety reassurances to YWS and the QCE whilst the improvement works were developed. As a result Butterley IRE was drawn down to 8.0m below top water level (TWL).

All interested parties were in agreement that work to the spillway was necessary but retention of the existing structure's key features had to be maximised. Permission was granted by the LPA to carry out additional investigation work in an attempt to retain as much of the existing structure as possible. Multiple cores of both the spillway walls and the spillway invert were carried out along with trial pits around the spillway to verify its condition. A section of the right hand side spillway wall was also taken down to allow a condition assessment of the concrete backing to be completed.

Figure 4 - Photograph of RHS spillway wall dismantling

The additional investigation work provided YWS, MMB and the QCE with confidence about the structural stability of the existing concrete backing, both under the invert and behind the walls. As a result, the wall raising design in the upper spillway was developed to incorporate the existing walls and the internal face of the raised wall section would now be finished in slim masonry cladding to tie in with the original masonry.

The physical model results were also revisited and the decision was made to remove the wall raising on the left hand side of the upper spillway. This was able to be implemented due to the curvature of the spillway, meaning water depths during PMF are highest on the right hand side. This allows for approximately 100m of the existing spillway wall to be retained unmodified.

Carrying out the additional investigation works allowed for more of the existing spillway structure to be retained, including both spillway walls in the upper section (Figure 5). However the cascade section would still need to be replaced with a more constant gradient invert to prevent flow leaping the cascades and causing serious damage as it impacted the flatter sections. With this revised solution YWS submitted an appeal in October 2014 against the rejection of the planning and listed building application.

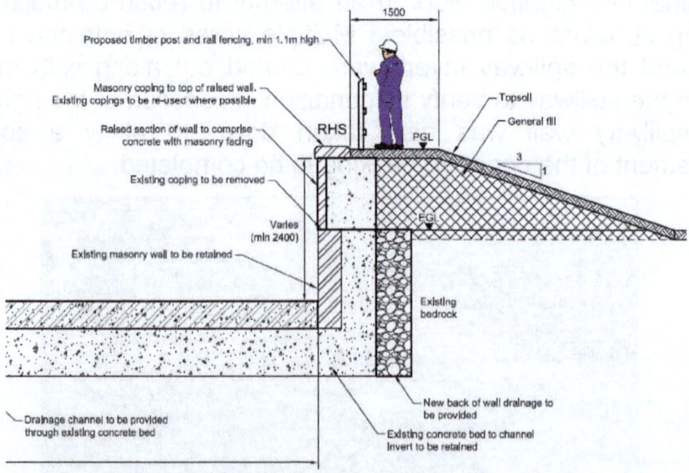

Figure 5 - Extract from drawing PR02-07-233 (P1) Proposed new spillway RHS wall section (Planning Appeal)

In January 2015 an appeal inquiry was facilitated by a Secretary of State Planning Officer. YWS, the LPA and a local interested third party group were represented. Over a one week period the three parties provided evidence and were cross examined on key topics including: engineering, heritage, landscape and environmental impact. Again the substantial harm caused to the Grade II listed spillway by the proposed work was discussed. All parties understood that work of any variety would cause harm; however those in opposition believed that the harm had not been sufficiently mitigated.

In March 2015 planning permission and listed building consent were granted, with no less than 30 conditions. The conditions ranged from the type of masonry to be used to method statements for the safe removal and replacement of key heritage features.

Following a period of detailed design, a request for discharge of all conditions was submitted in November 2015 and approved by the LPA in January 2016.

FINAL SOLUTION

Through the planning process the final solution was modified to best meet the needs of the interested parties. This involved retaining as much of the existing structure as possible while also meeting the requirements set by the QCE and maintaining the safety of the reservoir. As a result the final design became somewhat of a hybrid between the existing structure and a new spillway.

Use of Masonry

Instead of the originally proposed formliner, masonry cladding will be used on the new and raised walls of the spillway to retain the look obtained from the original Victorian design. The masonry that will be used in the new modifications will be sandstone, as per the existing structure, and will also be of similar size and arrangement as the existing masonry blocks. It was suggested that the new masonry could be stained to better replicate the existing stone but this was rejected during the planning process in favour of the masonry obtaining a natural worn colour over time.

The existing masonry coping stones on the right hand side wall and in the lower section of the spillway are to be reused on the new and raised walls where possible and are to be in the same stepped formation as per the existing coping stone layout.

Lime mortar will be used to best replicate the narrow joints in between the existing masonry courses of the new spillway walls.

Wall Raising

The original design included the raising of both walls in the upper section of the spillway in order to retain the PMF. However through the course of the planning process this was reduced to raising only the right hand side wall of the spillway.

The newly constructed walls on both sides of the lower section will be to the same raised height as the right hand side wall in the upper section.

New Concrete Invert

In the upper section, i.e. between the crest bridge and the stepped cascades, the existing spillway invert is of stepped masonry block construction and divided into bays with the existing masonry keystones separating them. These masonry blocks between the keystones are to be removed and replaced with fibre-reinforced concrete in a stepped formation to best replicate the existing arrangement. The existing keystones, which are in a more satisfactory condition than the invert blocks, will remain.

The concrete invert is to be fibre-reinforced in order to prevent the need to construct a complicated steel reinforcement arrangement on site. Fibre reinforcement will also reduce the risk of break-off of the corners of each step.

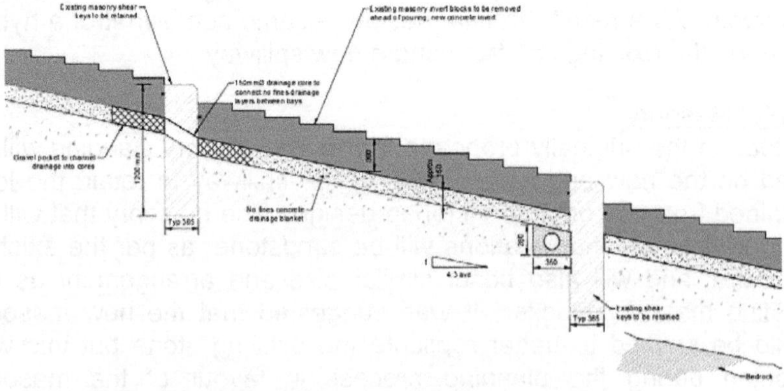

Figure 6 - Section through spillway invert of upper section

<u>Reconstructed Lower Spillway</u>

The existing lower section of spillway will be replaced with a reinforced concrete channel at a renewed profile to remove the cascades and prevent a similar failure to that which occurred in 2002. The channel invert will also be in a stepped formation, with masonry 'mock' keystones placed to replicate the appearance of the upper section.

Due to the re-profiling of the invert the existing walls will need to be demolished and new reinforced concrete walls built in their place. These walls will be clad in masonry with lime mortar between the joints as per the raised walls in the upper section. Existing coping stones will be modified and reused where possible; however some new coping stones will be required on the new lower section walls as the overall length of the spillway will be extended.

<u>Pillar Modifications</u>

The existing masonry pillars protrude into the spillway about 200mm from the walls and as a result can cause plumes of water up to 20m high which then falls onto the adjacent embankment during a PMF event. To prevent this it is necessary to provide a smoother profile to the inner face of the spillway channel walls. The pillars are to be retained and therefore their faces within the spillway channel will be made flush with the adjacent wall faces.

Where the spillway walls are to be raised it will also be necessary to increase the height of the associated pillars. It is intended to raise the pillars with similar large blocks of sandstone as per the existing pillar construction and place the existing capping stones back on top.

Drainage

Due to the hybrid nature of the upper spillway invert, i.e. the remaining masonry and the new concrete, it is likely that some water may penetrate underneath the slab invert and therefore drainage will be required to reduce the risk of damage to the structure.

The drainage solution comprises a 150mm layer of no-fines concrete beneath the new concrete invert to allow water to pass underneath the slab. Cores will be drilled through each keystone to allow the water to pass through to the next bay. At every third keystone a collection pipe will catch all the water from the three bays above and carry it out to the left hand side of the spillway, where it will flow into a new back-of-wall drainage pipe. The back-of-wall drainage will run down the left hand side of the spillway and will be discharged back into the spillway downstream of the terminal pillars.

Flood Mitigation Measures

In addition to the drawdown programme within the valley, the QCE instructed that other mitigation measures should be in place during the construction works of the spillway improvements to prevent damage to any of the existing structure or to the embankment if a flood event greater than 1 in 1000 year return period occurs.

Upper Section

The mitigation measures during the construction works will include leaving approximately one third of the spillway channel available to pass flow at any one time. The construction of the spillway improvement works has been programmed in such a way that work will begin on the left hand side and the central bays of the upper section of the spillway, leaving the right hand side as per existing so that it can safety pass flow.

Bulk gravel bags will be placed down the length of the spillway to divert the flow of water into the right-hand side of the spillway. Once the left-hand side and the central bays have been completed then the bulk gravel bags will be repositioned to divert the flow into the newly constructed left hand side and central bays in order to allow the right-hand bay to be constructed.

Lower Section

During the reprofiling works the same arrangements will be made in the lower section as per the upper section.

Once the lower section has been reprofiled, bulk gravel bags will also be placed outside of the spillway walls until the new spillway walls are constructed, in order to dissipate any flow and prevent it causing damage to the surrounding ground and the embankment.

Figure 7 - Visualisation of the Final Solution

CONCLUSION

Developing a listed structure can be a challenge even if only minor modifications are required. In the example of Butterley IRE the improvement works were significant and did require significant refinement throughout the lengthy planning process.

The final solution allows the existing structure to be retained in operation with many of the key features retained. The planning challenges that have been faced by the multidiscipline team have led to the production of a solution that may never have been realised if the structure had not been Grade II listed and so highly regarded by the local people. The bespoke design of the new spillway at Butterley IRE integrates the old and new providing a solution that mitigates some of the substantial harm to the grade II listed spillway.

The spillway at Butterley IRE is a magnificent example of Victorian Engineering and will hopefully be celebrated for many years to come once the much needed improvements are completed in 2017.

REFERENCES

Environment Agency (2010). *Guidance for the Design and Maintenance of Stepped Masonry Spillways - Project: SC080015.* Environment Agency, Bristol

HMSO (1975). *Reservoirs Act 1975.* HMSO, London.

HMSO (1990) *Planning (Listed Building and Conservation Areas) Act 1990.* HMSO, London

ICE (1996). *Floods and Reservoir Safety, 3rd Edition.* Thomas Telford, London, UK.

Dams – Benefits and Disbenefits; Assets or Liabilities?
ISBN 978-0-7277-6133-0

doi: 10.1680/dbdal.61330.361

The Hampstead Heath Ponds Project – achieving dam safety in a highly sensitive area

Dr A K HUGHES, Atkins, Epsom, United Kingdom
B C JONES, Atkins, Epsom, United Kingdom
T E CREED, City of London Corporation, London, UK

SYNOPSIS This paper will describe the works currently being undertaken on Hampstead Heath. There are three 'statutory reservoirs' on Hampstead Heath but two chains of more than 12 reservoirs. This paper will describe the process followed which identified a significant deficiency in spillway capacity down the two chains, the way in which the flood study calculations were carried out, the risk assessments made, and the design decisions made.

The works include a new embankment and a raised dam to provide additional flood storage capacity. The paper will describe the significant amount of consultation undertaken and the concerns voiced by the stakeholders.

INTRODUCTION

On Hampstead Heath, North London, there are several ponds of which eleven are in two pond chains, fed by two tributaries of the River Fleet. The ponds, including three "bathing ponds", are retained by earth embankments believed to be around 300 years old. There are three statutory reservoirs, one on the western (Hampstead) chain, and two on the eastern (Highgate) chain. However, the chains, of five and six ponds respectively, are closely linked.

The Panel Engineer, Dr Andy Hughes, raised concerns that the existing spillways, which generally consisted of pipes of 200mm to 450mm diameter, were inadequate, and the dams were therefore at risk of being overtopped in a flood event.

Most of the dams had very little grass cover on the downstream slopes to provide any erosion protection, and so the Panel Engineer judged that overtopping of the dams was not acceptable.

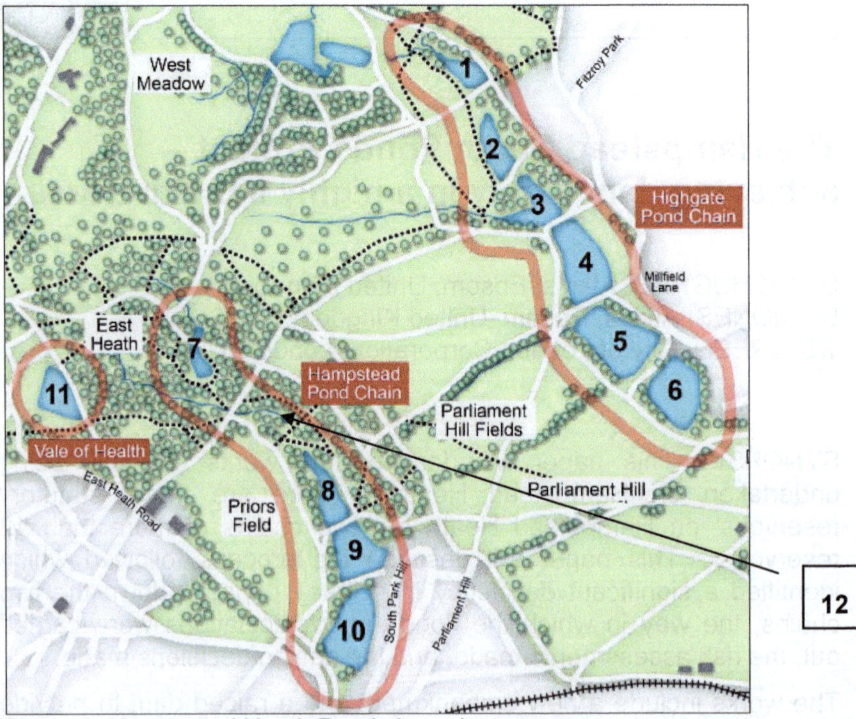

Figure 1: Hampstead Heath Ponds in project scope
(© City of London Corporation)

Table 1. Pond data

Pond No. in Fig 1	Pond name	Height (m)	Length (m)	Capacity (m^3)
1	Stock	3.0	67	6,400
2	Ladies' Bathing	3.7	24	14,200
3	Bird Sanctuary	0.7	101	13,000
4	Model Boating	4.2	94	46,000
5	Men's Bathing	3.1	140	55,000
6	Highgate No 1	3.8	132	42,800
7	Viaduct	4.3	46	5,000
8	Mixed Bathing	0.9	66	11,900
9	Hampstead No 2	5.2	103	25,400
10	Hampstead No 1	4.4	95	50,600
11	Vale of Health	4.8	131	17,800
12	Catchpit (silt trap)	0	10	0 (no dam)

The condition and level of the dam crests, the uneven downstream faces and the size of trees on most of the downstream slopes of the dams, meant that the volumes and speeds of flow overtopping the dams presented a significant risk that overflowing flood water would erode the dam fill material causing uncontrolled breach and release of stored water. A scheme concept was proposed whereby additional flood storage capacity was to be added to the middle of each chain, to reduce the amount of works required to make the downstream dams safe. These works would consist of increasing spillway capacity and/or raising crest levels. The ponds in the centre of the Highgate chain had the largest surface area and on the Hampstead chain there was space to build a flood detention embankment.

The owners and custodians of the Heath, the City of London, recognised the need to virtually eliminate the risk of dam failure by overtopping, and commissioned Atkins to carry out a flood assessment and evaluate options for achieving dam safety. These options needed to be considered in the context of each pond chain as a system, as well as identifying the best solution for each pond.

It was also recognised that the dam safety works would provide opportunities to improve water quality in the ponds, e.g. by desilting ponds, opening culverted sections into natural channels, and creating wetland habitats, but done in a way sympathetic to the setting of the Heath.

Duties of the City of London

Having established a risk of dam breach, the City recognised that it had to comply with the Reservoirs Act 1975 (HMSO, 1975), (where this applies to the three large statutory reservoirs on the Heath) and must also take into account the Flood and Water Management Act 2010 (HMSO, 2010), which may have an extended remit to include cascades of smaller reservoirs coming into effect in the next few years.

In carrying out works to reduce the risk of dam failure, the City of London, as the custodian of Hampstead Heath, was obliged to comply with the Hampstead Heath Act 1871 which requires the City to "...at all times preserve, as far as may be, the natural aspect and state of the Heath..."

ASSESSMENT OF DESIGN FLOOD

Studies carried out by Haycock Associates (Haycock Associates Ltd, 2006; 2010) suggested that during 'extreme rainfall events,' the earthen dams retaining the ponds on Hampstead Heath could not be relied on to store the additional volume of water. Excess flood water would flow over the top and round the sides of the dams, possibly leading to a breach.

If the dams were breached, the water normally stored in the ponds would also be released and combine with the flood water – very quickly and in a completely uncontrolled way – with risk to life and property downstream. The Haycock studies used bespoke methodologies raising concern that the results were not consistent with using accepted industry standard methods – for instance the magnitude of the floods could have been over-estimated.

To address these concerns the designer undertook a fundamental review to assess whether the dams could withstand the design flood, which was the Probable Maximum Flood (PMF) as the dams were all considered to be Category A, as defined in the 3rd edition of 'Floods & Reservoir Safety' (ICE, 1996), because of the close proximity to the highly populated areas of Camden and Kentish Town. A range of smaller floods were also considered, down to the 1 in 100 year return period event.

This fundamental review of storm events and resulting flows through the ponds was carried out using industry standard methods, based on established guidance from the Department for Environment, Food and Rural Affairs (Defra, 2004) and the Institution of Civil Engineers

The flood calculations showed flood peaks were generally 30% to 50% lower than those estimated by Haycock. However, even at these smaller floods the dams would overtop and breaches were possible, with risk to life and property.

The detailed analysis included the development of hydrology for the catchment area using FEH and FSR methods, and the creation of a hydraulic model to calculate the flood return periods at which the dams were overtopped. The hydraulic model was created using industry standard ISIS TuFlow software, with a combination of 1D elements for the reservoir units, spill units, channels and pipes, and 2D grid elements to model overland flows of excess floodwater between ponds and downstream as far as the River Thames.

All dams were overtopped in a PMF. The return periods causing overtopping varied from a low of 1 in 5 at Stock Pond to 1 in 100 at Highgate No.1 Pond at the end of the Highgate pond chain, up to 1 in 10,000 at Hampstead No.1 Pond at the end of the Hampstead Pond chain. The results are summarised in Table 2 below:

Table 2. Summary of Current Standards of Protection as calculated in Assessment of Design Flood Report, 2013

Pond (Hampstead chain)	Smallest return period flood overtopping dam crest	Pond (Highgate chain)	Smallest return period flood overtopping dam crest
Vale of Health	1 in 1,000 years	Stock	1 in 5 years
Viaduct	1 in 1,000 years	Ladies Bathing	1 in 20 years
Catchpit	n/a	Bird Sanctuary	1 in 20 years
Mixed Bathing	1 in 100 years	Model Boating	1 in 20 years (auxiliary spillway)
Hampstead No 2	1 in 100 years	Men's Bathing	1 in 50 years
Hampstead No 1	1 in 10,000 years	Highgate No 1	1 in 100 years

Using the results of the model, such as overtopping flow rates, heights and durations, the velocities of floodwater flowing down the dam slopes were calculated in order to compare with Figure 12 in 'Floods and Reservoir Safety' (ICE, 1996). This comparison assumed poor cover for most of the dams on the Heath, as these were covered in trees that prevented good grass coverage. In all cases the peak flow velocities and durations exceeded the erosion capacities of the dam slopes, and supported the Panel Engineer's judgement that overtopping of the dams was not tolerable.

OPTIONS EVALUATION

The options considered had to meet the key objectives of the project:

- improve dam safety on all the dams in the chains.
- preserve the Heath as a natural open space.
- maintain (or increase) the standard of protection downstream.
- do not increase the rate of flow discharged from the last dam in any flood event, compared to the existing scenario.

Design Principles and Design Philosophy

The project design principles and design philosophy were developed to balance dam safety requirements, with feedback from engagement with stakeholders and the wider public, while having regard to the environmental considerations of each pond and the 'natural aspect and state of the Heath'.

Figure 2: View across Men's Bathing Pond to the dam at Model Boating Pond

Figure 3: View across Mixed Bathing Pond to the dam

These considerations included: retaining existing pond water levels and the distinctive character of the Heath and key views, minimising the scale of intervention and impact on visual amenity and the use of the Heath for all users – including swimmers, anglers, walkers and nature enthusiasts.

Environmental management was an integral part of the project. In addition to improving water quality the project had to ensure that, following construction work, reinstatement of the Heath's natural aspect would take place. The collaboration between technical specialists ensured that none of the options being considered would preclude pond and terrestrial habitat reinstatement and restoration. The use of appropriate and natural materials and minimal intervention was designed to retain the natural aspect of the Heath.

Design principles that applied to all of the preferred options included:

- Each chain of ponds was considered as a whole system, so that any significant increases in storage capacity are focused in the least sensitive locations, at ponds with the largest surface areas, limiting tree loss around ponds and reducing the residual works required elsewhere.
- Each dam had to be able to pass the PMF safely.
- Each option was designed as a passive system to improve the resilience of the dams without reliance on any mechanical system (such as valves or pumps) or human intervention. The passive system of each option has been designed to pass excess flood water at each dam following these principles:
- Where the overtopping of the dam crest is not tolerable, which applies to the majority of the dams in the pond chains (due to the number of trees on the crests and on the downstream slopes), some works to raise or restore the dam crests and creation of natural open spillways were proposed, to pass the PMF in order to minimise risk of dam failure.
- Where overtopping of the dam crest is tolerable (which only applied to the dams at Mixed Bathing Pond and Bird Sanctuary Pond), and excess flood water up to the PMF still needs to be passed over the dam crest or the downstream slope, reinforced grass on the downstream face may be required to allow flow over part or all of the width of the dam crest.

There was therefore a trade-off at each pond between the amount of dam crest raising, and the width and depth of the spillway required to pass the PMF safely. The width of the spillway was often limited by the space available, avoiding specific trees and going around the dams as much as possible.

OPTIONS EVALUATION: HIGHGATE CHAIN

The groups of options were evaluated, using the hydraulic model, by varying the amount of additional flood storage capacity in the middle of each pond chain then assessing the works required downstream to meet the above principles.

On the Highgate chain, this meant varying the height of the raising of the dam at Model Boating Pond, between 1 and 3 metres. This would be done by adding fill to the upstream face of the existing dam. The results were shown in the form of a flow chart which reduced in size at each stage as the options were reduced from 6 to 2 preferred combinations.

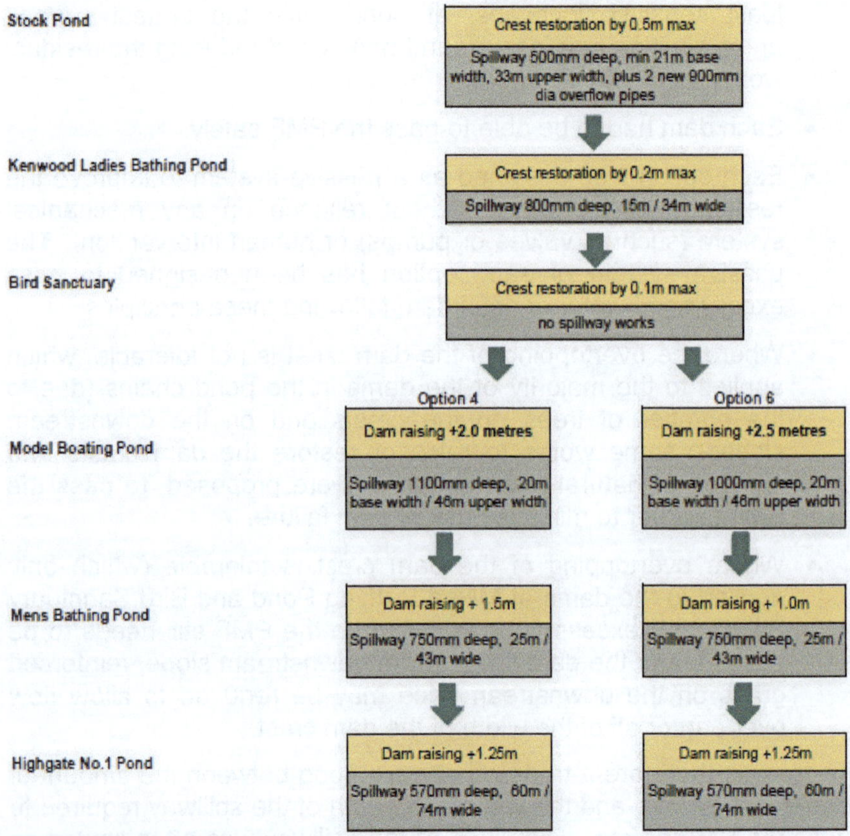

Figure 4: Preferred options on the Highgate chain

The option of raising Model Boating Pond by 3 metres would have reduced the required raise of the downstream dams to 0.5 metre, but this was discounted by the client and the stakeholders. There was a desire to limit the number of lorry movements in the earthworks to win fill on site and dispose of the silt within the borrow pits.

The proposed works at Model Boating Pond also included excavating the natural ground slope above the west side of the pond, to widen the surface area of the water and remove the sheet piles on that side to create a softened edge. This excavation was intended to provide material for the dam and so could be shaped to avoid trees, by leaving an island around the group of lime trees half way along the west bank. The upper slope of the west bank would be cut from the existing slope of around 1:10 to 1:5.

A borehole into the pond bed found that the depth of silt was around 3 metres, which meant that more clay fill material was required for the dam to start at hard bed level. A borrow pit on the top of the hill west of the pond was investigated for the material, and found to contain good sandy London Clay.

Figure 5: Visualisations of the altered view downstream across Model Boating Pond, with an island formed on the right (west) bank

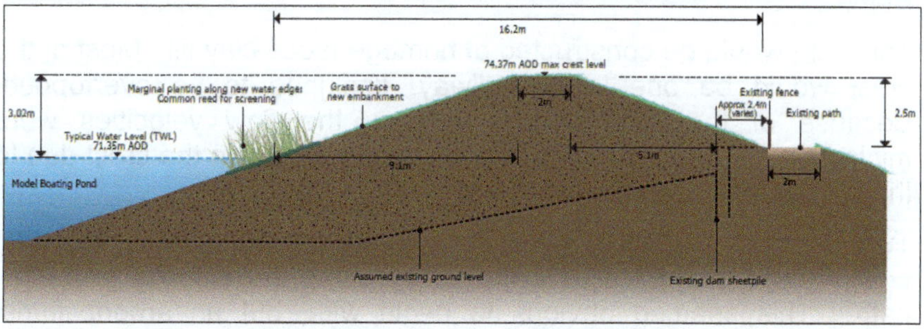

Figure 6: Cross section of the raising dam at Model Boating Pond

OPTIONS EVALUATION: HAMPSTEAD CHAIN

Options for the Hampstead chain involved adding storage capacity by raising the dam at Mixed Bathing Pond (the third of the five dams) by 1m to 2m, and by constructing a new embankment dam across the valley upstream of Mixed Bathing Pond.

The focus of the options evaluation on this chain was an avenue of highly valuable London Plane trees on the crest of the dam at Hampstead No.2 Pond. To minimise the loss of these trees to two, a drop shaft was proposed to increase the driving head and therefore the discharge through a new concrete box culvert overflow. This was combined with fine tuning of the pipe through the catchpit dam and the introduction of a labyrinth weir spillway on the last dam (at Hampstead No.1 Pond) in order to balance the tree loss with the other objective of maintaining the standard of protection to the area downstream of the ponds.

Figure 7: Visualisations of the effect of removing two London Plane trees at Hampstead No.2 Pond (before on left, after on right)

<u>Catchpit Dam</u>

The primary extra flood storage capacity on this chain was to be created by a new embankment dam, 5.6m high above the valley bottom, normally dry but storing approximately 12,600m^3 during a PMF.

This dam would be constructed of homogeneous clay fill. Most of the crest would be one large spillway, designed to be overtopped. Because of the size of the spillway, the flow velocities were minimised so the only reinforcement required was on the crest due to the erosion by walkers.

Two possible positions were considered for the dam. The first position was straight across the valley along the existing clearing / path. This seemed obvious to begin with, but an arboriculturist survey highlighted the high value of the trees bordering the clearing, compared to lower quality trees further upstream.

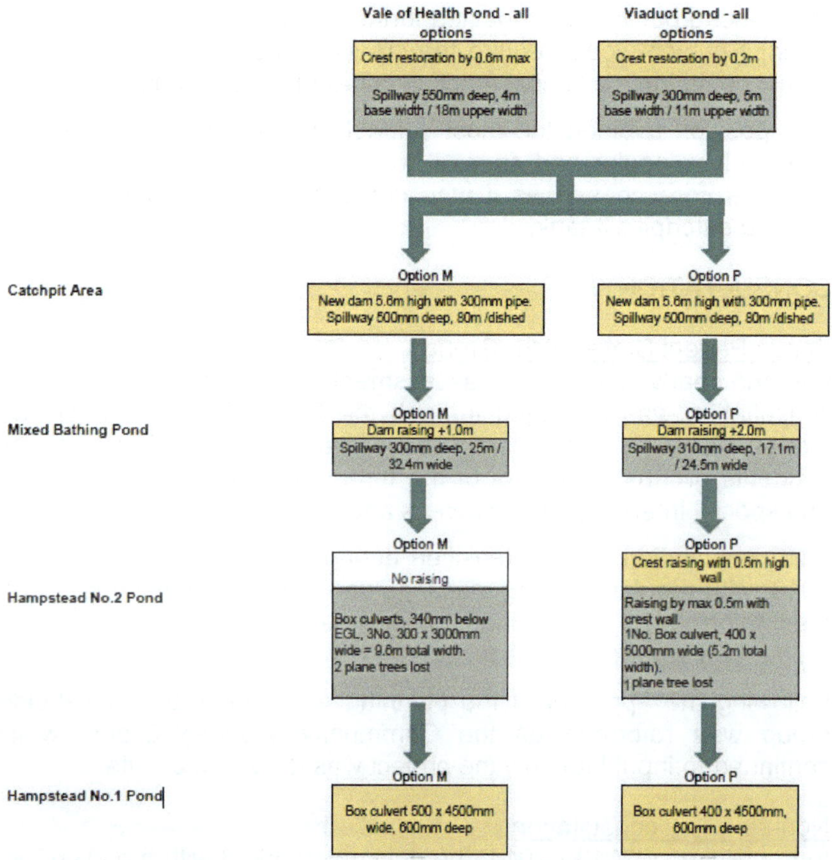

Figure 8: Options flowchart for Hampstead pond chain

Figure 9: Location of preferred dam position upstream of the clearing (stream flows to Mixed Bathing Pond on left of photo)

The second position was located further upstream above the existing catchpit (which would require the creation of a new wetland habitat which would have a similar function in trapping sediments).

This position avoided the most valuable trees (for example a large hybrid black poplar and several hornbeams). The second position was therefore chosen as it was considered easier to remove the existing catchpit silt tank.

CONSULTATION

Ponds Project Stakeholder Group

For the early stages of assessment and design, stakeholder consultation was managed through the Ponds Project Stakeholder Group (PPSG). This comprised representatives of the various residents groups, members of the three bathing pond associations, and special interest groups such as anglers.

This consultation stage intensified during the spring and summer of 2013, with regular evening meetings and weekend workshops which then informed a constrained options report, a shortlisted options report, and a preferred options report.

Following the approval of the planning application, the stakeholder group was reformed as the Community Working Group, which continued to input into how the project was delivered on site.

Non-statutory consultation

In the winter of 2013 / 2014, the designer worked with the client and specialist communication consultants to produce a range of materials to inform and engage the wider public. As this was before the formal consultation phase of the planning application, this was known as the non-statutory consultation. Around 80,000 postcards were posted to the residents of the Hampstead, Highgate and Camden areas, thousands more leaflets were handed out in and around the Heath, inviting the public to view one of two exhibitions. This involved a storyboard of 14 A1 sized posters with selected visualisations, and visual illustrations such as the volume of a PMF shown as multiples of Lido swimming pools. Further A1 posters with 'before and after' visuals for each pond were fixed to A-frames at the ponds.

PLANNING APPLICATION

The planning application was submitted to the local authority, Camden Metropolitan Borough Council, in July 2014. As well as the standard statutory elements of a planning and design access statement, environmental impact assessment, and outline designs, the planning application process was supported by a parallel independent review, and the planners' decision making was influenced by the outcome of a judicial review.

Judicial Review

Planning permission was not granted until a judicial review, launched by the Heath and Hampstead Society, had been heard in court.

The case was taken to the Royal Courts of Justice and a ruling given by the Honourable Mrs Justice Lang DBE.

Independent Review by AECOM

Due to the specialised nature of the project case, the local authority commissioned an independent review of the project by AECOM, paid for by the client. The scope of the review was to answer the five questions summarised below:

1. *Whether the proposed project is technically appropriate and necessary in the context of legislation and current best practice guidance,*
2. *Written confirmation that the modelling methodology and assumptions underpinning the project are suitable, reasonable and have been applied appropriately,*
3. *Whether a sufficient range of alternatives to address the dam safety hazard have been assessed to an appropriate level of detail to justify being discounted,*
4. *Whether the applicant's claims that the scheme will neither alter the ponds interaction with the Thames Water drainage network nor increase the potential for surface water flooding downstream under all operating conditions (including overtopping and spillway activation) are sufficiently evidenced and reasonable,*
5. *Whether the comments made by the third party identify any reasonable concerns about the technical content or considerations of the submission which should be addressed by the applicant by way of further submission, prior to planning permission being able to be recommended to be granted.*

The designer assisted in this review by providing the assessment reports and by modelling of the pond chain outflows in the smaller return periods (such as 1 in 5 up to 1 in 50) with which Thames Water were concerned.

Answering this question involved meetings with Thames Water, whose models proved that the existing sewer system would be filled by a 1 in 30 flood and so using the existing system was not an option in the design floods being considered for dam safety.

The AECOM report on the review concluded that the Panel Engineer and designer had followed industry standard guidelines in their assessments and design decisions, and that the proposed works would not adversely affect the residential areas downstream.

Verified views

A significant component of the planning application was the production of verified views. These were developed from the visualisations created during the pre-planning consultations, but were formalised by using photographs at surveyed locations superimposed on a 3D landscape model. The viewpoints were agreed with the local authority to include key strategic views such as the view across Hampstead No.2 Pond to the London Plane trees and the view from Parliament Hill down to Men's Bathing Pond.

Following the judge's decision in the judicial review in favour of the City of London, which clearly vindicated the Panel Engineer's judgement, and the supportive conclusions of the independent consultant's review, the planning application was approved in late January 2015.

Site clearance commenced in February 2015 and enabling works began that spring. The programmed project completion date is October 2016, and at the time of writing this is still the target.

CONCLUSIONS

The Hampstead Heath Ponds Project is an example of how a large scale dam safety works can be successfully designed in a sensitive location. The philosophy of considering the pond chains as a whole, and the central design concept of additional storage capacity, minimised the impacts on downstream ponds, and was inherently sustainable. By using clay from the Heath, and partly filling in borrowpits with silt taken from six of the ponds, long term benefits to water quality and the natural environment will be achieved.

Every effort was made to consult with stakeholders on the design options, and a range of hydraulic engineering techniques were used to find ways to optimise the solutions within a uniquely difficult set of constraints. The justification for the scheme was vindicated and independently verified.

REFERENCES

Defra (2004): *Floods and Reservoir Safety: Revised Guidance for Panel Engineers*. Defra, London

Haycock Associates Limited (2006). *Hydrological and Water Quality Investigation and Modelling of the Hampstead Heath Lake Chains and Associated Catchments.* Haycock, Pershore, Worcs.

Haycock Associates Limited (2010): *Hydrology Improvements Detailed Evaluation Process (HiDEP): Hydrology and Structure Hydraulics.* Haycock, Pershore, Worcs.

HMSO (1975) *Reservoirs Act 1975*, Her Majesty's Stationery Office, London

HMSO (2010). *Flood and Water Management Act 2010.* The Stationery Office, London

ICE (1996) *Floods and Reservoir Safety. Third Edition*, Thomas Telford, London.

Dams – Benefits and Disbenefits; Assets or Liabilities?
ISBN 978-0-7277-6133-0

doi: 10.1680/dbdal.61330.377

Challenges of design and construction for reservoir safety improvements within an historic estate

D E NEEVE, Arup, Leeds, UK
M JENKINS, Arup, Leeds, UK
S PORTER, Chatsworth House Trust, Chatsworth House, UK
T KILBANE, Fox (Owmby) Limited, Lincoln, UK
P KELHAM, Arup, Leeds, UK

SYNOPSIS Detailed design and construction of any reservoir safety and improvement works can be challenging. The team successfully overcame the constraints of three historic dams with protected ecology all set within a National Park and a Grade 1 listed parkland with a remote moorland access.

The chosen solution provides compliance for the two high risk dams and peace of mind for the third within a limited budget for the charitable owner. The works included new top water level and auxiliary weirs with associated spillways to three reservoirs and crest raising on two high risk (Category A) dams.

INTRODUCTION

Chatsworth House Estate includes an interconnected series of lakes and a Grade I Registered Historic Park. The oldest (Swiss Lake) dates from 1710 and they are all still actively used to provide water for the famous fountains and garden water features along with the firefighting system, toilets and hydroturbine. Two of these lakes (Swiss and Emperor) fall under the Reservoirs Act 1975 (HMSO, 1975) and required significant improvement works to increase freeboard and overflow capacity following a statutory inspection in 2011.

The two smaller reservoirs, Mud Pond and Ring Pond, do not currently fall under the provisions of the Act (each has an approximate volume of 10,000m³). The original aspiration was to design all the works but only construct those required in the interest of safety at Swiss Lake and Emperor Lake. The remedial works at Mud Pond and Ring Pond would be constructed at a later date. However, due to design efficiencies it was possible to bring forward the Mud Pond works.

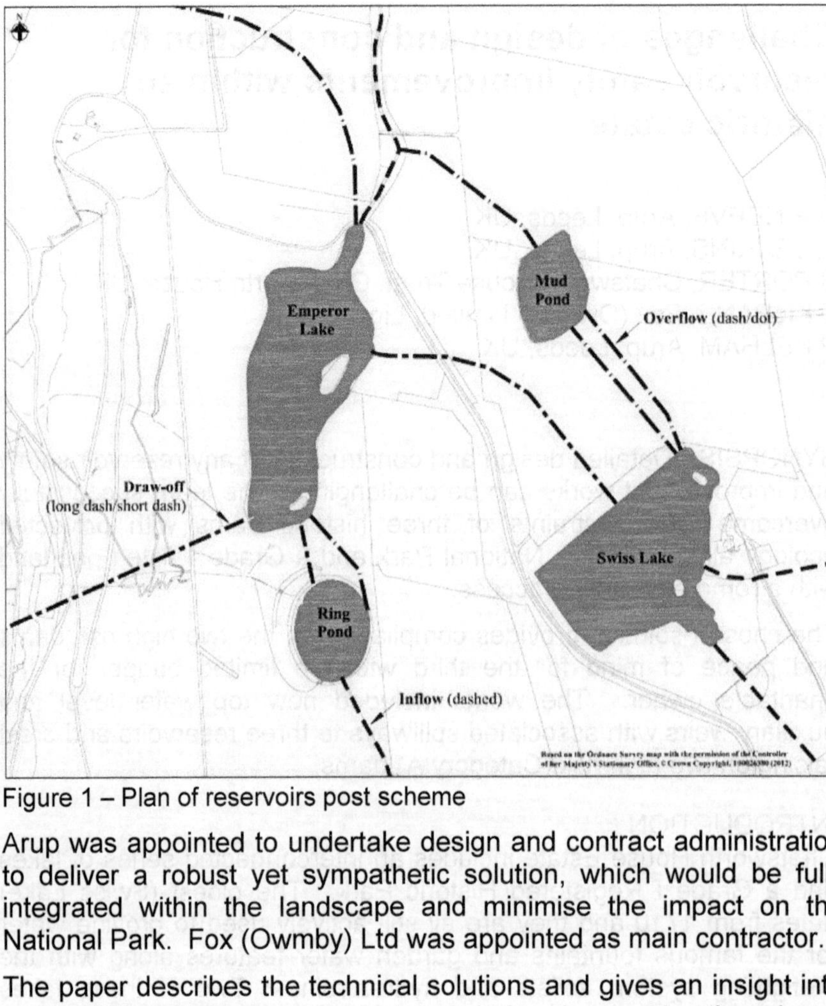

Figure 1 – Plan of reservoirs post scheme

Arup was appointed to undertake design and contract administration to deliver a robust yet sympathetic solution, which would be fully integrated within the landscape and minimise the impact on the National Park. Fox (Owmby) Ltd was appointed as main contractor.

The paper describes the technical solutions and gives an insight into some of the challenges encountered during the construction phase. A previous paper (Neeve *et al*, 2014) discusses the navigation and concessions required to achieve an acceptable planning solution with a diverse range of drivers and constraints that can be time consuming and costly for the client.

SOLUTION

The client's primary concern, beside compliance, was ensuring that the scheme was sympathetic to the landscape and surrounding woodland, whilst minimising expenditure to the charity. By reducing the amount of construction the design minimised the impact to the benefit of the estate, visitors and planning authority.

The pre-scheme reservoirs had inherent problems as the level difference between the weirs and the embankment crests were insufficient, resulting in widespread embankment overtopping during flood events. At Emperor Lake this occurred during a 1 in 5 year event.

Figure 2 – Emperor Lake pre-scheme overflow weir level very close to embankment crest level

The technical aim of the remedial works was to ensure the design flood could be conveyed through the reservoirs and safely discharged to the River Derwent in the valley below. This was achieved by increasing the weir capacity at each reservoir and raising the crests at Emperor Lake and Swiss Lake.

Table 1. Dam and Reservoir Information

Reservoir	Dam Height (m)	Dam Length (m)	Capacity (m^3)	Pre-scheme outflow capacity (m^3/s)	Design Flood (m^3/s)
Emperor Lake	6	420	72,000	1.7	15
Swiss Lake	3	320	56,826	1.6	4.5
Mud Pond	1.5	190	9,500	2.6	7

The pre-scheme outflow capacity in Table 1 is based on a static water level at the embankment crests and therefore overtopping from wave action would be occurring. Even with no wave allowance the table shows the pre-scheme overflow capacities were inadequate compared to the design flood (Aecom, 2013).

Crest Raising

Limited information was available from the dam's construction, some 170 to 300 years ago. Hence the pre-design investigations were to establish basic design data rather than confirmation of existing knowledge. The ground investigation revealed that the topsoil was not removed before Swiss Lake construction commenced in 1710. The potential slope stability failure mode of a soft layer at the base of the dam was reviewed using the increased loading from the crest raising.

Figure 3 – Site Investigation trial hole at Swiss Lake

The embankments were most likely constructed from locally won weathered bedrock, a mixture of sandstone and mudstone, making a good clay. The approach of using locally won material was also utilised for the remedial works for probably the same reasons as the original construction, ease of transport, convenience and cost. Keeping with these principles the scheme's major excavation for the Emperor Lake spillway was utilised to provide the embankment raising material for both Emperor Lake and Swiss Lake, avoiding a separate borrow pit.

To contain the flood rise and wave allowance the works on Swiss Lake and Emperor Lake included embankment raising of approximately 400mm and 600mm respectively. The wave protection is different at both lakes, with Swiss Lake having an inclined wall and Emperor Lake a rip-rap slope with a notional low wall at the crest. The design criteria were to ensure that good erosion protection was provided up to the level of the predicted maximum wave height using guidance within Floods and Reservoir Safety, 3rd Edition (ICE, 1996). The solution was to increase the protection level by using locally sourced sandstone blocks with grassed topsoil above.

Spillways

The works to the reservoirs included new auxiliary weirs and associated spillways and discharge channels. Reinforced grass auxiliary spillways have been designed to minimise visual impact at Mud Pond and Swiss Lake.

Swiss Lake

The new auxiliary weir was located on the north east abutment to utilise the existing topography. This ensured major earthworks were avoided through the open sections of fields, which contains historic ridge and furrow patterns, between Swiss Lake and Mud Pond. The location contained a number of mature trees that were removed, including their stumps and large roots to reduce the likelihood of future leakage paths.

Figure 4 – Swiss Lake auxiliary spillway looking to Mud Pond

Mud Pond

A large mature oak tree is located on the crest of the dam (see Figures 5 and 11) which dominates the character of the immediate vicinity. A risk assessment was undertaken to consider whether reservoir safety would be best served by retaining or removing the tree. It was concluded removal of the tree could lead to future leakage along the substantial root system and the best outcome would be to retain and manage the tree. To increase reservoir safety the embankment has been raised over the adjacent section to prevent overtopping and the subsequent risk of a breach.

The pre-scheme spillway, located immediately to the left of the oak tree, had no formal channel, which had allowed erosion of the dam to occur and had exposed the root system. To avoid this risk the top water level weir and spillway have been moved to the southern auxiliary weir and spillway to align with an existing drainage sough. To prevent continuous use of the reinforced grass auxiliary spillways the top water level weir also acts as a low flow channel, which is set 100mm lower than the auxiliary weirs.

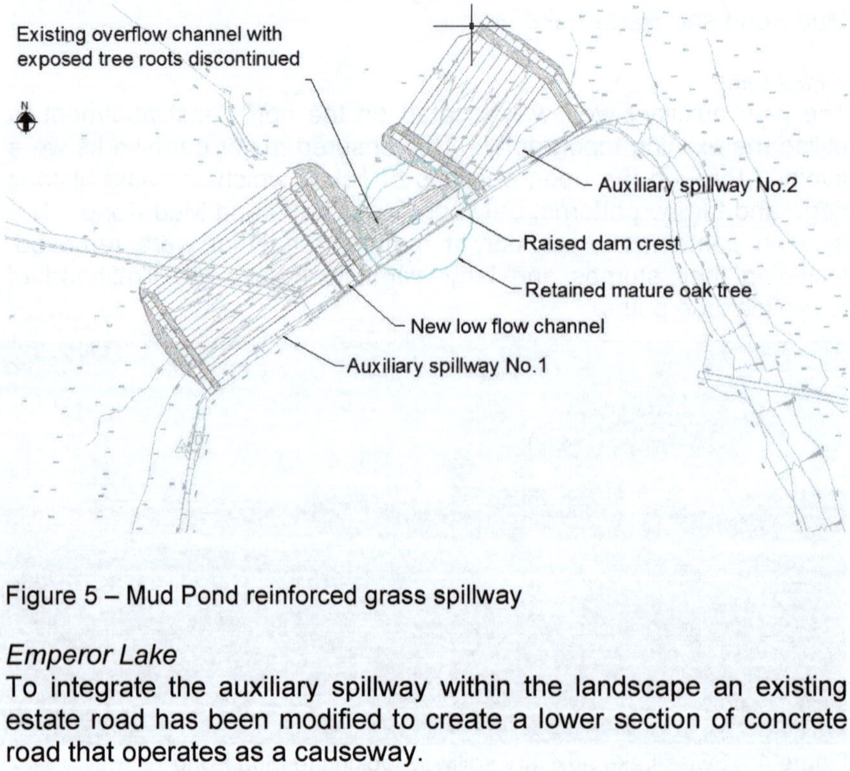

Figure 5 – Mud Pond reinforced grass spillway

Emperor Lake

To integrate the auxiliary spillway within the landscape an existing estate road has been modified to create a lower section of concrete road that operates as a causeway.

In normal operation a box culvert beneath the causeway allows flows in and out of the reservoir. During flows above a 1 in 5 year event the causeway conveys flood water out of the reservoir.

The new top water level weir, formed from a 60m long reinforced concrete structure, creates a new pond area that acts as a stilling area for the flood flows from Emperor Lake and the Emperor Stream catchwater.

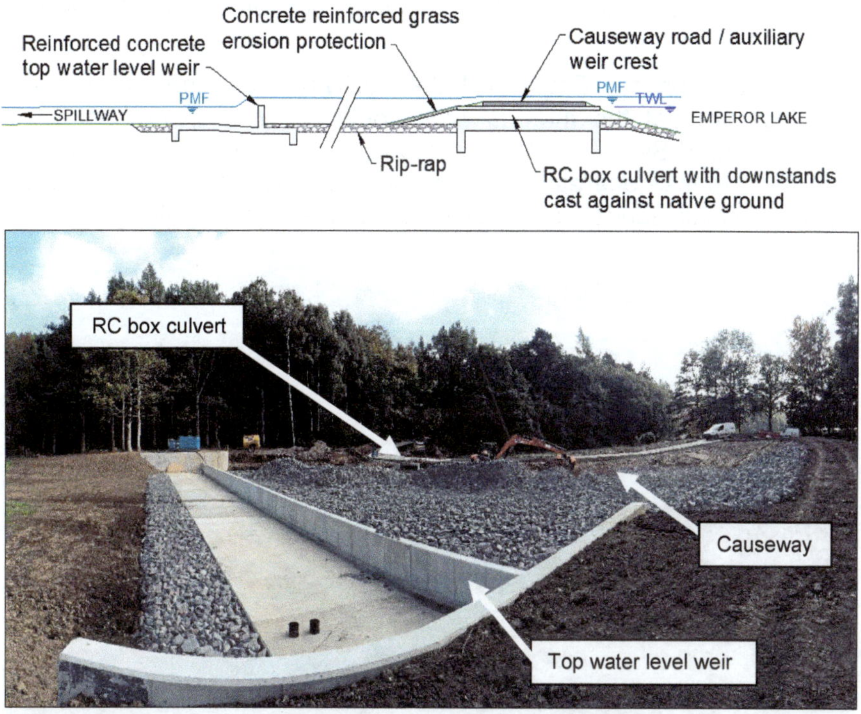

Figure 6 – Emperor Lake top water level weir and causeway weir

CONSTRUCTION

The construction phase of the project took over approximately six months between spring and autumn 2015. Logistics played a large part to enable a wide variety of materials and resources to reach the working areas at the right time. This was further complicated by the remote moorland access and numerous interfaces with the visiting general public and estate activities. To minimise questions and frustration the work was widely promoted and signage used throughout the works.

The works occurred during the estate's main visitor season that meant it was essential to provide a water supply to the garden water features, as well the firefighting system, toilets and hydroturbine. A strategy was agreed to accommodate the Construction Phase Flood, which was estimated as a 1 in 33 year event, whilst maintaining the estate's supply. The strategy was implemented without incident throughout the works by the estate team. To further guard against flooding the work was sequenced to minimise the risk of embankment damage by leaving existing weir structures operational for as long as possible and carrying out embankment raising activities in sections.

The excavation for Emperor Lake spillway was one of the first activities so that the material won could be used in embankment raising around Swiss and Emperor Lakes. No material was imported and the excess generated from the spillway excavation was used to sympathetically landscape an area to the west of the spillway.

Figure 7 – Emperor Lake spillway © The Devonshire collection, Chatsworth.

Emperor Lake Weir and Causeway

The most complex civil engineering aspect of the improvement works was undertaken in two distinct phases. First, the new weir structure, with cut-off and integral wing walls was built offline, forming an extension to Emperor Lake with Emperor Stream over pumped.

The second phase involved connecting the new weir to Emperor Lake by breaching the existing embankment and installing a new culvert and causeway structure. This phase could only begin once the initial phase had been inspected and approved by the QCE.

Particular challenges included the sequencing of the work to ensure that the upstream and downstream cut-off trenches beneath the weir were dug and cast against virgin ground with the minimum of time exposure. Concrete deliveries had to be carefully planned and managed, with it taking about 30 minutes to navigate through the multiple field gates on the construction access route off the A619.

The route was chosen to minimise impact on the visiting public and villages.

Figure 8 Emperor Lake Weir construction © The Devonshire collection, Chatsworth.

Emperor Lake Embankment Raising

The embankment was raised by approximately 600mm by stripping the topsoil to a suitable formation (verified by hand shear vane test and visual inspection), then building up the embankment in layers with Class 2 cohesive material from the spillway excavation. Tree stumps left from pre-commencement felling were removed prior to filling.

Large sandstone blocks were chosen to match the local stonework, support the raising works and provide continuity of the existing wave wall/riprap protection. This presented a particular challenge to the construction team as one of the planning requirements imposed did not permit the use of any cementitious material for this element (due to the setting and historical perspective within the National Park). Blocks were instead bedded onto competent material, which could be partly on the existing uneven blocks/riprap and partly on embankment clay material which was individually excavated and shaped to form an even bed. Vertical joints were kept tight and gaps at the base of the blocks were covered with locally recovered riprap.

Other issues encountered included dealing with a Japanese Knotweed infestation and topsoil quantities being less than expected. To compensate, surplus topsoil was recovered from the spillway channel excavation.

Figure 9 – Emperor Lake embankment raising works completed

Swiss Lake Spillway

A concrete cut-off wall was constructed behind the existing dry stone boundary wall to effectively act as the weir. This was deeper than anticipated in order to key into good material. When excavating for the spillway, an old drainage sough was discovered, presumably pre-dating the construction of the lake. Before its removal, an accurate historical record was made of its construction and location by an archaeologist. The contractor worked closely with the archaeologist to ensure that proper records could be made of historical discoveries, without unduly compromising the programme.

Swiss Lake Embankment Raising

The embankment was raised by 400mm by the same method as described for Emperor Lake. The difficulties with modifying 300 year old structures and tree stumps were highlighted during this part of the project. Whilst is desirable for long term stability to remove tree stumps entirely, this can cause other issues, as was demonstrated when a section of wall collapsed on the southern shore following removal of a stump that was very close to the wall. This was sympathetically re-built by the client who had particular experience in maintaining such structures.

The contractor managed third party and public interfaces in a sensitive manner throughout the site works, such as accommodating canoeing activities for an international scout event.

Figure 10 – Archaeological record of drainage sough at Swiss Lake

Mud Pond

An iconic mature oak tree was retained, with two grass reinforced spillways, one on either side. The level was locally raised around the tree and a previous weak spot in the embankment reinforced. The original intent for a concrete cut-off in front of the tree was not possible due to the number and size of roots. Instead the cut-off extended as far as possible beyond the spillways with the remaining section being was hand dug around the roots, then backfilled with bentonite pellets.

Figure 11 – Completed Spillway at Mud Pond

A drainage sough was discovered under the northern auxiliary spillway, which was archaeologically recorded and removed to prevent a future spillway weakness. During construction of the southern spillway, a leak was discovered. A scour hole had developed behind the pond wall with water draining along the footings of a connecting dry stone wall. The wall was removed, the dam repaired and the void filled.

Damage was caused to the completed structure when livestock were unexpectedly released into the surrounding field and the sheep took a surprising liking to the plastic Erosamat material. The damaged areas were subsequently repaired and stock proof fencing erected.

CONCLUSION

The team successfully delivered an efficient design within a limited budget. The solution integrated well within the landscape and balanced the needs of maintaining and enhancing the historic setting without compromising on reservoir safety standards.

Figure 12 – Completed Causeway and Weir Structure at Emperor Lake

REFERENCES

Aecom (2013). *Flood Study* January 2013

ICE (1996). *Floods and Reservoir Safety, 3rd Edition.* Thomas Telford, London, UK.

Neeve DE, Porter S, and Marchant L, (2014). Challenges of planning significant reservoir safety improvements within an historic estate. In *Maintaining the Safety of our Dams and Reservoirs; Proc of the 18th Biennial Conference of the British Dam Society, pp 35-47.* ICE Publishing, London, UK

HMSO (1975). *Reservoirs Act, 1975. Chapter 23.* Her Majesty's Stationery Office, London, UK

Dams – Benefits and Disbenefits; Assets or Liabilities?
ISBN 978-0-7277-6133-0

doi: 10.1680/dbdal.61330.389

Waun Pond - New Overflow

K M H BARR, Fairhurst
M HAYWARD, Fairhurst
V FORD, Fairhurst

SYNOPSIS Following a periodical inspection of Waun Pond, Blaenau Gwent County Borough Council appointed Fairhurst to address safety recommendations. The overflow capacity was found to be grossly inadequate. The solution involved design of a new overflow which accounted for significant physical constraints including services laid along the crest of the dam below a public road and conflicting stakeholder interests in terms of amenity and flood risk downstream.

The adopted solution was a box-type overflow, with the potential for flow to enter on all four sides, discharging through culverts formed by tunnelling below the crest of the dam using pipe jacking, into a downstream stilling basin and channel. The paper covers the flood study, design and construction aspects of the new overflow.

BACKGROUND

Waun Pond is a redundant industrial Category A impounding reservoir in Blaenau Gwent, South Wales, now used for wildlife and fishing. It is situated near the head of the Afon Ebwy Fach, between the towns of Brynmawr and Nantyglo, at an elevation of about 340m above sea level.

History

Waun Pond is first recorded on Ordnance Survey mapping in 1880, when it is shown as one of a complex of small reservoirs associated with the Nant-y-glo Iron Works. From the late 1940s the reservoir was used as a water source for another nearby industrial site, the Brynmawr Rubber Factory, owned latterly by Dunlop. The factory was built to the north-east of the reservoir for the manufacture of rubber goods including floor tiles. A large stone faced concrete pump house was added to the south side of the reservoir to abstract water for the factory.

The original surface area of the reservoir was about 8ha. Over succeeding years its surface area has been progressively reduced by infilling for development land around the northern and western shorelines and is now a little over half of the original area.

Following the closure of the Dunlop factory, the reservoir passed through various hands. It eventually came into the ownership of a local trust, its intention being to preserve it as a local angling amenity and wildlife refuge.

Description

The reservoir is impounded by a main embankment about 200m long and 7m high. This merges at its eastern end with ground around the same level as the crest over a length of around 150m. At the eastern end of the reservoir the downstream level falls away and the reservoir is impounded by a secondary embankment about 75m long and 2m high.

Around the reservoir the land to the north and west has been redeveloped for a retail park and housing. The area downstream to the south is occupied by the site of the former Nantyglo School, which closed in 2011, industrial workshops and a leisure centre. The downstream land, including parts of the embankment, is owned by Blaenau Gwent County Borough Council (the Council).

A public highway, Pond Road, runs along the crest of the main embankment from a roundabout on the A467 adjacent to the toe of the east secondary embankment.

Prior to reconstruction the overflow was located in the right abutment. It was formed by screened letterbox openings into a covered reinforced concrete chamber, which discharged at right angles into a pipe running below the crest of the dam. The pipe discharged to the open channel of the Afon Ebwy Fach running along the toe of the main embankment. A drawoff discharges near the dam toe and joins the channel immediately downstream. The pump house that formerly abstracted water is now disused and is a Grade II listed structure.

A short distance downstream of the dam the watercourse enters a culvert adjacent to the former school site. The culvert extends downstream for a considerable distance.

Previous inspections

There is no record of any inspection being carried out under the 1930 Act. The first periodical inspection under the 1975 Reservoirs Act was carried out in 1986. A flood study was carried out, resulting in the conclusion that the overflow capacity was grossly deficient. The

Inspecting Engineer recommended improvements to the overflow to convert it to a siphon, and protection of the eastern secondary embankment to allow it to act as an auxiliary overflow. No action is believed to have been taken.

Successive periodical inspections were carried out from 1997 onwards, initially appointed by the Council as Enforcement Authority, and then by the then owner. Further flood studies reinforced the earlier findings.

A scheme to discontinue the reservoir by reducing the water level and subdividing it was reportedly proposed in 1998, but was not taken forward. Subsequently measures to address the overflow capacity deficiency were proposed, but again did not proceed.

Following the transfer of enforcement responsibilities to the Environment Agency in 2004, enforcement notices requiring recommendations to be carried into effect were issued in 2005. Due to a dispute over the identity of the Undertaker no action was taken.

<u>2010 periodical inspection</u>

By 2010, the dispute over the identity of the Undertaker had not been resolved, but the Council appointed Mr J C Ackers to carry out a further periodical inspection. The principal safety recommendation made in the Inspecting Engineer's report was that the existing overflow arrangements should be upgraded so that the appropriate Category A design incoming flood from the catchment could be safely routed through the reservoir and conveyed downstream.

Subsequently, Fairhurst was appointed by the Council to carry out the studies recommended in the Section 10 report and design the resulting works. Mr Ackers was appointed as Qualified Civil Engineer to supervise and certify the works.

STAKEHOLDER CONSULTATIONS

Consultation with stakeholders including the pond owners, the Council and Natural Resources Wales (NRW) revealed potentially conflicting aspirations. The owners wished to minimise any reduction in water levels to maintain the function of the reservoir as a local amenity. NRW was unwilling to accept increases in discharge rates over a wide range of return periods due to downstream flood risk to property.

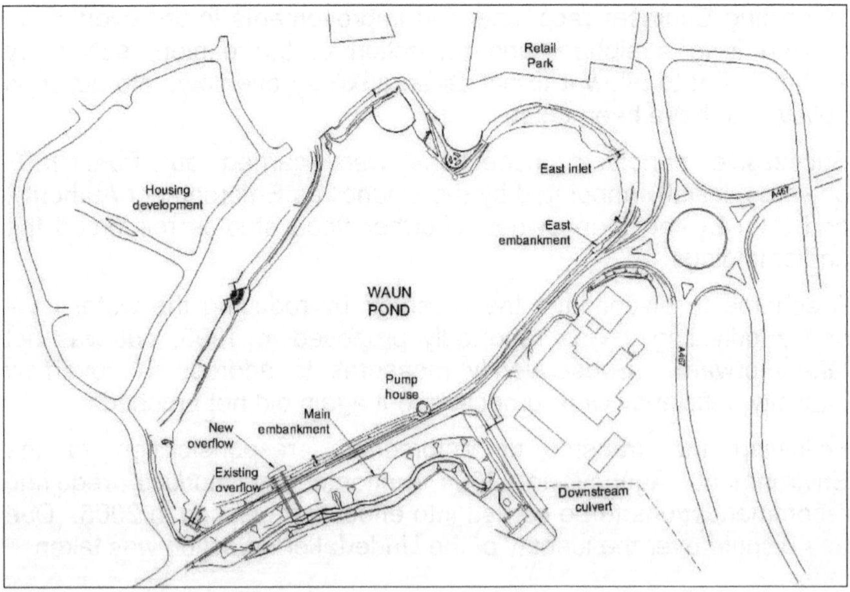

Figure 1. Location plan

FLOOD STUDY

The previous flood studies had resulted in inconsistent conclusions. Review of the previous studies established that they had each relied on assumptions to a greater or lesser degree due to incomplete information.

The agreed remit for the flood study covered the following work:

- Re-assessment of the outflow rating of the current arrangement.
- Re-assessment of the catchment contributing and hydrological assessment of flood flows entering the Pond.
- Identification and assessment of options for alterations to the overflow arrangements.
- Estimate of flood lift, maximum discharge rate and water level under existing and proposed conditions for a range of events.
- Following consultation with stakeholders, recommendation of a preferred option to be progressed and outline hydraulic design of the preferred option.

Data collection

An initial data gathering exercise was undertaken.

Topographic Survey
Survey information available from previous studies was incomplete or carried out prior to changes made to the dam and reservoir margins. A detailed topographic survey was therefore undertaken to provide information for the flood study. The survey covered the reservoir margins and the upstream face of the main and eastern embankments above normal water level, the crest and downstream faces of the embankments.

The lowest point on the crest of the main embankment was found to be around 341.24mAOD to the east of the valve tower. The lowest point on the crest of the east embankment was about 340.81mAOD. The sill level of the overflow letterbox was at 338.70mAOD.

Catchment
The reservoir is fed via several inlets from a catchment of around 2km², partly direct and partly indirect. There were inconsistencies in the catchment area adopted in undertaking the previous studies. The catchment is highly urbanised and drainage patterns are affected by spoil from historic mining activity.

For this study the catchment area was re-assessed using information available from past studies, Ordnance Survey mapping, aerial photography, online photography from Google Streetview, sewer network records and a catchment walkover. Following review of available information, a catchment area of 2.82km² was adopted.

<u>Analysis of existing conditions</u>
Overtopping of the existing eastern embankment was deemed to be acceptable in an extreme flood event. The 10,000yr flood was therefore adopted as the design standard.

The reservoir system was represented in a numerical hydraulic model using ISIS software (now Flood Modeller). At low flows the inlet letterbox sill was found to control outflows. At higher flows, discharge was controlled by the culvert inlet within the chamber.

Natural Resources Wales (NRW) had identified downstream flood risk as a key constraint in the development of any remedial solution for the reservoir. Lower return period events were therefore simulated, in addition to the 10,000yr flood, to establish a baseline for this assessment. The flood study concluded that the secondary east embankment would overflow in about the 30yr event, flooding the A467. The predicted discharge through the existing overflow in the 10,000yr event was only 2m³/s. The predicted flow spilling over the eastern embankment in the 10,000yr event was 25.4m³/s. It was concluded that a new overflow was required.

Option appraisal

Three basic options to provide sufficient overflow capacity were modelled, together with a fourth option representing complete removal of the impoundment. It was assumed that altering existing road levels along the crest would be undesirable because of the expense and disruption, and that the existing top water level (TWL) should be retained, or reduced as little as possible.

Option 1 – Auxiliary overflow.

This required no alteration to existing dam crest levels. Predicted downstream flow rates were increased for all return periods except the 10,000yr design event. The 100yr plus 20% climate change peak flow was increased by around 33% relative to the existing conditions.

Option 2 – Auxiliary overflow and minor embankment works

This option required a minor raising of the east embankment and provision of a low wave wall. The downstream peak flow rate in the 100 year plus climate change event was found to increase by 13% over the existing flow rate.

Option 3 – New overflow and lowering of TWL

This option involved a modest lowering of the reservoir together with the replacement of the existing overflow with a new structure. It was found to be possible to match or improve on the existing flow rate in the 1 in 50 year event and greater.

Option 4 – Discontinuance and removal of storage

Complete removal of the existing storage was not considered acceptable in terms of its effect on downstream flood risk, but was included to provide a worst case comparison.

Selected option

The conclusion of the Flood Study was that Option 3, involving a lowering in TWL and construction of new overflow facilities was the best option. Further refinement was then undertaken to optimise the solution.

The adopted solution involved lowering the reservoir TWL by 700mm to increase flood storage whilst minimising the impact on the aesthetic and amenity value of the reservoir.

A compound overflow arrangement was developed, with hydraulic inlets set at various levels to control outflows to rates similar to the existing outflow rates for various return periods while limiting water levels within the reservoir. This included the following elements:

- Low level orifice inlet with cross-sectional area $0.4m^2$ at an invert level of 338.0mAOD.
- High level weir length 11m at a level of 340.3mAOD.
- Secondary high level weir at a level of 340.6mAOD.
- New drawdown facility.

The final design reduces flows by over $1m^3/s$ in the 100yr plus climate change event. Increase in outflow compared to the existing outlet is limited to the 25yr event. The increase in this scenario is less than $0.3m^3/s$ and NRW confirmed that this was acceptable.

Water is allowed to spill over existing crest of eastern embankment. The maximum depth of overflowing predicted in the 10,000 year event is 200mm and duration 2.2 hours. Following a site visit by the QCE, he determined that this was acceptable.

OVERFLOW DESIGN

It was ascertained that a new overflow structure was required as the existing structures could not be modified and a new outfall conduit through the face of the dam would be necessary. The apparent solution at concept stage was to install the new overflow by open cut. This would have necessitated the closure of Pond Road, which the Council was not keen on as it was a local bus route. The length of the closure, if timed correctly, would not have caused excessive traffic disruption.

As part of early due diligence for the proposed works a full services search was undertaken as the road across the crest was public highway. The results showed that there were gas mains, water mains, recently installed telecoms and, in addition to low voltage and high voltage electricity mains, two bundles of three 132kV electricity cables located within the crest of the dam. The services were spread across the highway with drainage also being evident but no records were available. The viability of suspending the services over a 10m span was investigated and found to be viable, albeit with substantial structures.

Discussions were held with the electricity supply company for the area as to the nature of the cables, what they fed and restrictions in undertaking excavations near them. It transpired that they were strategic cable routes and a back-up so it was not possible to turn both off. For health and safety reasons any works utilising mechanised plant would need to be kept at least 2m away from them in any direction. It became evident that suspending all services and excavating under them was not a viable option, and early discussions with contractors reinforced this conclusion.

Pipe jacking solution

After assessing the proposed works and restrictions it was concluded that the only viable alternative was pipe jacking. The general concept was discussed with a pipe jacking contractor who confirmed that, based on the proven make up of the dam and the jacking length, the proposals, on paper, were feasible. The initial design was based on a rectangular culvert but the contractor stated that pipes would be much easier to jack and therefore would be more cost effective. The maximum standard size of pipe which could be jacked through the dam in the desired location based on the site restrictions was 1.8m.

The soil investigation had shown granular made ground being present up to approximately 5m below Pond Road level which also contained perched water. Working back from the existing level of the outfall channel being mindful of minimum channel and pipe gradients it was proven that the pipes should be able to pass approximately 1m below this layer. This would negate the need for expensive pre-grouting of the ground in advance of the jacking.

The soil investigation and design was sent to a specialist pipe jacking contractor to verify that the design, as produced, was feasible. They confirmed this and produced a budget cost.

The client was appraised of the proposal and concurred that, based on the restrictions highlighted, pipe jacking was the best option.

Hydraulic design

Hydraulic requirements and construction constraints mean that flows are conveyed through the dam in twin 1800mm diameter pipes. The upstream inverts of the pipes are offset to provide a low flow route.

The downstream spillway required a return along the toe of the dam at 90° to join the receiving watercourse. The concrete walls of the channel were curved to facilitate the change in direction. The walls were built-up due to high downstream ground levels and to account for super-elevation of the water surface in moderate flood events. A low-height baffle weir with low flow notch was also installed in the spillway downstream of the bend to dissipate energy in moderate return period floods, prior to flows discharging to the watercourse.

Water levels in the spillway are controlled by the capacity constraint introduced by a downstream twin 1m diameter pipe culvert in more extreme flood events. The influence of this control extends beyond the low height baffle weir in the spillway in the 100yr flood. In the 10,000yr flood the backing-up effect of the downstream control extends as far as the culverts through the dam.

STRUCTURAL DESIGN

From concept stage it was acknowledged that a concrete structural solution was the optimum way forward. Early discussions were held with precast concrete manufacturers about off-site production of panels or entire sections which were then to be installed on site to minimise construction times and potential disruption due to inclement weather. The general feedback was that the units would be large and heavy. Installation of precast concrete units was also considered and it soon became apparent that due to the relatively restricted width of Pond Road and the possible loadings onto it that only a relatively small crane was viable, therefore limiting the potential. After further negotiations with the Council and the local highways officer it became apparent that closure of Pond Road would be resisted even for a short length of time.

Therefore a full reinforced concrete structural design was undertaken. Various different loading scenarios were modelled including the upstream box being full of water with no external restraint, and the upstream box being empty and earth and water pressures imposed externally to the top of wall. These scenarios resulted in 500mm thick walls and base. The walls were wide enough to allow the top grillage platform to be securely fixed to it.

The letterbox low-level overflow and the two outfall pipe locations were designed to be boxed out. The two outfall pipes were designed to be cast into the structure. Due to the exposed conditions and desired design life RC50 concrete was specified with 50mm cover to the reinforcement. Water bars were specified for all joints to ensure a watertight structure.

The base of the upstream box was designed to have a screed which would be laid to falls which would initially channel water to the most westerly outfall pipe. This would take “everyday” flows. Higher flows through the letterbox would result in flows being allowed to utilise the second pipe.

A secure access to the platform on the upstream structure was required and this was designed in expanded galvanised mesh. For security reasons a lockable gate with side fencing was specified. A set of cast in-situ concrete steps afforded access to the concrete apron in front of the letterbox overflow.

The penstock was a proprietary design fixed to the outside face so that water pressures would assist in providing a watertight seal.

The downstream structure was designed as a "U" shaped channel with a nominal depth of 4m and an initial width of 7m where the twin pipes exited the dam, tapering to 5m after the channel swung 90° along the toe. The worst case loading criteria was external backfill up to the top level. The channel had a sloping invert both longitudinally and to a lower central point along which low flows would run, keeping the sides dry.

A stone-filled Reno mattress was installed at the downstream end of the reinforced concrete channel to act as erosion protection.

The Council's Health and Safety officer insisted on a 2m high non-climbable fence around the inlet. Two different means of access and egress are provided by a gate in the downstream grill of the westerly overflow pipe and a ladder via the platform on the upstream box.

CONSTRUCTION

After competitive tendering, local contractor Alun Griffiths (Contractors) Ltd was appointed to undertake the works. The first challenge was to access the upstream working area and install the temporary works. Whilst sheet piling had been suggested as the preferred option for constructing a cofferdam, concerns had been expressed at how they would be removed after the works. Cutting them off was not an option due to the potential uses of the reservoir. It was decided that an access ramp was to be installed along the face of the dam down to water level to enable a cofferdam bund to be constructed. Both were built using locally quarried stone. The bund had a core of smaller finer stone to produce a relatively watertight structure with a small layer of silt dredged from the bottom of the reservoir being placed on the outer face to provide further waterproofing. The bund was taken up to the 1 in 20 year estimated water level which generally concurred with the highest recorded water level over the previous few years. A pump was utilised to remove the small inflow through the cofferdam.

The downstream working area was restricted and the existing outfall flowed along the toe from the western shoulder eastwards through the area. The new outfall was positioned such that the scour valve was downstream of it, which meant the scour could be utilised as the main discharge point for the reservoir and keep the works area dry. As the scour valve had a much reduced capacity in comparison to the 750mm diameter main overflow, the overflow was not sealed so that it could act if required. The reservoir had its water level reduced to aid construction but it could not be totally drained due to the local wildlife and fish stocks in the reservoir. The top level of the forebay in the scour valve assembly was utilised to keep a minimum level of water in the reservoir and fortunately this coincided with the minimum water level ecologists required.

It was decided to jack from the downstream side so that if the upstream temporary works were overtopped the actual jacking tunnels would remain dry. A 4m by 4m by 8m long mass concrete thrust block was constructed and the downstream face excavated as near vertical as possible to take the jacking pipes and shield. The shield was pushed in advance of the pipes and the material excavated by compressed air hand held spades. Each 26m long jack took two weeks to complete. Works were delayed by water-filled gravel lenses being encountered, but these drained sufficiently to allow works to proceed within a day or so. The downstream works were halted for three days by the former main overflow being activated due to significant inflow into the reservoir.

The annulus around the pipes was pressure injected with a cementitious grout.

While the pipe jacking was proceeding from the downstream side the excavation for the upstream box structure was being undertaken. The box was approximately 13m long by 4m wide and 6m deep. The base was nominally 1.5m below existing bed level and was cast on lean mix blinding with upstand kickers to enable the wall reinforcement to be tied in. A waterstop bar was installed between the kicker and the vertical walls. Two buildouts were incorporated to accept the jacked pipes and the letterbox was similarly formed with a boxed out section. A proprietary formwork panel system was used and all the walls were poured together.

The jacked pipes entered the upstream box on line and level, then the remaining walls were completed. The rear and sides of the box were backfilled with material from the bund and the penstock, steps and access platform installed.

The downstream structure was constructed in a similar manner to the upstream box with the base and kicker walls cast first then the walls cast. Proprietary curved formwork was used to form the bend which produced a neat finish.

When the overflow was practically completed the flow was allowed through it which meant the existing overflow and the two scour valves could be grouted up. A land drain was laid along the western section of the ditch along the toe to remove any water which entered it.

In order to negate the requirement for the offsite removal and disposal of the rock which had been used for the bund and access ramp, the material was utilised to backfill around the downstream structure and the area of the toe between the new and old overflows. The Council's ecologists were very positive about this as it would provide habitat for a local lizard species. The remainder was left as an island to provide a habitat for overwintering birds safe from dogs.

The original four month programme was extended to six months due to inclement weather; the original pipe jacking contractor withdrawing from the project just before they were due to start, and delays due to inundation of the downstream works.

Figure 2. Completed overflow inlet

CONCLUSION

The spillway replacement works were completed and a Section 10(6) Certificate issued in 2015.

The overflow has functioned well in the time since completion with no issues with regard to the relatively small letterbox inlet being restricted by pond weed, which was a major problem with the previous overflow.

Dams – Benefits and Disbenefits; Assets or Liabilities?
ISBN 978-0-7277-6133-0

doi: 10.1680/dbdal.61330.401

Castle Irwell Flood Detention Reservoir - Design and Construction

A COURTNADGE, Jacobs
J GOSDEN Jacobs
T AMBLER, Jacobs
R LEWIS, Jacobs
E ADALI, Jacobs
A RADLEY, Axis P.E.D. Ltd

SYNOPSIS Castle Irwell Flood Detention Reservoir is a new offline flood storage reservoir, located in Salford, approximately 2km upstream of Manchester city centre. The reservoir sits in the lower reaches of the River Irwell catchment on the former site of Manchester Racecourse. The project client is the Environment Agency. The reservoir is fully bunded within a meander on the River Irwell by a 2.3km long, 3m high, zoned earth embankment. Inflows into the reservoir are controlled by a grass reinforced inlet weir at the upstream end and outflows are controlled through twin, automatically actuated, 1.5m square penstocks. Construction commenced in February 2015 and is due to be completed in 2016.

INTRODUCTION

In the early 1990s, the Environment Agency (EA) constructed a flood defence scheme through Salford to provide a Standard of Protection (SoP) of 1 in 75 Annual Exceedance Probability (AEP), comprising an upstream storage basin with a capacity of 650,000m³ and downstream defences. The scheme reduced the risk of flooding to 6,500 properties within the Salford floodplain and was completed in the mid 2000s.

In February 2013, the EA secured funding from Defra to develop and construct a second phase of the scheme, namely the Castle Irwell Flood Detention Reservoir, to raise the downstream SoP to 1 in 100 AEP. The funding came with the condition that scheme needed to be substantially started on site by 1 April 2015. Additional funding of £5M was made available from Salford City Council.

Jacobs UK Limited, with the support of Axis P.E.D. Limited ("the designer"), was commissioned in April 2013 to commence the Project Appraisal Report to confirm the preferred approach and develop the outline design. A design and build strategy was subsequently adopted for the earthworks element so as to exploit innovations and cost efficiencies from the EA's framework contractor.

At present, Galliford Black and Veatch ("the contractor") has significantly advanced works on site with the majority of the earthworks complete and the remaining ancillary structures to be completed in Spring/Summer 2016.

The project presented a number of specific challenges requiring some innovative solutions from the Project Appraisal Stage through to construction. These included:

Table 1. Design features

Design constraint / issue	Design feature/resolution
Restrictions on import/export and need for excavated flood storage capacity.	Landscaped knoll to achieve a cut/fill balance
Relatively rapid lowering of reservoir to enable follow on floods to also be attenuated	Programmed logic control to control outlet penstocks relating to reservoir and downstream water levels
Area prone to vandalism	Vandal resilient design e.g. vulnerable elements buried.
Environmental improvement	Creation of wetland habitat area
Tight programme necessary to gain grant funding	Parallel working on planning application and reference design. Design-build approach for embankments

ROLES AND RESPONSIBILITIES

The designer was required to produce a reference design for the earthworks, and a detailed design for the landscaping, ancillary structures and drainage systems. The reference design showed a possible embankment arrangement and specified the criteria for the contractor's detailed design giving them flexibility to maximise the re-use of available materials.

The contractor was responsible for the detailed design of the earthworks. This initiative allowed an earlier start date on site, which was a condition of the Defra funding, and was also intended to generate efficiency savings on the basis that the contractor had most control of the materials derived from the borrow pit.

SCHEME OUTLINE

Castle Irwell Basin is situated on a meander in the lower reaches of the River Irwell catchment bounded by the Salford University Student Village to the south. The site covers approximately 30 hectares. The site provides sports playing pitches for Salford University and amenity parkland for the local community, however access to the site is limited and the poor quality of the amenity area has meant that the site is not used to its full potential.

The fundamental principle of the scheme was to supplement the flood peak attenuation provided by the existing storage area (approximately 1km upstream) by adding a second storage basin, thereby increasing the SoP downstream to 1 in 100 AEP.

In doing so, the scheme sought to:

- Incorporate habitat enhancements and deliver a strategic semi-natural greenspace;
- Replace the existing sports pitches disturbed during the creation of the new storage area;
- Retain as much of the mature tree cover to the river corridor as possible; and
- Enhance an existing Sculpture Trail;

The final design was a reservoir, offline from the river, located within the low-lying, natural meander. The reservoir is retained by a 2.5km long embankment which runs around the perimeter of the site. Additional storage volume was also created by lowering the ground levels inside the reservoir by between 0.8m and 2.6m depth. This ground lowering provides approximately 20% of the storage capacity during the design flood with 1 in 100 AEP.

The principal components of the reservoir are described below and illustrated on Figure 1, with the key statistics shown in Table 2.

- **Embankments** - These are discussed in detail later
- **Knoll** - The surplus of cut material from within the reservoir basin would be used to create a knoll landscape feature in the northeast corner of the site in order to achieve a cut/fill balance.
- **Inlet structure** - At the southwest corner of the reservoir the crest of the embankment is lowered to form an inlet spillway. During significant flood events, with an AEP of 1 in 8 or rarer, water will spill from the river over the spillway into the reservoir. As the flood recedes water will initially return to the river over this structure.

- **Outlet control structure** - Following an impounding event the impounded water below the level of the inlet spillway will drain back into the river via an outlet structure in the northeast corner of the site. The outflow from this structure is controlled by two automated penstocks.
- **Reservoir Basin** - The reservoir basin is divided into two areas as shown in Figure 1. The slightly larger area to the south comprises sports pitches for the University of Salford to replace those previously on the site, whilst the lower lying area at the northern end has been designed as a new area of permanent wetland habitat with an elevated knoll at the end.

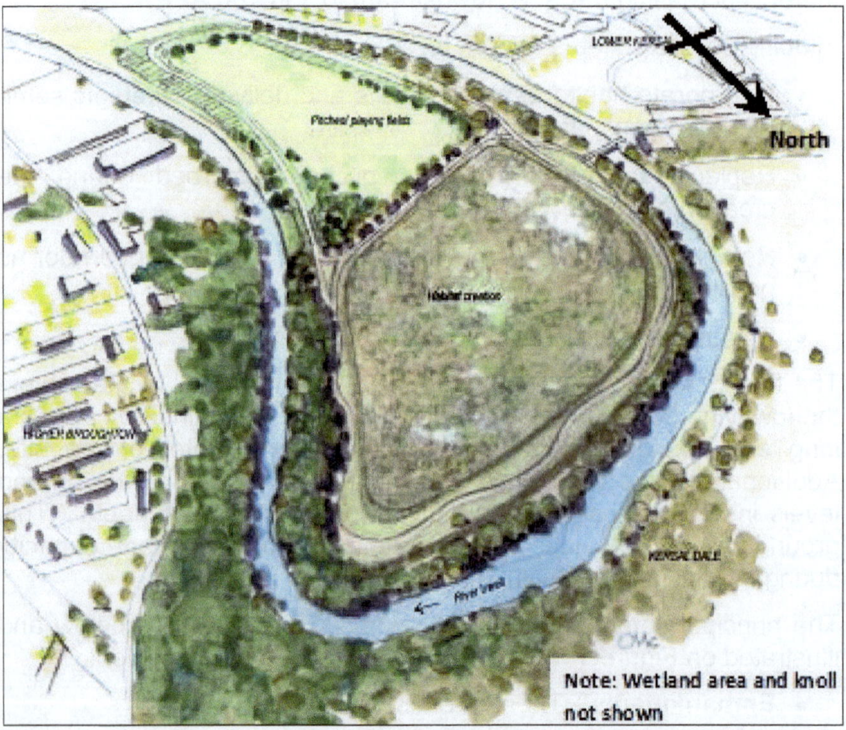

Figure 1. Birds-eye view artists impression of the Castle Irwell Flood Basin

Table 2. Principal Statistics

Parameter	Value
Embankment length	2.5km
Embankment height	3m to 5m
Design flood	1 in 100 AEP
Original ground level	28.0 to 30.5mAOD

Parameter	Value
Typical finished basin level: Sports pitches Wetland area	 26.3m AOD 26.2m AOD
Inlet spillway crest level	30.4mAOD
Reservoir storage level	30.98mAOD at design flood
Embankment crest level	33.0mOD
Storage volume: At design flood storage level: At embankment crest level:	 580,600m³ 1,004,000m³
Inlet spillway dimensions	100m (with transition slopes at 1v:10h)
Outlet culvert size	2 no.1.5m diameter culverts
Drainage orifice size	300 mm square
Outlet invert level	26.0mOD (upstream) 25.9mOD (downstream)

PROJECT APPROACH

Interactive planning was crucial in the development of conceptual designs at the onset to the project. Engineers, planners, and landscape architects recognised the challenges faced by three very different disciplines. This approach subsequently led to a documented set of 'Design Input Parameters' which evolved as better information became available but ensured buy in from all parties on key design parameters and particularly user requirements.

Leading up to contractor appointment, the programme was further compressed to meet Defra targets by running the planning process in parallel with engineering detail design. Key milestones of the project are summarised in Table 3.

Table 3. Delivery Timescales

Key Milestones	Timescale
Project Appraisal Report	April 2013 - March 2014
Design Development	March 2014 – October 2014
Planning Application	April 2014 - October 2014
Detailed Design	September 2014 –December 2014
Tender	December 2014 – January 2015
Construction	February 2015 – October 2016

The total volume that could be excavated for the wetland area was agreed in advance between engineers and landscape architects, ensuring storage capacity could be realised whilst delivering an acceptable sports pitch layout and landscape scheme.

RESERVOIR OPERATION

Under normal conditions, the reservoir basin will be empty and the penstocks closed. Normal rainfall events will be drained via a 300mm square orifice situated in one of the penstocks. This orifice was sized to pass rainfall events below 1 in 10 AEP, to prevent the sports pitches from flooding in smaller events. By incorporating the orifice into the main outlet culverts the need for an additional conduit passing through the dam was avoided.

Flood modelling showed that a flow control device in the drainage orifice inlet was not required since inflow from the river in non-impounding events, and outflow in impounding events, would be minimal and would not affect the operation of the reservoir.

The inlet spillway was designed to overtop in a 1 in 8 AEP event. The outlet penstocks are designed to remain closed until after the flood peak, when the water level in the basin begins to recede by flowing back over the inlet spillway. The penstocks are automated based on an operating philosophy which has three main criteria:

- Before a flood event the basin should have as much capacity as possible;
- After a flood event peak the basin should be emptied as soon as possible, to provide maximum capacity for the next flood peak; and
- Drawdown of the reservoir should not cause flooding downstream

Triggers within the control philosophy are based on data from on-site telemetry, including the water level in the basin and the rate it is receding, and the downstream water level in the River Irwell.

A passive system, using flap valves or similar was found to be unacceptable as it would not provide optimal use of the sport pitches. In addition, initial modelling showed that a passive system at Castle Irwell basin would reduce the effectiveness of the original flood storage reservoir located 1km upstream at Littleton Road.

In order to maximise reservoir drawdown the outlet structure was located as far downstream as possible, thereby increasing the head difference between the basin level and the river level. Due to the river's meander, the outlet structure was located 1.7km downstream of the inlet spillway, which allowed the required drawdown rate to be provided by smaller culverts.

EMBANKMENT DESIGN

There were a number of challenges to overcome in the design as discussed below.

Three dimensional geometry

The layout of the reservoir was complicated by a number of constraints including:

- Planning restrictions severely limited the volume of material which could be imported or exported from the site, so the volume of excavation needed to balance volume of fill;
- Practical limits on the size of the knoll to accommodate surplus material without compromising storage capacity;
- The flood storage capacity, including the depth/storage relationship, needed to achieve the flood risk objectives of the scheme as modelled in the hydraulic model;
- Several factors constrained the embankment slope angle such as slope stability, overtopping flow velocities and long term maintenance;
- The alignment of the inner toe and the basin level was firstly constrained by the need to reinstate seven sports pitches to an acceptable standard, aligned in accordance with Sport England/ governing body guidance and with no overall loss of playing field area. Secondly, a planning requirement to meet the aspiration from the Local Planning Authority/ community/ EA for creation of a new wetland habitat within the storage basin;

Developing a solution within these constraints required an iterative approach to the design, using 3D CAD modelling software to check the cut and fill balance. In order to achieve the flood storage characteristics required from the hydrological modelling a surplus of material needed to be excavated from the reservoir basin. This was accommodated within the design by building the embankments higher than otherwise required and by the creation of the knoll.

Suitability of fill material

The site is overlain by superficial deposits of Alluvium, over Glacial Till with Sherwood sandstone bedrock. The approximate quantities of the various materials available from the dam foundation and basin excavations are summarised in Table 4.

Table 4. Contractor's estimated material quantities at the outset of construction

Material	Quantity (m³)
Topsoil	60,000
Made Ground	43,000
Alluvium	158,000
Sand and Gravel	38,500
Sandstone (rock)	2,500
Total	**302,000**

The most abundant material for the embankment construction was alluvium, which ranged from a silty sand to a very sandy silt (with silt content ranging between 20% and 80%). Whilst such silty material is not normally favoured for dam fill, due to its susceptibility to internal erosion, limited strength and variable permeability, it was considered viable given the relatively low height embankments. Indeed an advantage of silty material for flood detention dams is that it is less prone to desiccation cracking which is often an issue for such structures built using more cohesive materials.

The material excavated was significantly wetter than its optimum moisture content (OMC) (the mean moisture content was 21%, compared to the OMC which ranged from 11% to 21%). The undrained shear strength of the in situ material was typically around 25kN/m² which meant it was unsuitable for trafficking or compaction without treatment.

The embankment cross section was zoned in order to make the best use of available materials as illustrated in Figure 2. The central zone of alluvium provided the required water tightness whilst the outer sand/gravel shoulders acted as a filter layer to mitigate the risk of internal erosion, protected the core from desiccation cracking and also enhanced the slope stability.

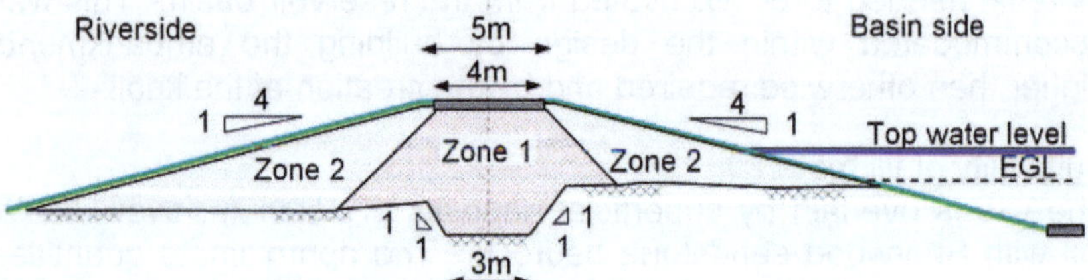

Figure 2 Embankment cross section

Extreme floods over spilling the crest

Given its location in a highly urbanised area the reservoir needed to be designed as a Category A dam in accordance with Floods and Reservoir Safety, 4th Edition (ICE, 2015), which meant it had to withstand a PMF flood. Analysis showed that in extreme floods, with an AEP greater than 1 in 10,000, the embankments would be overtopped and the river would effectively short-cut the whole meander. In such extreme conditions most of the reservoir embankments would be submerged but there would be a head of water flowing across some sections of the eastern embankment, flowing from the reservoir to the river. In a PMF flood it was estimated that water could overtop parts of the dam crest by between 0.5m and 1m depth creating high flow velocities and the potential for erosion and breach. To mitigate this risk the outer face of the eastern embankment was protected using open mesh reinforced grass.

OUTLET CONTROL STRUCTURE

The outlet control structure consists of two 1.5m diameter concrete culverts. These culverts were sized to be capable of emptying the reservoir in 24 hours following a flood event. This was to enable the reservoir to be ready to operate for any further flood events, and also allowed the reservoir to be drained quickly in the unlikely event of a structural problem occurring with the dam.

Vandalism was a key concern at the site. The nearby Littleton Road flood storage reservoir showed heavy damage to both the structure and the embankment, primarily due to vehicle impact and motorbike erosion. Table 5 discusses how the risk of vandalism for each key design feature was mitigated.

Table 5. Control Structure Design Approach

Design Feature	Design approach
Trash screens and access hatches	Impact resistant trash screens and locked access hatches.
Penstocks	Access to penstocks prevented by trash screens. Penstocks located below actuator chamber to reduce visibility.
Stilling tubes and pressure transducers	Where possible stilling tubes were located within the trash screens. Elsewhere they were located in chambers buried behind the structure sidewalls. Access to these was via locked access hatches.

Design Feature	Design approach
Downstream security screen	Two grilles were designed preventing downstream access into the culverts. These grilles included: • A locking mechanism to allow opening from above, located within a secure box. • Hinges with locking nuts. • Failsafe shear pins if the grilles become blocked with trash. • Dummy fixings on not-hinged edge.
Electrical actuators for penstocks	A concrete chamber was created above the penstocks to protect the actuators from vandalism. A heavy duty steel security door provided access to the chamber. Removal of heavy concrete planks forming the roof of the actuator chamber allows the penstocks to be craned out for maintenance. Holes in the planks allow emergency manual operation of the actuators using T-keys.
Earth retaining wall around outlet structure	A concrete wall was designed surrounding the outlet structure. A local graffiti artist with the assistance of local schools will create a mural on the face of the walls.
Motorbike erosion of the embankment	Desire lines along the inlet spillway were paved with asphalt to reduce damage to the Enkamat.

The final design of the outlet structure is shown schematically in Figure 4.

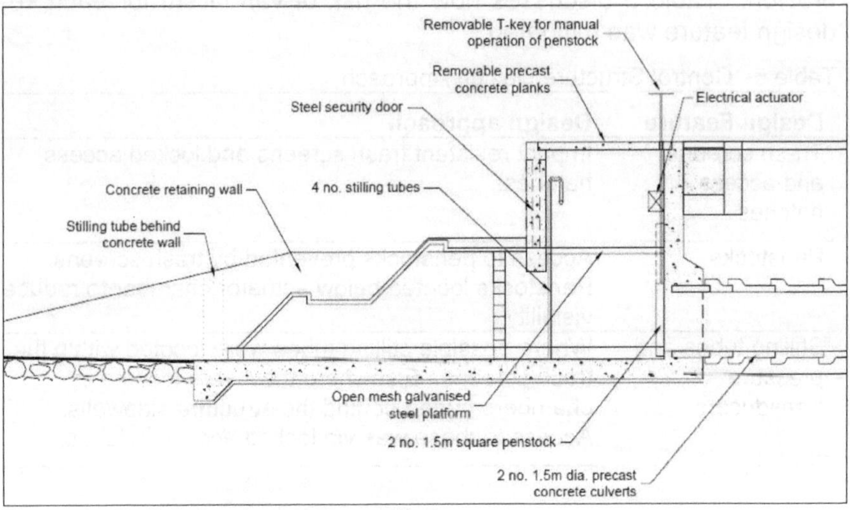

Figure 4. Outlet structure cross section

CONSTRUCTION

The specification for the earthworks was a conventional end-product specification requiring the compacted fill to have a field dry density of at least 95% of the maximum dry density (MDD) achieved in a laboratory compaction test, and a moisture content between 2% drier and 4% wetter than the optimum moisture content (OMC) (or -0% wetter to +6% drier where the material was stabilised with quicklime). The high in situ moisture content of the alluvium was always anticipated to be a problem and to overcome the problem treatment with lime stabilisation was initially trialled during placement of the core trench. Although these trials were successful in treating the material, the process was costly and potentially hazardous. As construction progressed effective techniques were developed to work the borrow pit, taking advantage of the large, flat site and the good weather to adequately dry the material prior to placement and compaction, thereby avoiding the need for further lime treatment for the majority of the core. Lime stabilisation was again introduced for the months of November and December when the weather conditions prevented the material drying naturally.

During construction it became apparent that the available volume of Zone 1 core material had been over-estimated at the design stage and the embankment cross section was accordingly redesigned with a slightly narrower core. Analysis was carried out to demonstrate that this would not adversely affect the risk of seepage, internal erosion or slope instability.

Construction commenced in Feb 2015 and was due to be complete on programme in April 2016. Good progress has been helped by dry weather through much of 2015, although the site was flooded in the heavy rains over Christmas 2015 when the river spilled over the inlet spillway into the reservoir (Figure 5). Due to the temporary lack of vegetation, the flooding caused erosion damage to parts of the river banks, including the area near the inlet spillway, and deposits of silt on the sports pitch area, which has delayed completion. However, no significant damage was caused to the dam structure.

Non-native invasive species were widespread throughout the site, including significant quantities of Japanese knotweed and Giant hogweed. Management of this issue posed considerable challenges and warrants a paper in its own right. In summary off-site disposal was too expensive so contaminated material was instead encased in a sealed lined cell and buried within the knoll. A root barrier membrane was installed around the perimeter of the embankment where it interfaced with the existing river bank.

Figure 5. Inlet Spillway, Boxing Day Floods 2015

CONCLUSIONS

Salford FAS has proved to be a successful project despite a number of challenges in the design and construction. The first major success was in achieving the ambitiously short 20 month programme for planning, design and tendering the contract to achieve a construction start in February 2015. This success is attributed to interactive planning at the outset by the various parties involved, developing the detailed design in parallel with the planning application and fixing deign parameters with the client at an early stage. A disciplined approach to managing drawing XREF layers also helped to expedite the drawing production with inputs from multiple design disciplines.

The construction stage has also progressed well, without major problems. This is in part due to adopting a relatively conservative and flexible reference design at the outset which allowed alterations to be made during the construction to accommodate unforeseen problems when they arose.

ACKNOWLEDGEMENTS

The authors of this paper would like to acknowledge their gratitude to Paul Robertshaw from the Environment Agency, Salford University. Wildlife & Wetland Trust, Galliford Try and Black & Veatch.

REFERENCES

ICE (2015). *Floods and Reservoir Safety, Fourth Edition.* ICE Publishing, London

Dams – Benefits and Disbenefits; Assets or Liabilities?
ISBN 978-0-7277-6133-0

doi: 10.1680/dbdal.61330.413

ice
Institution of Civil Engineers
publishing

Design and Construction of Mitford Flood Storage Reservoir

J G PENMAN, CH2M
A D MYATT, Environment Agency
A B CARR, CH2M
V S COVERDALE, CH2M
D M LATHAM, JBA (formerly CH2M)
I G EDMONDS, CH2M

SYNOPSIS Following severe flooding from the River Wansbeck in 2008 a flood alleviation scheme for Morpeth, Northumberland, UK was designed and constructed which includes one of the largest flood storage reservoirs the Environment Agency has commissioned in England.

The dam comprises an earth embankment 14m high with a crest length of 370m, a spillway capacity of 760m^3/s and a storage capacity of 1.4Mm^3.

Construction of the scheme was completed in 2015. The paper will describe the design of the dam, the construction work and the implementation of the environmental mitigation measures.

INTRODUCTION

Morpeth is the market town of Northumberland located approximately 15 miles north of Newcastle upon Tyne on the River Wansbeck. The Wansbeck catchment area is 292km^2 and its two major tributaries join the Wansbeck within a 10km reach upstream of the town.

Morpeth has a long history of flooding dating back to at least 1839, with 21 flooding events have been recorded during the last 175 years. The most significant recent flood events were in March 1963 and September 2008. The latter was the largest recorded in Morpeth's history when over 1000 properties were flooded.

The 2008 event posed a serious risk to life due to the rate of rise of the flood peak, the number of vulnerable people living in the floodplain and the difficulty in evacuating these people. Around 400 people were rescued from their homes, and many hundreds more left their homes voluntarily. The flooding caused devastation to much of the centre of Morpeth and the impact of the event was felt throughout the region.

In response to the 2008 event the Environment Agency sought to develop a scheme that had community support and utilised the existing defences where possible. The scheme identified was a combination of a flood storage reservoir on the River Wansbeck and raising / reusing of the flood defences within the town. This paper covers the flood storage reservoir (FSR).

SITE SELECTION AND SCHEME LAYOUT

Initial studies identified a suitable reservoir site on the Wansbeck about 1km upstream of the centre of the village of Mitford. However, this site met with local opposition so an alternative site some 500m further upstream was adopted.

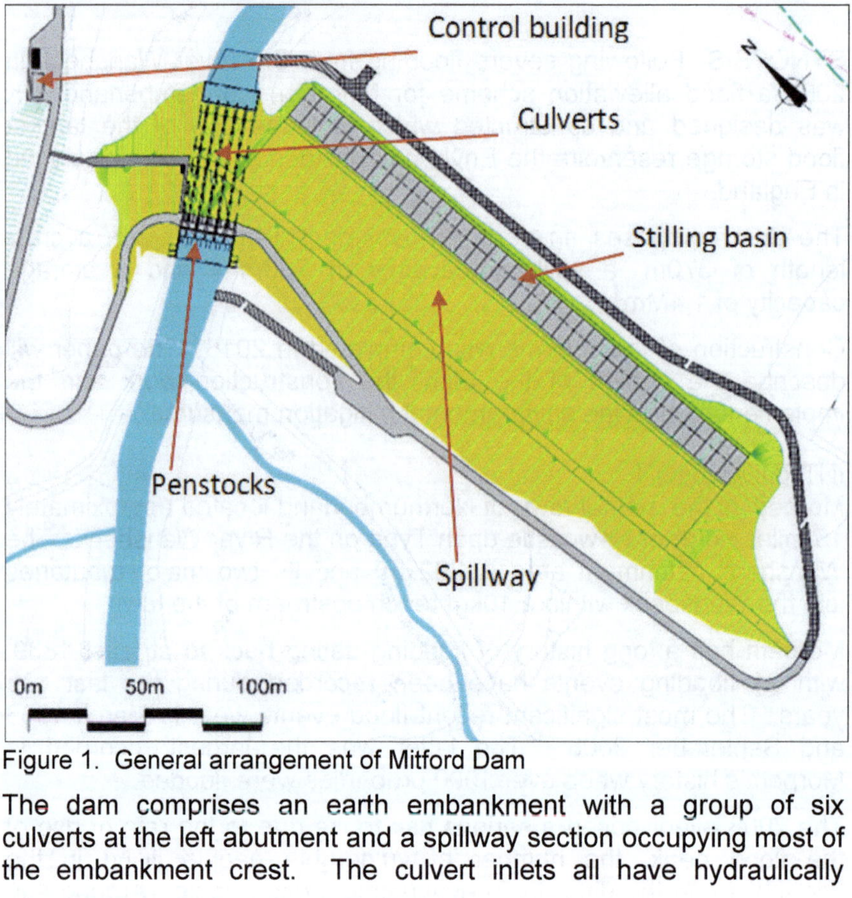

Figure 1. General arrangement of Mitford Dam

The dam comprises an earth embankment with a group of six culverts at the left abutment and a spillway section occupying most of the embankment crest. The culvert inlets all have hydraulically actuated penstocks. The embankment is 370m long with a maximum height of 13.8m and 1v:4h side slopes. The storage volume is 1.4Mm³ making it one of the Environment Agency's largest flood storage reservoirs.

The dam axis deviates in the upstream direction at the left abutment. This innovative arrangement allowed the culverts to be accommodated whilst maximising the length of the spillway.

Figure 2. View of Mitford Dam from left abutment

HYDROLOGY AND RIVER MODELLING

The river network upstream of Morpeth comprises the Wansbeck itself, the Hart Burn and the River Font. The dam is located approximately 2.5km upstream of the River Font confluence – this is significant as the flood storage reservoir only attenuates the flows in the Wansbeck.

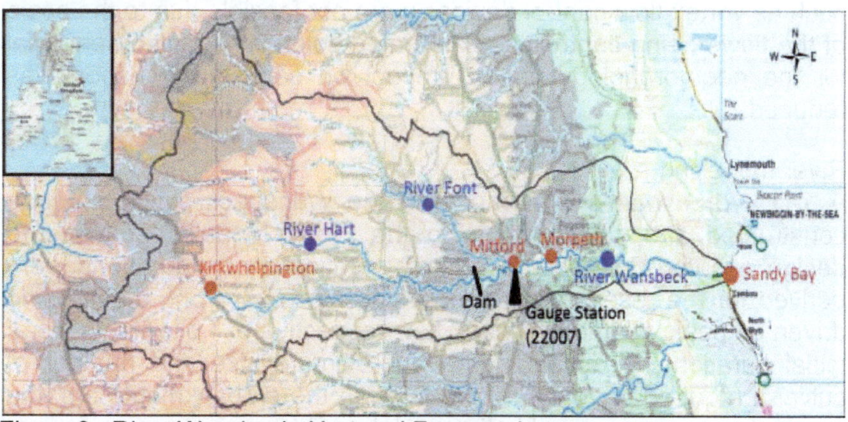

Figure 3. River Wansbeck, Hart and Font catchments

A hydrological assessment of the catchment was initially undertaken in 2005 as part of the Environment Agency's Morpeth Strategy Study by JBA Consulting. Peak flows were estimated using the FEH statistical method, incorporating gauge data from Mitford, historic events, and pooling group methods. Design hydrographs were derived using the FEH rainfall runoff method and a semi-distributed routing model. This assessment was largely adopted for the detailed design of the dam after confirmation that the application of the more recent FEH guidance now available would not impact on the assessment.

September 2008 Flood event

The occurrence of the September 2008 flood event significantly changed the scope of the FSR. The hydrological assessment was not updated to incorporate the event because updating immediately after the event could bias flow estimates upwards. However, it was agreed that the scheme should protect against this event, which had an estimated return period of 137 years. The economic appraisal for the scheme (completed before the 2008 event) determined the most cost effective standard of protection (SOP) to be 1 in 115 years. A passively controlled flood storage reservoir could have provided sufficient attenuation to protect against this event but not the 2008 event.

A gated solution was deemed necessary to enable the maximum tolerable flow to pass downstream without starting to impound the reservoir. This would preserve as much as possible of the available storage volume for attenuating the peak of the flood hydrograph, ultimately reducing the total flood storage needed and therefore lowering the peak water level in the flood storage reservoir to an acceptable level. Other means of providing optimisation of storage such as vortex flow control devices were not feasible due to the scale of the flows being controlled. The gated solution was also beneficial for the geomorphology as the impact on sediment transport was reduced.

River modelling

A 1-D hydrodynamic river model of the Wansbeck catchment was constructed and calibrated using the Flood Modeller software to design a gated flow control regime. Critical storm durations were derived for the reservoir catchment and the River Font, these being driven by peak volume and peak flow respectively. Following several initial iterations to ascertain the approximate size and number of culverts required to convey flows through the embankment dam, the reservoir was represented in the model as follows:

- LIDAR generated reservoir unit to define reservoir stage / surface area (and hence volume) relationship
- Five gated culverts, comprising sluice gates controlled by logical rules and conduit units

Note: a smaller crayfish and eel culvert was omitted from the model since it would close at relatively low flows.

This model was used to investigate four possible flow control regimes with various gate operating regimes and flow monitoring techniques. They are summarised in Table 1.

Table 1. FSR flow control options

Flow control regime	Dam height+ (m)	Gate operation	Flow gauging required
Fixed orifice	14.8	N/A	N/A
1	14.3	3 of 5 gates close simultaneously when reservoir trigger level hit.	Reservoir level measurement used in conjunction with culvert discharge curve to pass a prescribed target reservoir outflow.
2	13.8	3 of 5 gates close one by one at defined trigger levels. The staged closure of the gates provides some optimisation of the storage.	Reservoir level measurement used in conjunction with culvert discharge curve to pass a prescribed target reservoir outflow.
3	13.2	All gates close partially and move incrementally to maintain the desired outflow to further optimise the storage.	Flow gauge immediately downstream of the outlet used to pass a prescribed target reservoir outflow.
4	12.6	All gates close partially and move incrementally to maintain constant flow into Morpeth town, providing maximum optimisation of the storage	Flow gauges immediately downstream of the dam outlet and on the River Font used in combination to pass a prescribed target flow into Morpeth town.

+ Above outlet structure invert level

These flow control regimes are illustrated by the following hydrographs generated by the hydraulic model, which plot reservoir inflow, outflow and stage, as well as flow at Mitford (i.e. flow into Morpeth town).

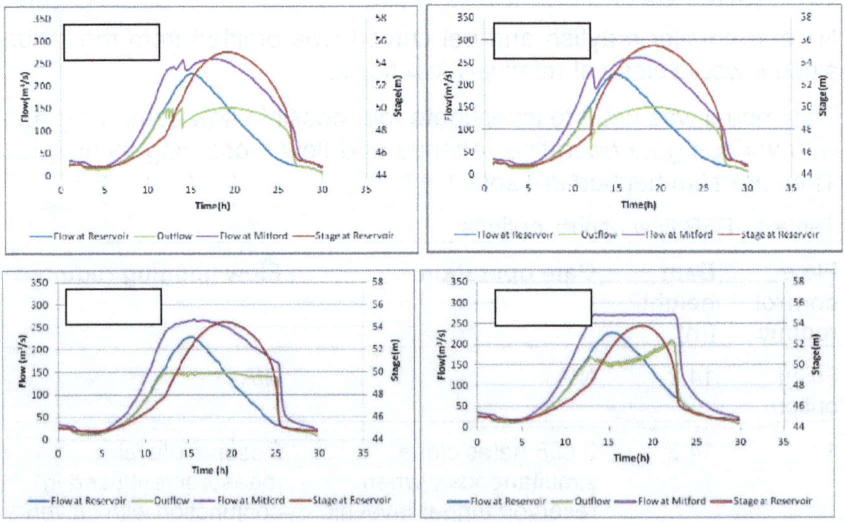

Figure 4. Potential operating regimes for the FSR

The arrangement adopted was Regime 2 as the reductions in reservoir volume provided by Regimes 3 and 4 were outweighed by the increased complexities of flow gauging and gate operation. The maximum pass forward flow for Regime 2 was 150m³/s.

CULVERT / PENSTOCK DESIGN

The control structure comprises a 1.8m x 1.8m culvert and five 3.0m x 3.0m culverts, all 53m long.

The 1.8m x 1.8m culvert (Culvert #1) is on the left side and is designed for the passage of crayfish and eels. It is set lower than the other culverts such that it maintains a flow in dry weather conditions.

The five 3.0m x 3.0m culverts (Culverts #2 to #6) are identical with the exception of the left hand unit (Culvert #2) which has an invert level set 150mm lower than the other four and a rectangular low-flow channel to enhance passage for salmon and trout.

The culverts are all formed from precast units. There are in-situ inlet and outlet structures at either end of the culverts. The wingwalls on the inlet and outlet structures are extended to river bed level with gabion retaining walls. There are coarse debris screens immediately upstream of the culvert inlets.

The penstocks are located at the culvert inlets and are powered by hydraulic actuators which can be submerged. They are controlled automatically according to operating rules enforced by a programmable logic controller (PLC) located in a dedicated control building overlooking the dam. A hydraulic power pack located in a plant room on the embankment crest provides power.

Figure 5. Culvert inlet structure and penstocks

Culvert #1 closes when the flow in the river exceeds 30m³/s to maintain the stability of sediments present within the culvert. The culvert does not play a part in the attenuation of floods.

The operating rules for the remainder of the culverts are shown below in Table 2. The closure sequence is assigned to ensure the desired culverts remain open and reduce the risk of erosion to the outlet left bank. Re-opening of the penstocks occurs at the same trigger level less a nominal depth.

Table 2. Summary of penstock operating rules

Penstock number to close	Combined flow through culverts (m^3/s)	Flow through one culvert (m^3/s)	Upstream level (m AOD)
1	150	30	48.00
2	150	37.5	48.80
3	150	50	50.30

Pressure transducers located at the inlet and outlet structures provide live water level information to the control panel. Sensors installed on the penstocks themselves monitor the position of the penstocks and will relay information to the panel. Should a penstock fail to close or close partially, it will re-open and the panel will assign a different sequence to accommodate the failed penstock remaining open. This process provides a high level of redundancy.

Redundancy to the power supply is provided through a back-up generator. Hand pumps to close the penstocks are a secondary back-up. Manual control of the gates can be assumed in the event the operating rules cannot be communicated by the PLC.

<u>Outlet energy dissipation</u>
A CFD model was used to design the dam outlet energy dissipation measures. The use of empirical design methods for the scour protection detailed design had been considered, but none was found to be applicable to the specific design situation where the reservoir is full and two culverts are discharging full bore. The model provided robust hydraulic data including flow velocities and depths, which were used to design energy dissipation and scour protection measures comprising:

- A 1m deep, 15m long stilling basin to dissipate energy by allowing the jets from the individual culverts to coalesce, and significantly reduce the risk of extensive scour downstream of the outlet
- A fibre-reinforced concrete slab to withstand the hydraulic conditions within this basin, and a heavily-graded riprap revetment to protect the river bed and banks respectively

Loose riverbed material was used to fill the stilling basin and hence provide an environment amenable to fish and native crayfish, and this material will scour out when subjected to high flows. The benefits of the basin and the hydraulic conditions predicted by the CFD are shown in Figure 6.

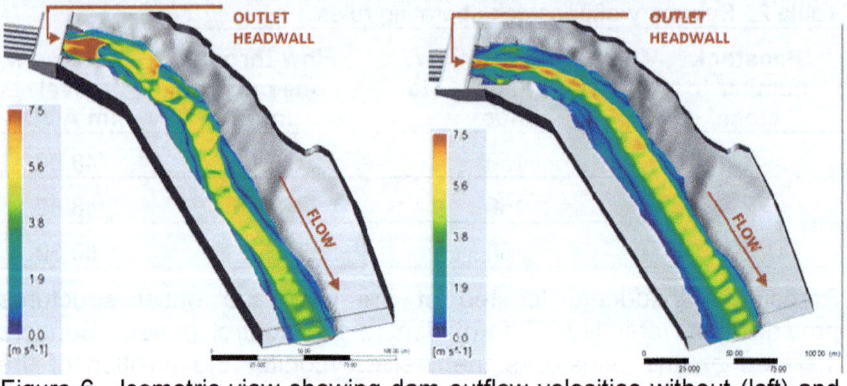

Figure 6. Isometric view showing dam outflow velocities without (left) and with (right) stilling basin in place

SPILLWAY DESIGN

The dam was designed, in combination with the culverts, to pass a Probable Maximum Flood of 932m³/s. The design concept for the spillway adopted was to discharge all overtopping flows over a single 240m wide spillway protected from erosion by an articulating concrete block revetment system. The revetment was covered with sacrificial topsoil and grassed. This solution helped to secure planning permission and landowner acceptance, providing a finished appearance in keeping with the local landscape.

The estimated peak velocity on the downstream face of the spillway was slightly in excess of 9m/s. This is greater than the limit recommended in CIRIA 116 (Hewlett *et al*, 1987) but was considered acceptable as the blocks specified are some 40% heavier per unit area than those used in the CIRIA 116 trials.

Careful detailing and installation of the revetment to achieve inter-block friction was important to ensure the revetment acts monolithically and hence performs to its capacity. This was enhanced by specifying a deep-rooted grass seed and controlling topsoil depths such that the roots penetrate into the joints between the blocks.

A reinforced concrete stilling basin 16.5m long and 2m deep was provided to contain the hydraulic jump at the toe of the embankment. The edge of the stilling basin doubled as an edge beam for the base of the concrete blocks.

GEOTECHNICAL DESIGN

Ground investigation

The geology of the site comprises alluvial deposits overlying glacial till. The left abutment comprised a pre-existing landslip and the right flank included a possible glacial moraine. A ground investigation targeted these features as well as characterising the dam foundations and identifying sources of fill material.

Ground and groundwater conditions

The landslip failure was found to be due to an over-steepened slope due to river erosion and high ground water pressures caused by a thick water-bearing sand lens outcropping just above the proposed dam crest level. The drainage of the sand lens appeared to be restricted by detritus which mantled the spring line. The shoulder of the dam is at the edge of the unstable zone as the sand lens appears to terminate beneath the upstream mitre channel. The solution to this was to remove the slipped material within the dam footprint and install drainage to allow the sand lens to discharge into the river without the groundwater reaching the slope face.

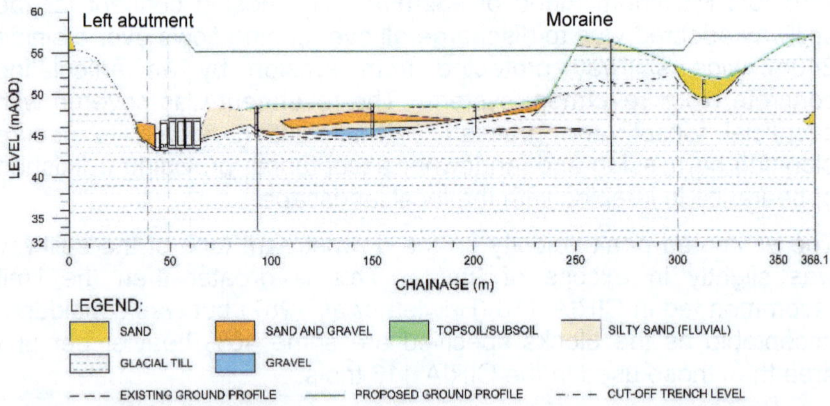

Figure 7. Longitudinal section of dam site

The glacial moraine was found to comprise an upper layer of very gravelly clay, with occasional lenses of sand. This was underlain by a high strength grey sandy clay. The mound contained occasional water strikes. It was determined that the upper portion of the mound was insufficiently impermeable to be incorporated into the dam, although the lower part could remain in-situ. The properties of this horizon were similar to the underlying glacial clay.

The floodplain comprised several former river channels beneath a more recent fine or sandy alluvium. Perched water tables were identified in the former river channels suggesting that they would provide potential conduits for flow beneath the dam. The presence of the buried channels required the inclusion of a cut-off trench beneath the dam foundation.

Design

Stability analysis of the embankment was not a governing criteria as the 1v:4h slopes, required for maintenance reasons, ensured an adequate factor of safety. A cut-off was included to limit seepage, and the risk of piping, through the dam foundations. Long term settlement was predicted to be of the order of 100mm.

Construction materials

The specification for the construction materials was in accordance with Specification for Highway Works, Series 600, and a modified class 2C was specified, which incorporated a limit on coarse soils to control permeability, a limit on clay sized particles and plasticity to control the risk of shrinkage, a limit on permeability and a limit on dispersivity. Whilst importing construction materials was considered, the preferred option was to use a borrow pit close to the site and thus minimise vehicular movements on the local roads. A borrow pit was identified in the field immediately north of the left bank of the dam and testing was undertaken in the field. Whilst the majority of materials were compliant, the upper clay layer in the field was found to be dispersive. It failed the crumb test, but not the pinhole test, therefore this material was determined suitable for use within the less sensitive parts of the embankment (i.e. areas remote from the crest and upstream face). The lower clay horizon was initially compliant and was used in the cut-off channel, but as the construction works progressed, it was found to have a higher clay content (>30%) suggesting that it could be prone to shrinkage. However, the clay was not high plasticity and therefore the risk of shrinkage was acceptable. The topsoil thickness on the dam was increased from 100mm to >300mm to mitigate against the effects of shrinkage.

EMBEDDED ENVIRONMENTAL DESIGN

Background

The natural hydrology / geomorphology of many of the rivers in the Northumbria River Basin District has been significantly changed by weirs, sluices, bridges and dams. The Wansbeck is considered to be one of the least affected rivers. The river is considered to be one of the most important watercourses in England for the native white-clawed crayfish. The distribution of this species across England has been significantly affected by the introduction of the American signal crayfish, but the Wansbeck is free of signal crayfish. The Wansbeck also has good densities of brown trout with a wide distribution. Salmon are present in downstream sections.

The environmental design of the dam and its operational function was fundamental to the acceptance of the upstream storage option by Natural England and compliance with the Water Framework Directive (WFD). The underlining principles were to maintain the geomorphological characteristics of the river and reduce the impact of the structure on fish and white-clawed crayfish populations, in particular the impact the culverts may have on upstream movements of these species.

Fish and crayfish passage

Research led by Halcrow Group (now CH2M) and Durham University (Louca *et al*, 2014) demonstrated that culverts *per se* were not obstacles for the movement of white-clawed crayfish; however, bed composition together with stream depth and velocity were a significant influence on the ability of crayfish to traverse such structures. Stream bed composition required areas of refuge and cobbles in a range in sizes. A target value of <0.5m/s at the Q_{50} flow ($0.76m^3/s$) was recommended (Lucas *et al*, 2010). Culvert #1 was added to maintain passage for crayfish and freshwater eels. The inverts of this and the adjacent culvert were placed lower than the rest of the culverts to provide variation in bed levels across the culvert structure. Culvert #2 also included a low-flow channel (1000mm x 300mm). These features aimed to provide a minimum depth of water (300mm) through the structure during low-flow periods for fish passage.

Offset baffles, fabricated within the culvert units, were provided through Culvert #1 using guidance from SEPA (SEPA, 2010), to manage velocities in the culvert. The features gave a measured average culvert velocity of 0.4m/s for a river flow of $1.2m^3/s$. Since the velocity will reduce with lower flows, the target value of <0.5m/s at Q_{50} (Lucas *et al,* 2010) will be achieved.

Figure 8. Baffles in Crayfish Culvert (Culvert #1)

Establishing the channel bed form was more challenging. It was not possible to retain the natural bed given the gate function and the need to prevent undercutting of the dam embankment. The finish of the bed of the low channel in Culvert #2 was roughened with an exposed aggregate finish, and bespoke 'fish block' baffles fabricated from recycled timber were fitted within the low flow channel at regular centres. Eel matting was placed through Culvert #1 to aid passage. The baffles and fish blocks themselves will encourage the accumulation of material through creating eddies and drop-out areas where sediment is likely to collect.

During construction strict biosecurity procedures were enforced to reduce the risk of signal crayfish and / or crayfish plague introduction. The existing natural channel was de-watered and diverted. During this time over 9000 white-clawed crayfish were removed and translocated within the Wansbeck Catchment. This work was carried out under licence from Natural England which was obtained once the method had been agreed.

CONSTRUCTION

During the construction of the dam the River Wansbeck was first diverted through a temporary channel which was designed to convey the 1 in 10 year flow. The channel was lined with 150mm stone sitting on a geotextile underlayer. A temporary bridge was placed over the diversion channel to provide access to the culverts and left abutment.

Figure 9. Diversion channel

With the diversion in place the culverts were constructed behind temporary bunds in the river channel. The culverts were placed with 500mm gaps between adjacent units across the width of the structure. These gaps were backfilled with mass concrete. There is an inclined mass concrete haunch on the exposed faces of the outer culverts to limit the potential for cracking to develop between the fill and the concrete structure. There is a sand filter collar around the culverts about 4m downstream of the dam centreline which is designed to intercept any seepage running along the contact of the fill / culvert and hence reduce the potential for piping failure along the contact.

The river was diverted back through the culverts once they were complete, and the embankment, spillway and ancillary structures were then constructed.

Figure 10. Culvert construction with diversion channel behind

The placement of embankment fill was all controlled by GPS which greatly reduced the need for operatives to enter the working areas and contributed to the excellent safety record of the project. The fill was placed without difficulty apart from a short suspension due to inclement weather in the winter of 2014.

OPERATION

The scheme has operated twice since construction was completed in the summer of 2015. In the larger event on 5 January 2016 two of the main penstocks closed and the reservoir impounded to a height of about 6m above river bed level. The flood storage reservoir fulfilled its purpose of protecting Morpeth from what would have been another severe flood.

REFERENCES

Hewlett HWM, Boorman LA and Bramley ME (1987). *Design of reinforced grass spillways. CIRIA Report 116*, CIRIA, London

Louca V, Ream HM, Findlay JD, Latham DM and Lucas MC (2014). *Do culverts impact the movements of the endangered white-clawed crayfish?* Knowledge and Management of Aquatic Ecosystems. 414: 14, 1-17.

Lucas MC, Ream HM and Findlay JD (2010). *Impact of culvert structures on the movement of white-clawed crayfish*. Unpublished report to the Environment Agency by Durham University.

SEPA (2010). *Engineering in the water environment: good practice guide. River Crossings. Second edition. November 2010.* SEPA, Stirling

Dams – Benefits and Disbenefits; Assets or Liabilities?
ISBN 978-0-7277-6133-0

doi: 10.1680/dbdal.61330.427

Ageing Service Reservoirs - an increasing burden or scope for innovation?

I HOPE, Severn Trent Water

SYNOPSIS Whilst the structural integrity of service reservoirs (SRs) is the key focus for Panel Engineers, other regulatory regimes can designate a failure of these vital structures. Root cause analysis of incidents for bacteriological failures has revealed causes ranging from physical deterioration of assets through to complexities arising from loss of knowledge of the way in which the asset should be operated.

This vital asset base is aging, with some structures dating back to Victorian times originally comprising brick-built open structures. Over the years SRs have been significantly modified and repaired. The current replacement rate for these tanks that house food-grade water could be up to 200 years. Individual component parts such as water bars, roof membranes and joint sealants have a limited life.

Engineers are challenged to become more engaged by overseeing construction quality and seeking opportunities for innovation, for example by challenging the convention to backfill against SR walls.

INTRODUCTION

The nation's stock of SRs is ageing. Confronted with low asset replacement rates combined with an increasing number of bacteriological failures, engineers are challenged to rethink their approach to SR asset management from monitoring through to exploration of appropriate methods for extending asset life. Furthermore, they are challenged to inventively engage with new SR construction projects.

ROLE OF SERVICE RESERVOIRS

Primarily SRs retain a bulk volume of food grade water for human consumption. They are of sufficient volume to supply diurnal and seasonal demands arising from the water distribution network. Internal water levels vary in response to these requirements.

Constructed in elevated locations, they provide sufficient pressure (hydrostatic head of 10m at tap) for customer supply.

Treated water reservoirs, located immediately post-treatment are synonymous with SRs for the purposes of this paper. However these are typically sited in lower-lying locations, even in the floodplain where treatment works are adjacent to river abstractions. They are subject to higher external hydrostatic pressures from ground water.

SERVICE RESERVOIRS AND THE RESERVOIRS ACT 1975

Currently 20 of Severn Trent Water's (STW) 59 reservoirs under the Reservoirs Act 1975 (the Act) (HMSO, 1975) are SRs. The overall total will increase by a further 60, 42 being SRs, when changes to the Act reducing the volumetric threshold down to 10,000m³ , are implemented. A proportionate increase is mirrored in most other UK Water Undertakings (Hawley, 2015), see Table 1.

Table 1. Distribution of Regulated Reservoirs across the UK

Country	SRs in Act (above 25,000m³)	SRs between 10 – 25,000m³	% increase	Total
England	153	360	235	513
Scotland	23	50	217	73
Wales	5	43	860	48
N. Ireland	10	30	300	40
Total	**191**	**483**	**253**	**674**

Analysis of STW's asset base on SR size and age reflects a continued dependency upon older structures many of which have been subject to repair over the years, shown in Figures 1a and b.

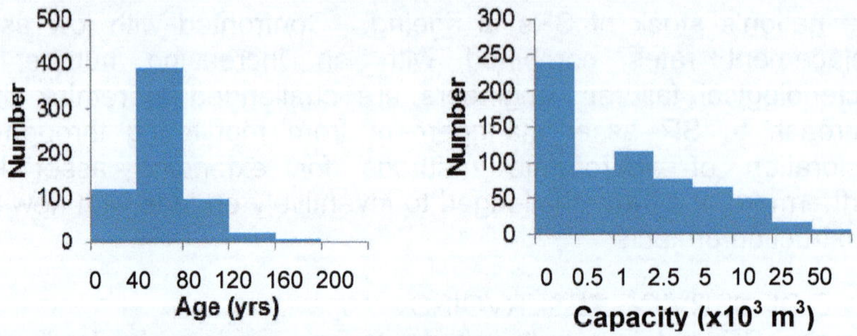

Figures 1a and b. Distributions of STW's SRs by age and size

OTHER REGULATORY REGIMES

Whilst the structural integrity of SRs is the key focus for Panel Engineers during their S10 and S12 inspections under the Act, other regulatory regimes play a pivotal role. For reservoir undertakers these can become competing drivers particularly when arranging internal inspection(s). An umbrella of legislation and governance covering SRs is shown in Table 2 (not exhaustive).

The Drinking Water Inspectorate's (DWI) regulatory role is to ensure that water at the customer's tap is safe to drink. Following a spate of coliform failures across the country, to ensure compliance with the DWI a number of Water Companies have implemented an increased frequency of SR cleaning. Similarly, as a result, STW is now working to a risk based cleaning programme as part of its Drinking Water Safety Plan (DWSP). This requires a drain down, clean and inspection of each SR which is a major intervention to the operation of the network, often taking comprehensive planning. Typically the cost of a drain down, clean, sterilisation and testing can be £50k-£75k, depending on SR size. Seizing on access opportunities, for statutory SRs it is planned to combine S10 inspections with these more frequent, DWSP driven, events, acknowledging that these are more frequent than legally required under the Act.

Table 2. Legislation and governance for Service Reservoirs

Criteria	Legislation / Process	Regulator / Overseeing Body
Water quality	Water Act 1945	Drinking Water Inspectorate
Safety	Health and Safety at Work etc. Act 1974	HSE
Security / resilience	Protection of National Infrastructure	CPNI / Defra
Protection against flooding	Reservoirs Act 1975	Environment Agency (England) NRW (Wales)
Emergency response to flooding	Civil Contingencies Act 2004	Local Resilience Forum
Funding	Asset Management Plans	Ofwat (economic regulator)
Customer Service	Continuity of supply	Ofwat (economic regulator)

In addition, a further external impact on operational strategies and safety standards can come from the requirements of company insurers which will vary for individual undertakers.

ASSURANCE OF RESERVOIR SAFETY

STW is split into two operating areas; East and West, reporting to the Director of Wholesale. Routine inspections of SRs are conducted by operational personnel based in the Wholesale directorate. Overseeing compliance with the Act, effectively operating in an assurance role, the Reservoir Safety Team reports into a separate directorate, headed by the Chief Engineer. This provides full independence of reporting and is a refinement, post re-structuring, to the previously reported structure (Hope, 2012).

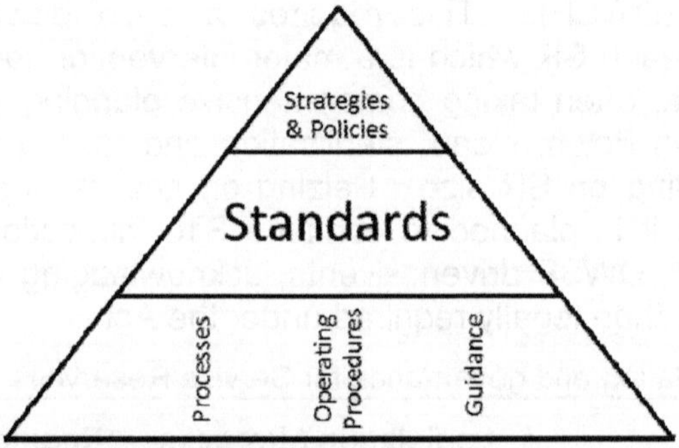

Figure 2. STW's Operational Strategy

IMPLEMENTATION OF OPERATIONAL STRATEGY

STW's strategy for asset management and operation is detailed in Figure 2 and expanded in Table 3. The goal is a clear line of sight from the strategic / legal intent through to how tasks are conducted by operating procedures across the company.

All training documents and codified procedures have been extensively illustrated with photos and explanatory diagrams. This has also served to bridge the knowledge management gap.

With its origins in three historical operating units, STW had an inconsistent approach to compiling SR inspection records. These are completed by operational personnel during routine inspections. The transcription of paper based information from a wet environment contributed to further inconsistencies. For example, certain discrepancies in reservoir volume were apparent. This initiated a major change project. Records of internal inspections have now been standardised across STW to consistently record asset condition, type, age etc. Beyond operational requirements, information is also captured to influence future investment needs

(AMPs). An App is currently under development and from summer 2016 it will replace paper-based records and further aid accurate record keeping.

Table 3. STW's Operational Strategy applied to Service Reservoirs

Tier	Driver	Explanation	STW Reservoir Team Input
Strategies and Policies	Legislation; British Standards; Government guidance	Legal drivers; Distribution Operational Maintenance Strategy (DOMS)	Compliance with Act Strategic direction Report to Board
Standards	Codified Procedures	Stating what must be done, defining systems of work, DWSPs	SR inspection procedures for operators. Defined support. Asset records and reports.
Operating Procedures	Systems of work	Operational up-skilling	Assessed surveillance training plus manuals
Processes	Operational tasks in SAP	Defines resources	Inspection and re-training tasks
Guidance	Consistent approach	Capturing the right information	New Inspection records, advice on repairs

With its origins in three historical operating units, STW had an inconsistent approach to compiling SR inspection records. These are completed by operational personnel during routine inspections. The transcription of paper based information from a wet environment contributed to further inconsistencies. For example, certain discrepancies in reservoir volume were apparent. This initiated a major change project. Records of internal inspections have now been standardised across STW to consistently record asset condition, type, age etc. Beyond operational requirements, information is also captured to influence future investment needs (AMPs). An App is currently under development and from summer 2016 it will replace paper-based records and further aid accurate record keeping.

RESERVOIR BACTERIOLOGICAL FAILURE RATES

Recent analysis of bacteriological failure rates on STW's SRs has exposed an average failure rate occurring at 28 years after construction (Leadbetter, 2016), illustrated in Figure 3. Resolution of these failures often involves detailed investigations and comprehensive remedial works following SR draining and cleaning.

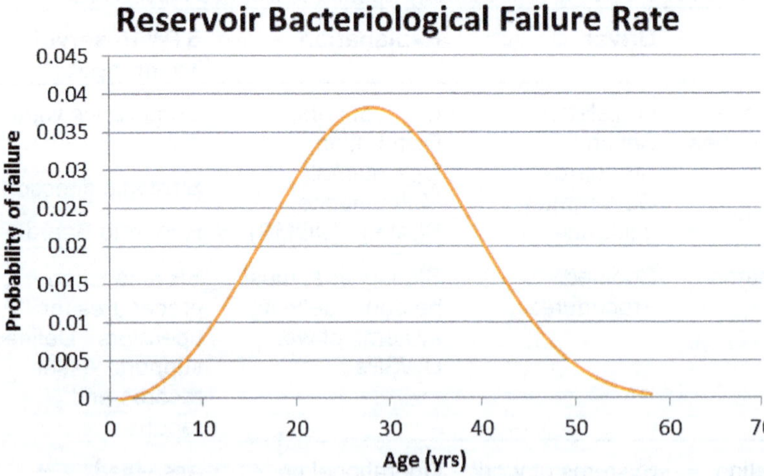

Figure 3. Analysis of bacteriological failure rates

Perhaps even more alarming is the revelation from the analysis that all SRs are likely to have suffered from bacteriological failure before they reach 60 years old, despite ongoing maintenance and repairs.

COMMUNITY OF PRACTICE (COP)

A COP has been established by the Reservoir Team to drive and embed continuous improvement and provide a hub of expertise and advice for operational practitioners. This has proven particularly useful for consulting on, establishing and embedding the new records for SR inspections. Incidents are readily reported, queries raised and learning shared across STW. For many operational teams, recent organisational change has led to new roles, (even moving from waste to clean water). The COP has served to help bridge the knowledge gap, promptly resolve day to day queries and openly share information and learning across the business, particularly for those new to the disciplines required when operating clean water assets.

SERVICE RESERVOIR CONSTRUCTION TYPES

Following research in 1988, CIRIA Report 138 (CIRIA, 1995) published a split of construction types. Whilst the proportion of reinforced concrete SRs has increased in recent decades it does provide an indication of the legacy of mix of construction types. This is contrasted against STW's 613 SRs in Table 4, below.

Older SRs, particularly brick and mass concrete, often rely on the passive restraint of backfill to support the walls. These require careful analysis prior to implementing low level intrusive works. After 1974 most reinforced concrete (RC) SRs were constructed to the CESWI (Civil Engineering Specification for the Water Industry) (WRc, 2015) specification which demands a full water test prior to backfilling. The rigours of a maximum defined loss of contents, namely 1/500th of the water depth retained over a 7 day period, and the ability to physically view and inspect the integrity and performance of the reservoir walls is essential in providing the asset owner with confidence of long term asset performance. A full reservoir with no passive restraint offered by backfill becomes a critical design parameter.

Table 4. Breakdown of construction type

Construction Type	National Proportion in 1988	STW SR assets as of 2015
Brickwork	19%	1%
Mass concrete	21%	8%
Reinforced concrete	57%	83%
Post-tensioned concrete	2%	1%
Others	1%	7%

Older SRs, particularly brick and mass concrete, often rely on the passive restraint of backfill to support the walls. These require careful analysis prior to implementing low level intrusive works. After 1974 most reinforced concrete (RC) SRs were constructed to the CESWI (Civil Engineering Specification for the Water Industry) (WRc, 2015) specification which demands a full water test prior to backfilling. The rigours of a maximum defined loss of contents, namely 1/500th of the water depth retained over a 7 day period, and the ability to physically view and inspect the integrity and performance of the reservoir walls is essential in providing the asset owner with confidence of long term asset performance. A full reservoir with no passive restraint offered by backfill becomes a critical design parameter.

Conversion of older open reservoirs

The UK's legacy of SR assets mirrors the development of public health engineering over the last 200 years. A typical history is illustrated by the Danes Castle SR in central Exeter. Built as an open reservoir in 1880, providing storage of filtered water for human consumption, it was covered over with a concrete roof in the 1920s to reduce contamination. Whilst avoiding a direct hit during Hitler's "Baedeker" bombing campaign in 1942, the SR walls suffered damage and could only be partially filled until its replacement in 1994 with a two compartment RC structure.

Older, adapted SRs generally have no side wall drainage, roof membrane or network of underdrains (to prevent uplift). This can result in the retention of elevated ground water levels exerting hydrostatic pressure on walls. Leakage of ground water into the reservoir will follow once a crack or defective joint is penetrated coinciding with reduction in internal water level. Retrospective construction of wall drainage can prove expensive, particularly if the side walls are not self-supporting, requiring an incremental approach to installation. The project to provide wall drainage to the Frankley WTW treated water reservoir at was implemented at a cost of £0.5m.

With the current limited rate of asset replacement (in some cases this could extend to 200 years), a better understanding of the way in which these assets degrade and the performance of individual components such as roof membrane, water bar and joint sealant is essential. Concrete structures will outlive the service life of their component parts. Over recent decades many of STW's SRs have been the subject of repair. Wall repairs include over-banding of joints through to the complete re-lining of the SR with butyl membrane (requiring numerous welds due to the presence of many internal columns supporting the roof). These repairs in turn will degrade and ultimately fail, raising the risk of bacteriological failure.

Over the past 18 months an increased focus for the Reservoir Team has been the detailed examination of construction features of non-statutory SRs and tanks in an attempt to trace potential sources of contamination. Crucially, a pioneering training package has been commissioned to support a thorough approach when tasked to analyse asset degradation and recommend remediation. It covers original approaches to design, e.g. considers whether roofs are tied to walls or allowed to slide through to causes of concrete degradation, approaches to historical repairs, understanding material degradation rates and includes specifications for investigations and remedial works. The package was compiled by Atkins using data, information and photographs from contractor Stonbury (Atkins & Stonbury, 2016).

Internal inspections can only reveal part of the condition of wall joints. A commonly encountered misunderstanding is the role of joint filler, (generally constructed as a 25mm fillet of polysulphide, visible on the internal walls). The filler itself does not contribute to the integrity /

water tightness of the SR. Leaking joints inevitably mean that the water bar embedded in the wall has failed. Modern water bar is made from uPVC, however on older tanks, water stops, if they exist, were constructed from steel or copper strip and later from rubber.

Water bar is supplied on rolls with junctions welded on site. Correct weld temperatures and a proper standard of workmanship are crucial to achieving a water tight joint. The condition of water bar, cast into a concrete joint or buried beneath a reservoir floor cannot be readily verified, potentially hiding poor workmanship. Water bar will move during casting if not adequately fixed, as shown in Figure 4..

Figure 4. An example of water bar movement after base and wall kicker pour requiring breaking out and re-setting. The process of repair can itself introduce potential flaws to tank integrity,

THE NEED FOR OPERATIONAL FLEXIBILITY

Taking a cell of a SR out of service for cleaning and inspection can be complicated by other system demands and lack of appropriate valving. For older reservoirs in particular, part height division walls impact on system operability by reducing available water storage by over 75% rather than 50% as in the case of a full height division wall.

Figure 5. An example of a part height division wall at Erdington SR.

EMERGENCY PLANNING AND RISK DESIGNATION

All of STW's reservoirs under the Act have an on-site plan (an emergency action plan) using the DEFRA template. This is updated annually by the Supervising Engineer (SE). The probability of failure of a SR is accepted as significantly lower than for impounding and non-impounding reservoirs. There are few if any documented failures following complete collapse of a SR wall. However, until the industry can confidently put figures to the probability of SR failure, the regulator in England will deem probability as unity and a reservoir will be designated as "high risk" if it is considered that there is a potential threat to human life in the downstream flood path should it fail.

SERVICE RESERVOIR FAILURE

Subsidence and foundation failure does happen, albeit rarely. In 1994 the new 20Ml Dunsford Hill SR in Exeter was subject to foundation failure during its water test. The reservoir excavation had been cut into a sloping hillside and the SR founded on a previously variably loaded strata. On first filling, the northern corner of the reservoir, where the shallowest excavation had occurred, started to rotate further outward (northward). The foundation in the area of the failure was successfully remediated prior to final commissioning.

The 45Ml Barr Beacon SR owned by South Staffs Water suffered pipework failure in 2012. Whilst not a reservoir failure as such, the impact of the flooding on the downstream estate was devastating causing in excess of £1m damage. The failure occurred at 05.00 and fortunately no-one was killed. The outcome could have been very different if this failure occurred later in the day, when the community was active. It is a stark reminder of the effects of stored energy (water at elevation), subject to uncontrolled release. Worthy of note is the emergency response by the Undertaker who banked social capital by their prompt and caring approach. Their openness to readily share lessons learnt with the wider industry is acknowledged.

CONTRACTOR SELF ASSURANCE

Towards the end of the last century construction industry practices were recognised as being inefficient and even adversarial. These systemic failings were highlighted by Sir Michael Latham (Latham, 1994) in his pivotal report "Constructing the Team". Over the last 30 years client / contractor initiatives have focused on collaborative approaches to contract management. In fully embracing project partnering, the majority of Water Companies have dispensed with the independent quality check provided by the Resident Engineer and Clerk of Works. Performance specifications rely on contractor self-certification. Arguably poor workmanship can quickly be buried by unscrupulous contractors. It becomes increasingly incumbent on all parties to exert greater controls in order to build to appropriate standards and thus avoid premature bacteriological failures.

Wider evidence of these concerns arises from the recently reported defects on Public Private Partnership schools in Scotland where tie bars were omitted in walls. However, in 2015, building contractor

Willmott Dixon undertook an intriguing review of the effectiveness of their self-assurance. A "Quality Delivery Audit" (Willmott Dixon, 2015) revealed an average compliance score over 50 sites of 71.8%. The report concludes *"very few sites are making a mistake that has not been made before elsewhere in the company"*. The company is to be congratulated for the openness in sharing these lessons learnt in striving to improve. They also report that they embrace the inclusion of a client's Clerk of Works on a project believing the outcome provides significantly improved quality of workmanship. Encouragingly, the £6m, 18Ml Outwoods SR currently under construction by South Staffs Water requires the sign-off by an independent engineer of key build stages.

Figure 6. A newly constructed reservoir, backfilled contrary to specification, without drainage membrane. The omission of a membrane allows the backfill to support an elevated ground water level with potential to introduce contamination if a movement joint fails or a wall cracks.

All the photographs in this paper were taken by the author in 2015 and are further evidence that all parties need to improve standards.

Figure 7. A further illustration of poor workmanship where the expansion joint, unusually employing rear guard water stop, has moved during construction. Furthermore the exposed detail allows ingress down the wall joint and into the reservoir.

PRECAST CONCRETE AND DESIGN FOR MANUFACTURING ASSEMBLY (DFMA) UNITS

Figure 8. The new Ambergate No. 1 reservoir under construction in 2015, employing conventional techniques.

As part of STW's £43m investment at Ambergate, an 87Ml SR (No. 1) is being constructed using conventional in-situ concrete base and walls.

Pre-cast concrete components are being used for columns, roof beams and roof soffit prior to concreting. Once commissioned the existing 110 year old reservoir will be taken out of service and a 50Ml reservoir (No. 2) employing DfMA wall construction will be constructed on its footprint. Final build costs employing DfMA wall units are closely comparable to conventional in-situ concrete construction. However, construction programme efficiencies are offered, typically saving £35k/week. Trials conducted using DfMA wall panels on valve chambers have provided essential learning for the site team. Aggregate size was reduced to 6mm to allow concrete to flow around the reinforcement during placing and compacting.

The £27m strategic treated water reservoir constructed for Anglian Water to improve resilience from the Grafham WTW network typifies the trend for walls to be constructed using pre-cast units. Wall joints incorporate hydrophilic water stop which swells following contact with water to provide a watertight joint. Fluctuations in water level may allow the joint filler to dry out and retract. The design life for the hydrophilic strip has been stated as being 120 years. However, there are significantly more joints using this methodology and joint replacement or repair of multiple joints will be costly and may well be inconclusive. Cambridge University's Centre for Smart Infrastructure and Construction has been commissioned to monitor the integrity of the roof using embedded fibre optic cabling. As the cost of these emerging technologies becomes more affordable, the potential applications open up. For example a credible option of wrapping pre-cast tank walls in fibre optic cable at various levels establishes the ability to monitor joint performance and importantly pinpoint deterioration.

Figure 9. Precast wall units erected prior to casting infill sections

The use of precast units for a batch of SRs is explained in more detail (Robson *et al*, 2012). However, remedial works were required to restore cover to reinforcing bars on faulty wall units in five of the first batch of seven SRs prior to going into service (Robson, 2016).

Figure 10. Wall joint detail incorporating hydrophilic strip

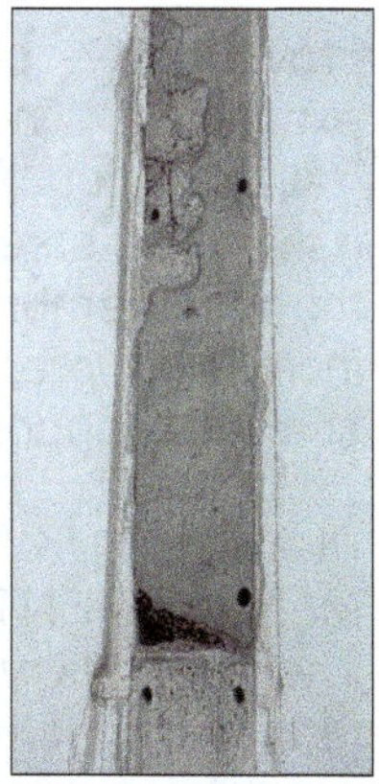

Figure 11. Infill casting between wall panels, completed in two lifts, creating further construction joints

FUTURE ASSET REPLACEMENT

What is the design life of a SR? How long is a newly constructed SR expected to last? For the water industry the current rate of asset replacement will lead to a 200 year cycle for demolition and re-construction. Component parts of the reservoir that contribute to the continued integrity of the process unit, in particular roof membranes, joint sealants and repairs such as banding, need to be considered. Currently under review for all new SRs is the installation of fibre optic / acoustic sensing technologies around the circumference of the wall to detect early signs of joint failure (leakage, stress, strain, etc).

A conventional concrete service reservoir will have a full backfill against the side walls and may even have a grass roof. Clients are encouraged to challenge convention. Why is it necessary to backfill against the walls of a reservoir? Design can accommodate thermal expansion cases. The presence of a fully backfilled wall encourages a positive water table. Any defective joint will allow potentially contaminated ground water to enter the tank. The case for backfilling against the reservoir is generally cost, occasionally driven by Town and Country Planning requirements. However, if we take the whole life costs the vulnerability of wall joints drives the need to challenge convention. A more robust approach to challenge the convention to backfill against service reservoirs is urged. Certainly excess spoil can be banked for landscaping purposes but why impose an increased operational risk by banking it against a new structure? Given that the design will be for a minimum 100 year life what component parts will have degraded, blocked, settled in that time? Furthermore how will defective component parts be repaired or replaced? A similar challenge applies to roof construction. After flood testing of a new concrete roof, membranes are applied and typically covered with 200mm of gravel or crushed stone in order to provide UV protection of the membrane and reduce thermal expansion. In providing an alternative to this convention, Hunter Valley Water, Australia, offer the option in their specifications of a galvanised steel roof over the concrete roof, facilitating inspection of the roof membrane. Whilst this may pose a planning challenge, with increasing pressures to install solar panels perhaps this offers a combined opportunity in appropriate locations?

PLANNED UKWIR PROJECT: SERVICE RESERVOIR ASSET MANAGEMENT TOOLKIT (RG05B207)

The author is working with water industry leads to oversee the above forthcoming UKWIR (UK Water Industry Research) project. The project is currently at tender evaluation and planned for completion in spring 2017. A key deliverable from the project will be a framework for expenditure decision-making and interventions, hopefully taking account of the contents of this paper.

CONCLUSION

Engineers cannot afford to lose sight of the fact that these tanks house food grade water throughout their extensive asset life.

The challenges are clear; increased regulatory scrutiny, higher customer expectations; a potentially degrading asset base and emerging skills shortage (not an exhaustive list!). Inventive application of new technology heralds the ability for improved monitoring and thus early intervention. The opportunity presents itself for the current generation of engineers to secure improved assets for the future whilst ensuring appropriate quality of workmanship. As industry leads reservoir engineers need to be more involved in all aspects of SR asset management, and crucially their replacement.

ACKNOWLEDGEMENTS

The author is indebted to STW for permission to publish this paper and to Messrs Hawley, Hughes, Robson and Warren for their comments. The views expressed are the author's own.

REFERENCES

Atkins & Stonbury (2016) *Service Reservoirs Inspection Manual* Severn Trent Water, Coventry

CIRIA (1995), *Report 138 Underground service reservoirs – Waterproofing and repair manual.* CIRIA, London

WRc (2015) *Civil Engineering Specification for the Water Industry.* WRc Swindon

Hawley (2015). *Research into skills shortage of Inspecting Engineers on SR panel and proposals for change.* ICE, London

HMSO (1975). *Reservoirs Act, 1975.* Her Majesty's Stationery Office, London, UK

Hope (2012). Implementing Severn Trent Water's People Plan to become the best in Great Britain at managing reservoir safety in *Dams: Engineering in a Social and Environmental Context – Proceedings of the 17th British Dam Society Conference, Leeds,* pp43-55, ICE Publishing, London

Latham (1994), *Constructing the Team*, HMSO, London

Leadbetter (2016). *Analysis of bacteriological failures on Service Reservoirs.* Severn Trent Water, Coventry

Robson & Bull (2012). Precast Concrete in Reservoir Engineering – New Construction Techniques in Water Retaining Design in *Dams: Engineering in a Social and Environmental Context – Proceedings of the 17th British Dam Society Conference, Leeds,* pp314-325, ICE Publishing, London

Willmott Dixon (2015) *Quality Delivery Audit.* Willmott Dixon

ice
Institution of Civil Engineers
publishing

Section 8:

Repairs and maintenance to ensure safety

Dams – Benefits and Disbenefits; Assets or Liabilities?
ISBN 978-0-7277-6133-0

doi: 10.1680/dbdal.61330.445

Asset or liability: stabilising an historic dam

A PETERS, Arup
N WALDING, Arup
D CROOK, Arup
M COOPER, Arup
P KELHAM, Arup
B COTTER, Dŵr Cymru Welsh Water

SYNOPSIS This paper describes the investigations and studies carried out to confirm the construction details of a Grade 2* listed dam, the leakage issues, causes of lateral displacements, and potential for instability. The conclusions of this work are summarised in relation to the requirements to achieve a safe asset, and the approach taken to achieve this. A description is provided of the remedial works designed and constructed to ensure the future stability.

INTRODUCTION

Upper Neuadd is a Grade 2* listed structure, situated in the heart of the Brecon Beacons below Pen y Fan in the Taf Fechan valley about 13km north of Merthyr Tydfil. The dam is the upper dam in a cascade including Lower Neuadd and Pontsticill reservoirs. It is formed from cyclopean concrete clad in massive masonry. The flanking sections to each side of the central overflow are supported by large embankments on their downstream faces which provide passive resistance to deflections of the relatively slender masonry structure.

The dam was constructed between 1896 and 1902 and has been plagued with leakage issues and safety concerns for many years. The leakage was extensive enough to raise concerns over the stability of the structure as early as 1965. Repointing of the upstream face and grouting of the body of the dam was subsequently undertaken, and measures made to monitor residual leakage. Results indicated a gradual increase in leakage. In 1983 stone drainage blankets were installed due to concerns over the seepage emerging from the left hand flanking embankment. Further repointing works were carried out in 1990 and horizontal displacement measurements showed the structure moved with higher water levels.

The dam was Grade 2* listed in July 2005 'for its special interest as an architecturally-designed dam of spectacularly massive construction and definite character'.

In 2002 a formal extension to the scour outlet pipework provided a lower level bellmouth overflow, which retained top water level 6m below the original spillway level under 'normal conditions'. However, the catchment is typically steep mountainous moorland and the reservoir can fill rapidly. The existing infrastructure was frequently found not adequate to maintain the lower reservoir level and this resulted in unacceptable leakage and stability concerns.

Presently the reservoir is not utilised for water supply, but Dŵr Cymru Welsh Water (DCWW) considers that the asset will be required in the future and therefore permanent discontinuance is not preferred as a solution. However any potential solution had to satisfy the following recommendations in the interest of safety given in an inspection report dated 2013 to either undertake:

a) measures to limit the loading on the dam, by preventing the flood level in the reservoir during the passage of a PMF rising above a level that is 1m below the present spillway crest; or

b) measures to ensure the long-term stability of the entire dam when the reservoir is filled either to the present spillway crest level (or to a slightly lowered spillway crest level) including during the passage of the PMF.

This paper describes the investigations and studies carried out to confirm the construction details of the dam, the leakage issues, causes of lateral displacements, and potential for instability. The conclusions of this work have been summarised in relation to the requirements to achieve a safe asset, and the approach taken to achieve this, including a description of the remedial works designed and constructed to ensure the future stability.

CONSTRUCTION DETAILS

The dam effectively comprises of three sections. The central section is a conventional masonry faced gravity dam with a hearting of masonry and concrete. The flanking sections on either side of the valley are also gravity dams, but of narrower section and rely on the earth embankments downstream for stability. Immediately adjacent to the central section there are transition sections where the footprint of the dam rapidly reduces from about 14m to 4.7m in width.

The supporting downstream embankments have broad crests which are some 6m below the crest of the concrete dam. An elevation of the central section of the dam and a typical section through the flanking dam are shown in figure 1 below.

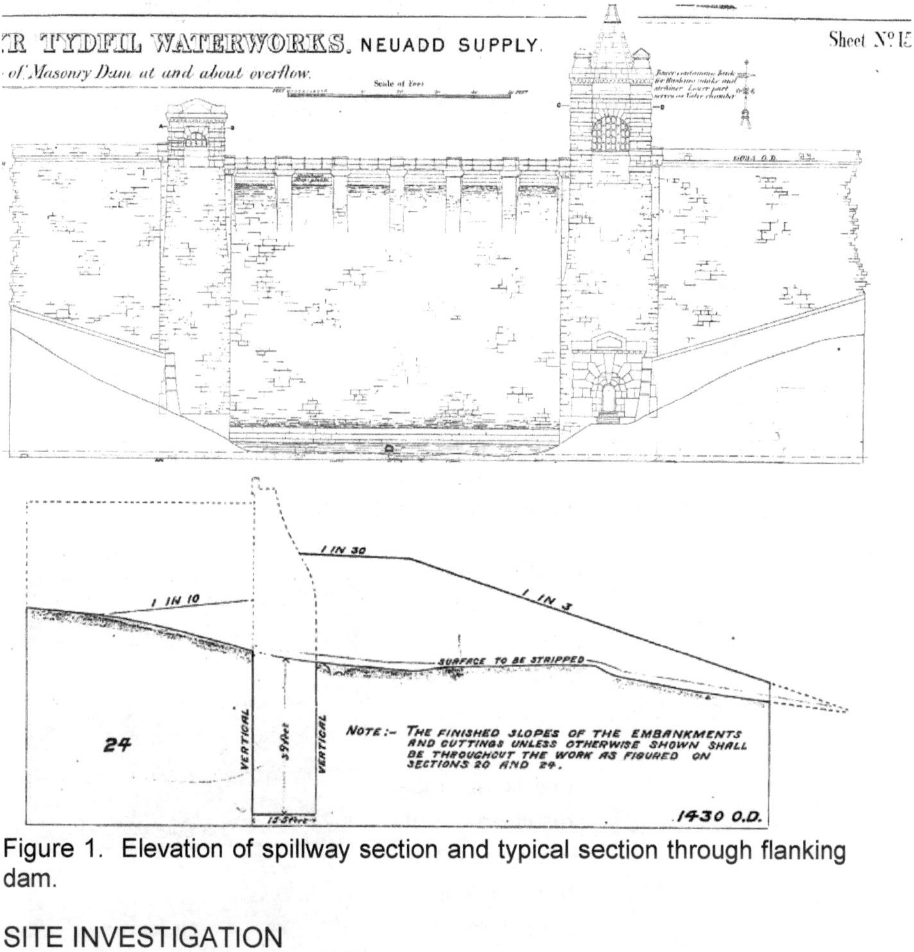

Figure 1. Elevation of spillway section and typical section through flanking dam.

SITE INVESTIGATION

Gaining a sound understanding of the likely ground conditions, and the form and composition of the dam though desk based studies was critical for the development of a targeted ground investigation.

The ground investigation was designed with the intention of achieving the following main objectives:-

- To confirm the geometry and form of construction of the dam as indicated on the 19th Century construction drawings.

- To identify primary seepage routes.
- To provide information on the geotechnical properties of the downstream flanking embankments for stability assessment.
- To provide additional instrumentation that may be used for future monitoring of the dam.

The site investigation confirmed that the cyclopean concrete core was no longer intact at multiple depths and locations along the length of the dam. Various samples were recovered as rounded concrete fragments, which had been eroded through long term seepage. The base of the cut-off was confirmed to lie approximately at the same level as shown on the 19th Century construction drawings, which provided additional confidence that the dam was constructed as per the drawings.

Exploratory excavations against the upstream face of the dam confirmed the presence of substantial timber formwork used to support the trenches during construction remained in place, and had degraded over time, see figure 2. Unfortunately the excavation works did not confirm the presence of the timber struts to the formwork, which were also shown on the construction drawings. It is considered likely that these struts could have been left in place and may provide seepage paths through the dam.

Figure 2. Evidence of timber formwork in trial pit, related back to record drawings

Standpipe piezometers were installed downstream of the dam to supplement those already in place. Together, these confirmed that relatively high piezometric pressures were being experienced in the formation soils downstream of the dam, particularly in response to increases in reservoir level.

PERFORMANCE OF THE DAM

The site investigation provided data on the condition and performance of the concrete hearting material and of the flanking embankments.

Two critical sections were identified for assessment of the stability of the dam. One section was taken through the overflow section and the other about 54m to the left of the overflow on the flanking section. The latter section was selected as it correlated to historic piezometers and V-notch monitoring installations. Data collated from the site investigation was used to define the concrete strength of the dam and the geotechnical parameters of the flanking embankments.

Initially the two sections were treated as purely gravity structures using the recommendations given in the CIRIA Report 148 (Kennard *et al*, 1996) and Design of Small Dams (USBR, 1987). The three principal load cases were usual, unusual and extreme, with the ice and seismic loads taken not to coincide.

Analysis based on past events

The analysis confirmed that although the non-overflow section was stable against sliding, albeit with a low factor of safety, it was found inadequate against overturning, with a factor of safety significantly less than unity. However, in January 2014 a storm event was known to have caused the reservoir to fill and overflow the spillweir. This suggested that the structure was stable under the loading applied during that event, although not necessarily with an adequate factor of safety. Consideration of this event led the team to conclude that there may be other factors also contributing to the stability of the flanking section.

The elements that could readily be examined included the density of the concrete; the magnitude of the hydrostatic uplift; skin friction on the vertical faces and enhanced earth pressure on the downstream face. Adjusting the concrete density made little difference to the stability. It was not possible to justify reduction of the uplift, as the piezometers indicated that uplift was present. The skin friction on the vertical faces made small differences. The biggest effect was therefore found to be through enhancement of the earth pressure on the downstream face.

The analysis gave an insight into the mechanisms that could have historically affected the performance of the dam. In order to generate the required earth pressure to act as a restraining structure, the flanking embankment would have to strain and the dam would have to deflect into the embankment.

During the January 2014 incident, there were significant and unacceptable leaks issuing from the downstream face of the left embankment. The evidence seemed therefore to suggest that the dam had indeed deflected towards the flanking dam section. The conclusion was that repeated deflections during storm events over many years had exceeded the tensile capacity of the upstream face, in discrete locations. This had resulted in cracks forming and water entering the hearting of the dam. Over time, the internal structure had deteriorated and preferential flow paths through the structure had formed (Figure 3).

This mechanism was backed up by the site investigation which had confirmed voids within the hearting and evidence of rounded aggregates, indicating long term and high velocity leakage paths.

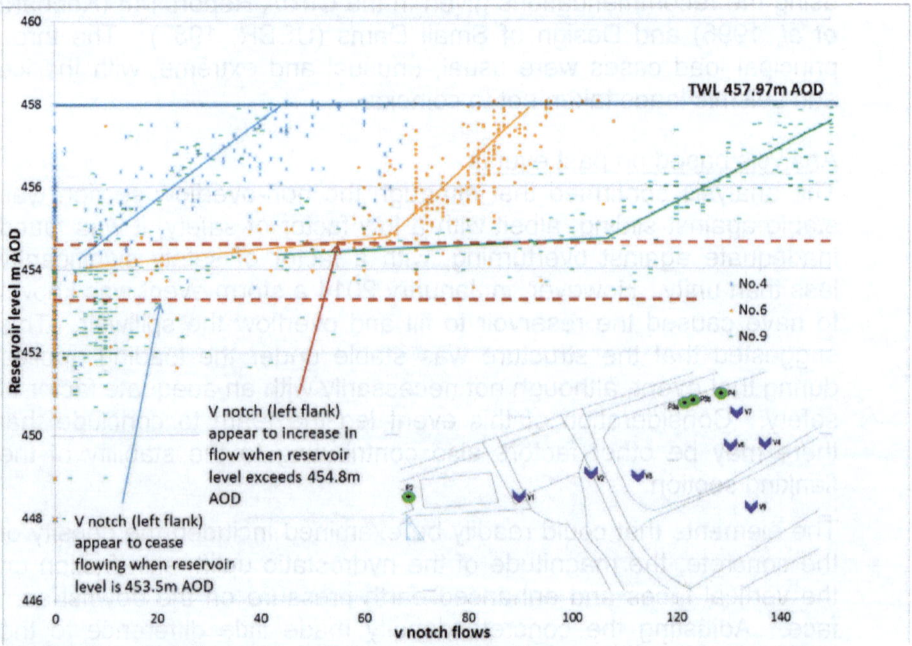

Figure 3. Relationship between leakage through dam superstructure related to reservoir level

Analysis based on site investigation

The construction records for the dam indicated that the flanking section had been constructed in a deep trench that was supported by timber shores. The site investigation on the upstream side of the dam found evidence to suggest the timber had not been removed.

The timber was in good condition, as would be expected where it was permanently immersed, and it was concluded that earth pressure could transmit through it to the structure. On the downstream side however, one of the boreholes recovered a small piece of rotten timber. This suggested that there could be a limit to the earth pressure that could be acting to stabilise the flanking section. The presence of the timber against both sides of the dam structure will result in full hydrostatic pressure acting with no effective dissipation. Therefore uplift could not be reduced.

Summary of analysis

Considering both the performance during the 2014 event and the findings of the site investigation it was concluded that the stability of the dam could not be confirmed at higher water levels. The rotten timbers between the concrete and earth structures meant that the dam was not performing as a composite structure and the lack of passive resistance available from the flanking embankment was resulting in excessive movement of the dam. Over time this had allowed cracks to form.

Short term management of the asset

The presence of the timber formwork used to construct the dam flanking sections raised safety concerns over allowing the reservoir to return to its operational top water level. The stability analysis had considered safe water levels for the usual and the PMF cases, which were both calculated as below the original top water level. The safe water levels to ensure structural stability were reviewed and a freeboard added to provide sufficient storage within the reservoir basin to ensure that the safe water levels were not exceeded during future storm events.

A short term solution was agreed with the QCE to open the lowest valve on the scour main in an attempt to maintain the reservoir at a lowered level whilst a longer term solution was developed. In the event of the reservoir filling above the safe water level, an enhanced inspection regime would be implemented and the QCE kept up to date on the dam's performance.

DEVELOPING A LONG TERM SOLUTION

It was agreed that further grouting works or installation of upstream membranes would not ensure a long term solution, due to the extent of the leakage issues and the concern over the stability of the structure at higher water levels. The team therefore considered options to ensure stability of the dam whilst managing leakage.

Options considered included:

- Construction of a reinforced concrete wall on the upstream face of the dam. This would allow the dam to be restored to its operational top water level.
- Vertical anchors with an upstream liner, or with grouting of the dam.
- Downstream buttresses built from the original dam foundation level with an upstream liner.
- Lowering top water level by cutting new openings beneath the spillway; lowering the spillway section; constructing an auxiliary weir through the flanking embankment; or removing the plug in the outlet tunnel.

Optioneering and outline design

The obvious solution to maintain a reduced water level is to reduce the overflow level. However, as the dam is a Grade 2* listed structure significant modifications to the overflow, which would change the appearance of the dam were not favoured by the Local Planning Authority and would be difficult to justify for listed building consent.

The next obvious solution is to strengthen the dam. However, the works required, particularly the temporary works, were very significant and the potential costs were found to be too high for the undertaker to justify, particularly as the storage volume of the reservoir is not expected to be needed to meet future water resources demands until 2030-35.

This left the option of draining the reservoir using the outlet tunnel through the dam on the left hand side of the overflow section. This contains the scour and supply pipework and would have originally been used for the river diversion during construction. The historic records indicated that the tunnel was plugged at impoundment with a 12 to 13 feet thick engineering brickwork plug on the upstream side. The tunnel is of the order of 3m in diameter.

Hydraulic modelling demonstrated that if the plug was removed the tunnel would have sufficient capacity to pass the PMF event with water levels in the reservoir basin not rising above the safe water levels determined from the stability analysis.

By carrying out works near the base of the dam, the appearance of the dam is not significantly affected and this option was therefore favoured by the Local Planning Authority in terms of listed building consent. The works are also reversible should DCWW wish to reinstate the dam to its original operational top water level in future years.

Detailed design

A CFD analysis was carried out to assess the flow conditions within the proposed outlet tunnel and downstream channel during the PMF and 150 year return period flood events. Typical outputs are given in figures 4 and 5.

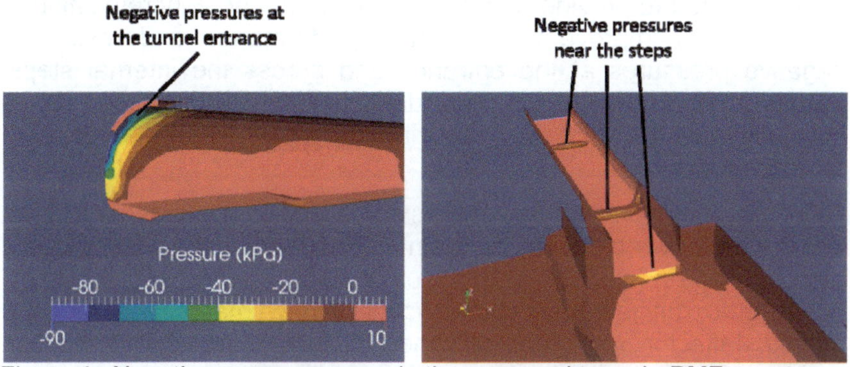

Figure 4. Negative pressure zones in the proposed tunnel - PMF event

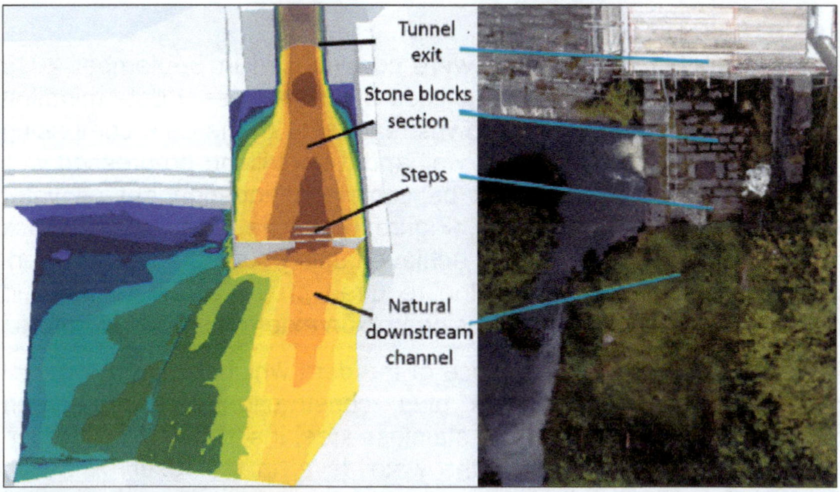

Figure 5. Variation of the maximum flow velocity above the downstream channel bed – 1 in 150yr event

The main findings of the analysis are as follows:

- During the PMF event negative pressures would be developed in the tunnel with peak values of -80kPa at the tunnel entrance.

- The flow rate through the tunnel is predicted to be $64m^3/s$ during the PMF event with a maximum velocity of 16.5m/s.
- Under the 150 year return period flood, the maximum velocities in the downstream channel are of the order of 14m/s

There was concern over the potential for negative pressures to 'pluck out' the masonry immediately above the tunnel inlet and the ability of the existing masonry that lined the tunnel to withstand the negative pressures and high velocities predicted. To mitigate these concerns it was decided to provide a 150mm thick reinforced concrete lining to the tunnel to improve the hydraulic performance and resist the negative pressures at the entrance and across the internal steps. Furthermore the entrance was designed to be protected by a stainless steel liner that would overlap the brickwork on the upstream face of the dam.

A further consideration was the risk of scour to the masonry apron and grassed river bank downstream of the tunnel outlet. The existing masonry apron was of similar construction to the dam spillway, however to mitigate the risk of scour a new masonry apron and curved deflector wall were designed to provide erosion protection and to deflect flows back into the existing stilling basin.

CONSTRUCTION

The main construction works were commenced in September 2015 following an extended planning/listed building consent determination process. The works on the downstream side of the dam comprising new masonry apron, retaining wall and tunnel lining progressed well as these works could largely be completed independent of water levels in the reservoir basin (Figure 6). The new 150mm thick reinforced concrete lining was achieved using curved steel formwork panels bolted together to fit the required profile of the tunnel and pumped concrete compacted using bolt-on external shutter vibrators.

The works on the upstream face of the dam which included removal of the engineering brickwork plug, construction of trash screen foundation slab, installation of stainless steel insert and trash screen were heavily constrained by the water level in the reservoir basin. The existing 450mm diameter scour main was kept open throughout the construction of the works, but as confirmed by a previous drawdown analysis, it had insufficient capacity to prevent the reservoir basin partially filling during the numerous storm events experienced throughout the winter of 2015/2016. A favourable weather window was finally achieved in February 2016, which allowed works on the upstream face of the dam to commence.

Figure 6. Construction stage: tunnel lining from downstream end

MANAGEMENT OF THE ASSET AND ITS FUTURE

Upper Neuadd reservoir forms part of the Taf Fechan water catchment and previously supplied water directly to the Lower Neuadd water treatment plant a few miles south of the site. More recently the resource was utilised in cascade with the Lower Neuadd reservoir and Pontsticill reservoir further south, and together provided water for the Merthyr Tydfil area and down to the city of Cardiff. In considering the desired solution for remedial works at the dam, DCWW had to take into account the possible future use for this water resource and balance this against the cost to the business and the risk the dam presented to the population downstream.

Careful consideration was given to full reinstatement of the original top water level through major structural works at the site, but on reflection and following a high level study of the catchment yield, present and future demand, it was decided that in the short to medium term it was not financially viable to consider this option.

<u>Risk Management</u>

To date, management of risk at the site has been carried out in line with industry best practice and through consultation with the QCE, involving three detailed visual inspections per week with weekly monitoring of the instrumentation at the site to record leakage and seepage through the superstructure. Additionally, settlement and horizontal displacement is surveyed twice yearly. All data collected is recorded and plotted to identify any potential changes that may indicate worsening conditions and increased risk of failure.

Future plans

Long term plans for the reservoir could potentially involve major structural works to re-instate the original operational top water level. In choosing the agreed solution to remove the tunnel plug and therefore prevent the reservoir from filling to unsafe levels, DCWW is well placed to initiate future works that would reverse this current arrangement without adding unnecessary cost.

Although the current solution effectively ensures the safety of the dam, DCWW will continue to monitor the behaviour of the structure through weekly visual inspections of the site, and will also continue to monitor the settlement and horizontal displacement to ensure the safety of the structure for many years to come.

CONCLUSION

Upper Neuadd is a Grade 2* listed dam in the heart of the Brecon Beacons. Detailed investigations, studies and analyses undertaken on the dam structure have enabled the team to understand the dam's performance in operation and explain the observed stability and leakage problems.

Following a review of options to abandon, rehabilitate or discontinue the asset, a solution was found that ensures the long term stability of the dam, whilst retaining the appearance of the listed structure and will enable DCWW to fully rehabilitate the dam to restore the original full operational top water level for the reservoir in the future.

ACKNOWLEDGEMENTS

The authors would like to thank DCWW for input into the paper and permission to publish. They would also like to acknowledge the input to the development of the solutions and consideration of risk and buildability by Skanska and Joseph Gallagher Ltd (JGL).

REFERENCES

Kennard MF, Owens CL and Reader RA, (1996). *Report 148. Engineering guide to the safety of concrete and masonry dam structures in the UK.* CIRIA, London

USBR (1987). *Design of small dams.* US Government Printing Office, Washington, USA

Dams – Benefits and Disbenefits; Assets or Liabilities?
ISBN 978-0-7277-6133-0

doi: 10.1680/dbdal.61330.457

Bollinhurst Impounding Reservoir – Spillway Apron Leakage Remedial Works

D R C HUGHES, J and D Geotechnical Services Ltd
K HARVEY, United Utilities
P J RIGBY, United Utilities

SYNOPSIS An overflow study undertaken in 2004 identified that the spillway of Bollinhurst Impounding Reservoir (IR) had insufficient capacity to pass the Probable Maximum Flood. An overflow model test was commissioned by the owner, United Utilities, which confirmed that additional overflow capacity was required. A replacement spillway was constructed between 2010 and 2011 which incorporated the original spill weir and adjacent wing walls. During refilling of the reservoir in 2012, leakage into the new overflow tumble bay under drainage system was identified. The leakage commenced at approximately 0.5m below overflow level and the flow, via the 160mm diameter connecting outlet pipe, into the connecting chamber was full bore.

Works were commissioned to investigate and remediate the observed leakage following a meeting in September 2012 which included AR Panel Engineer Dr Peter Mason, subsequently appointed as QCE for the remedial works.

This paper describes the history of the reservoir, investigation of the leakage source, and the determination of possible remedial options. It also discusses the observations made during the construction works and the design of the solution to fix the leakage.

CONSTRUCTION AND HISTORY OF BOLLINHURST RESERVOIR

Located approximately 1km south of the centre of Disley, Cheshire, Bollinhurst Impounding Reservoir impounds the Bollinhurst Brook and was constructed between 1871 and 1875. The Engineer for the works, which were for the purpose of water supply to Stockport District Water Works, was G H Hill.

Constructed of locally won Glacial Till and Coal Measures Sandstone, the embankment is 250m long and is dog-legged in plan with a maximum height of 21m. The embankment crest is approximately 4.6m wide and includes a tarmac access road. Bollinhurst embankment has a puddle clay core and cut-off trench which extends into the underlying bedrock. The clay core is located centrally within the dam crest for most of the embankment length. When the core reaches the spillway apron located on the right hand mitre of the dam, in front of the spill weir, it diverts upstream as a 'core arm trench' away from the embankment, across the masonry apron to proceed up the right hand mitre of the dam. Water tightness from the arm trench is maintained by a clay blanket to the overflow weir.

Flows from the spillway are discharged into Horse Coppice Reservoir located immediately downstream of Bollinhurst.

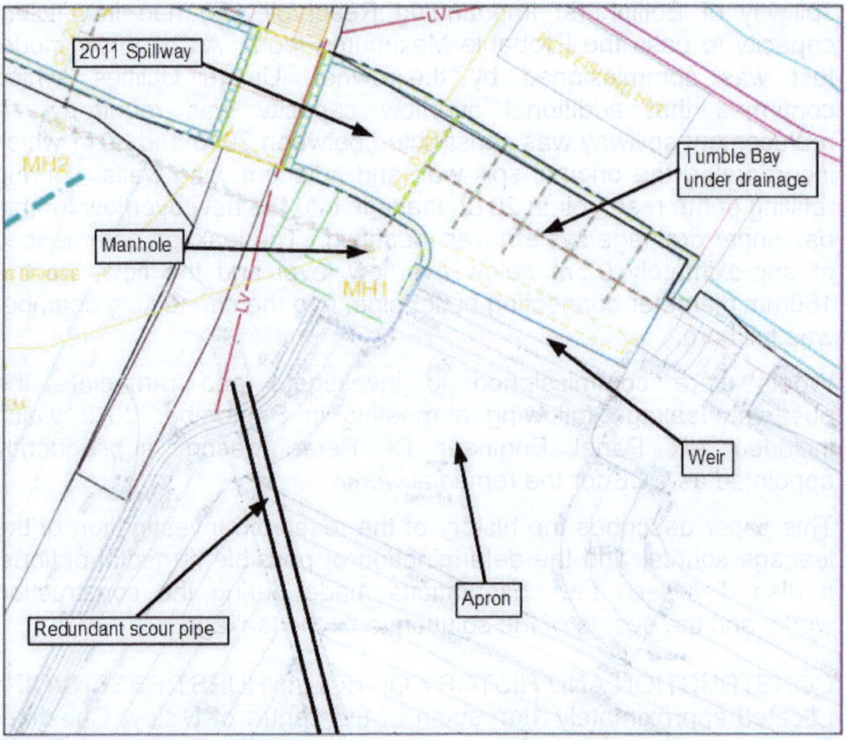

Figure 1. Bollinhurst IR Spillway and Tumble Bay Arrangement (extract UU drawing 7376_90021233_01_32_2107_AS BUILT)

During the 1950s improvement works were made to the embankment and spill weir. These works included raising of the dam crest and clay core to increase freeboard, construction of a new masonry wave wall, doubling of the spill weir length from 20 to 40ft (6.09 to 12.19m), widening of the spillway apron and demolition and reconstruction of the right hand spill weir wing wall and by-wash channel outlet.

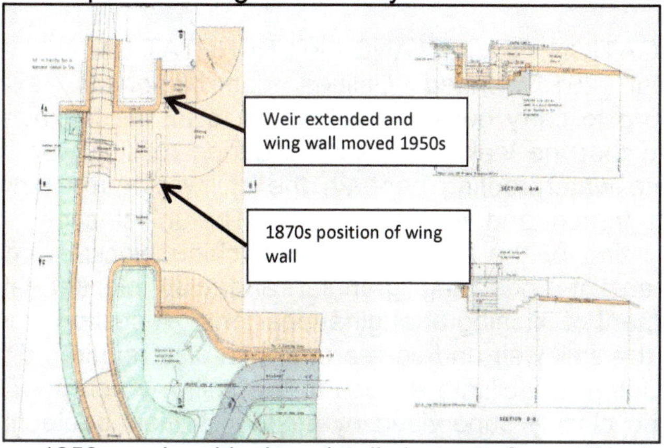

Figure 2. 1950s weir widening detail (extract from UU drawing ref BOL_t073-(F))

The reservoir is owned and operated by United Utilities. A 2004 study indicated the spillway had insufficient capacity during flood events with the embankment at risk of overtopping. Consequently, the reservoir level was lowered pending a review of the potential remedial options. Following an overflow model test, works to improve the overflow capacity of the reservoir were commissioned.

The principal items of work affecting the overflow facilities included construction of a new reinforced concrete spillway channel, tumble bay and stilling basin, including under drainage, down the right hand dam mitre. The previous overflow spillway channel was partially demolished and backfilled. The existing overflow weir, masonry steps into the tumble bay and bywash channel were kept and incorporated into the new works. A berm was also constructed to improve the stability of the downstream face of the embankment as part of the contract.

The spillway modification works were completed in 2011. During controlled refilling of the reservoir in 2012 leakage into the new overflow tumble bay under-drain system was identified. The leakage commenced when water levels in the reservoir were at approximately 0.5m below overflow level and the flow, via the connecting 160mm diameter outlet pipe into the collecting chamber No1, was full bore. The reservoir was immediately drawn down. Following a meeting on 4 September 2012, which included AR Panel Engineer Dr Peter Mason (subsequently appointed as QCE), works were commissioned to investigate and remediate the observed leakage.

United Utilities has good historic 'as built' records for the dam dating from both from its original construction in the 1870s and refurbishment works carried out in the 1950s.

INVESTIGATION, REVIEW AND INTERPRETATION

In August 2013 United Utilities commissioned Askam Civil Engineering to carry out remedial works to fix the leakage. It was suspected that the leakage observed during refilling was related to inadequate waterproofing beneath the spillway apron between the core arm trench and the spill weir. The initial scope of works included lifting of the spillway apron pitching blocks; removal and replacement of underlying granular and clay blanket layers and reinstatement of pitching at original positions. A section of wave wall between the spill weir and core arm trench also required dismantling and rebuilding to facilitate access to the spillway apron. The works were to be closely supervised by United Utilities' geotechnical and civil engineering staff responsible for site supervision, logging, measurements and sketches of exposures.

The removal of the pitching blocks commenced at the spill weir in September 2013 and progressed towards the reservoir basin. As the contractor removed the pitching each block was given a unique reference number with their location recorded on a plan for later reinstatement at the same location.

19th Century Foundation and Evidence of Historic Grouting

As the pitching was removed it became apparent that the weir extensions carried out in the 1950s and had not been carried out to a sufficiently high standard. On the right side of the spill weir it was observed that the foundation for the 1870s wing wall below the weir extension had not been removed as it was encountered directly beneath the pitching. Further removal of pitching revealed the absence of a clay blanket on the right hand side of the apron with sandstone bedrock observed beneath a thin layer of sandstone gravel. Rudimentary soakaway tests undertaken in this area confirmed the permeable nature of the bedrock.

Evidence of past grouting works, suggesting historical awareness of the leak, was encountered in this area, as indicated by drill holes through some of the pitching blocks and the presence of grout adhering to the underside of the pitching. Close observation of the grout showed evidence of dissolution in some areas, possibly caused by the passage of slightly acidic raw water.

Absence of Clay Blanket

On completion of pitching removal from the spillway apron it became clear that the clay blanket was absent from the majority of the area covered by the 1950s weir extension works. The core arm trench was exposed and found to closely follow the route shown on the historic drawings. Trial pits were carried out to prove the thickness and extent of the clay blanket across the apron. These confirmed the connection between the blanket and puddle arm trench. Evidence of an erosion sink hole through the clay blanket into the underlying sandstone was observed in one location. The extent and thickness of clay, at around 600mm thick, closely matched the 1870s G.H Hill as built drawings. The clay blanket was excavated from the spillway apron and removed from site to expose the sandstone beneath.

Figure 3: Exposed sandstone, wing wall foundation, puddle arm trench and syphon pipe

Widened Spill Weir

To confirm the condition of the bed rock below the extended spill weir the 1870s wing wall foundation was removed. The weir extension was found to have been constructed on poor quality mass concrete placed directly on fractured sandstone. Evidence, suggestive of seepage pathways through the weir foundation, was observed by of black staining on both the concrete and sandstone. In contrast, the Victorian weir had been built to high standard, with the dimensions matching the record drawings precisely.

Figure 4: 1870s weir and 1950s weir extension showing evidence of seepage

Geological Fault

As the investigation of the weir extension proceeded it became apparent that the formation beneath the pitching in the 1950s works had been constructed at a higher level than indicated on the as built drawings. As a result, The Coal Measures Sandstone bedrock was encountered almost directly beneath the pitching.

Figure 5: Geological fault beneath wing wall

The proposed remedial measures to install a replacement clay blanket across the entire apron necessitated the excavation of the bedrock to allow a 750mm thick clay blanket to be constructed. A geological fault, which passed beneath the wing wall, was encountered during the excavation of the bedrock adjacent to the right hand wing wall. This measured up to 0.5m in width and was infilled with sand and gravel. The presence of the fault and the fractured nature of the adjacent bedrock indicated the presence of a significant leakage pathway.

Redundant Siphon Pipe

A redundant siphon pipe was also encountered during pitching removal. This ran from the reservoir basin towards and through the spillway apron and adjacent embankment. This had been grouted when it was taken out of service. However, the pipe had originally been installed with a granular bed and surround through the arm trench and blanket which had also been left untreated post redundancy giving rise to another potential leakage path.

Spill weir wing walls

The original wing walls to the spillway had been retained as part of the spillway completed in 2011. During the leakage remedial works it was noted that water staining on both of the wing walls' masonry blockwork, and water levels in the back of wall drainage, suggested a potential leakage pathway between block work joints and through the wing walls.

REMEDIAL WORKS

Between November 2013 and February 2014, following the investigation phase of the works, the site works were temporarily suspended whilst a revised scope of works and design could be developed and discussed with the QCE. The agreed scope of works comprised the following elements.

Demolition of spillway wing walls

It was observed that the identified geological fault extended beneath the existing right wing wall structure, with a significant proportion of the wing wall founded on the fault line. A geological assessment of the fault concluded that the dip orientation, dip direction and the presence of low strength infill materials (clay and mica) on the sandstone bedding planes exposed within the excavation, represented a potential safety issue. The unfavourably dipping bedding planes, combined with very low frictional properties, might have resulted in sliding, leading to the failure of the wing wall structure into the excavation during construction.

It was considered that further excavation below or in front of the wing wall was not advisable without temporary support as it was likely that the fault gouge material would fall away resulting in the undermining of the wall foundations.

The fault also represented a favourable flow path from the reservoir through the highly permeable disturbed fault gouge material. It was proposed to locally over-excavate the fault and fill with concrete to ensure an effective water tight seal.

In addition, it was recommended that the wing wall and spill weir were dismantled and rebuilt to facilitate the replacement clay blanket works. Dismantling the wall and weir would also have the benefit of allowing the full exposure and treatment of the fault in order to remove it as a potential flow path.

Rebuilding of both the wing wall and spill weir foundations would remove potential seepage pathways through the existing masonry block work which were evident by the presence of black staining. The replacement foundations included a reinforced concrete wall, raised to the core level, to eliminate any further leakage pathways.

<u>Replacement wave wall</u>

A 25m-30m section of wave wall between the core arm trench diversion and the spill weir had been dismantled to facilitate access to the spill weir apron. Initially it had been intended to reassemble the wave wall in masonry. However, following the decision to replace the wing walls it was decided to continue the replacement of the wave wall using masonry clad reinforced concrete.

<u>Removal of redundant siphon pipe</u>

The redundant siphon pipe running from the reservoir basin presented a potential seepage pathway through the embankment and an obstacle to the construction of the new section of wave wall. The pipe was broken out, removed and the area made good by the contractor.

<u>Fibre reinforced concrete</u>

After the removal of the wing walls, section of wave wall, and the clay blanket the exposed sandstone foundation was lowered to allow a concrete sealing layer to be placed. The sandstone on the right side of the spillway apron at the 1950s extension was lowered locally by up to 400mm.

On completion of the formation preparation works a specialist contractor was employed to place a 150mm thick layer of blown polymer fibre reinforced concrete to seal fractures within the sandstone. Particular attention was given to the areas affected by the fault and the highly fractured sandstone beneath the extended weir. The fibre reinforced concrete extended across the entire spillway apron and provided an impermeable barrier which would also reduce the potential development of seepage sink holes which had been observed within the original clay blanket.

Figure 6: Fibre reinforced concrete installation

<u>Reinforced concrete structures</u>

The replacement concrete structures comprised a 30m section of wave wall, two spillway wing walls and the spill weir foundation extension. The design of the replacement structures was carried out United Utilities' civil engineer. Reinforcement design was undertaken by the contractor.

The lower sections of the structures incorporated steeply sloping rather than vertical faces to receive the rolled clay blanket thus accommodating future settlement of the clay without creating a potential leakage pathway.

Concrete was delivered to site by wagon and transferred to less accessible areas in skips to be poured into the formwork. Vibrating pokers were used to reduce the presence of voids within the concrete. In order to provide an aesthetic facing to the new concrete structures the existing masonry block work was reused.

Placement of rolled clay blanket over blown concrete connected to existing core arm trench

The remedial works specified the installation of a 750mm thick layer of rolled clay to be placed across the spillway apron and connected to the clay core at the embankment crest and also the puddle arm trench.

The clay was obtained from a quarry in Warwickshire as it complied with the requirements of the specification and the same source material had been used successfully on a number of previous United Utilities reservoir projects. The as-excavated clay comprised cobble and boulder sized lumps of stiff to very stiff clay which required processing on arrival to be made suitable. The clay was placed into a skip where it was broken up in to smaller pieces using a riddling bucket and sprayed with water in order to reduce its shear strength and assist fragmentation of the clay lumps.

The earthworks specification from the owner required the clay to be placed within undrained shear strength limits between 35kPa and 65kPa. A method compaction specification was used for the placement of the clay blanket. The clay was initially placed in 150mm to 200mm thick layers using an excavator. A remotely operated 1.5 tonne vibrating sheepsfoot roller then performed the required 20 passes, compacting the clay to between 100mm and 150mm layer thickness. The shear strength of each layer was checked using a hand shear vane to confirm that it fell within the shear strength limits specified. Clay cores were also cut into each clay layer and examined to ensure that no voids or horizontal fissures were present within the compacted clay or between each layer.

A short section of clay core excavated during removal of the wave wall was reinstated in a similar fashion to the clay blanket. Sheet piles were installed along the downstream face of the core to minimise the extent of excavation in that area by providing support to the embankment. The sheet piles were left in place in order to minimise potential damage to the core. The clay blanket and core were protected from desiccation between shifts by covering with plastic sheeting.

Granular regulating layer

Following installation of the rolled clay blanket a non-woven geotextile separator layer was placed beneath a layer of Class 6N granular material. Crushed sandstone, from a local supplier, was used as it was considered to be more resilient to dissolution than limestone.

Figure 7. Clay blanket and core reinstatement

Reinstatement of pitching blocks to spillway apron and masonry facing to wave wall and wing walls:
The contractor appointed a specialist stone mason to reinstate the pitching blocks. The numbering system employed when the pitching was removed was used to reinstate the blocks in original positions allowing a good finish with tight joints to be achieved. The new wave wall and wing walls were also faced in sandstone reused from dismantled structures.

The works were completed in April 2014. The reservoir was refilled and the weir allowed to spill. No flowing water was observed in the spillway under drainage Chamber No.1, where the flows had been observed during first refilling. The absence of water indicated the remedial works to have been successful.

CONCLUSION

The somewhat archaeological approach to the investigation and remediation of the main suspected leakage pathway permitted the identification and subsequent removal of several other potential leakage routes. The successful completion of the project was greatly assisted by the availability of accurate record drawings dating back to the reservoir's construction in the late 19th Century which allowed the early identification and understanding of the dam structure and potential leakage pathways.

The investigative nature of the recent works unearthed a few skeletons left by previous works showing that historical drawings may not always be reliable, and that short cuts taken during construction works can create long term problems. The project benefited greatly from the close cooperation between United Utilities' design team and the contractor's site staff thereby allowing a robust solution to be designed and constructed.

ACKNOWLEDGMENTS
The authors wish to acknowledge the permission of United Utilities to publish this paper, to thank Dr Peter Mason for his guidance during the project and the staff of Askam Civil Engineering for their commitment and collaboration during the works.

REFERENCES

Briscoe A, (2012) United Utilities. *Reservoirs Act 1975 Supervising Engineers Report - Bollinhurst.* United Utilities, Warrington.

Carter IC, (2007) *Report on an Inspection under Section 10 of the Reservoirs Act 1975. Bollinhurst Reservoir.* United Utilities, Warrington.

United Utilities Water PLC (2013). *Bollinhurst Impounding Reservoir Spillway Apron Remedial Works. UU Project No. 90021233. Conformed Solution Scope Book.* United Utilities, Warrington.

United Utilities Reservoir Archive. *Drawing reference BOL_t073-(F) Stockport Corporation Water Works (1951). Bollinhurst Reservoir – Reconditioning. Waste Weir and Watercourse Construction Drawing No.6.* United Utilities, Warrington

United Utilities Reservoir Archive. *Drawing reference 7376_90021233_01_32_2107_AS BUILT. United Utilities / Askam Civil Engineering / MWH (2011) Bollinhurst IR Overflow Work. New Overflow Channel Drainage Plan.* United Utilities, Warrington

Dams – Benefits and Disbenefits; Assets or Liabilities?
ISBN 978-0-7277-6133-0

doi: 10.1680/dbdal.61330.469

Investigation of Voids under a Spillway using GPR and its Resolution

J R FOSTER, Mott MacDonald Bentley
B DOOLEY, Mott MacDonald Bentley
R ROBSON, Mott MacDonald Bentley
A HOBSON, Yorkshire Water Services

SYNOPSIS This paper describes the findings of a project to refurbish an existing spillway on a Category A dam, in particular focusing on the investigation methodology used to identify voids beneath the spillway. The paper presents a summary of the investigation findings and the remediation works completed.

The paper explains how Ground Penetrating Radar (GPR) was used to investigate the presence of voids beneath the spillway. This was used together with more conventional ground investigation as verification. The paper reviews the benefits and limitations of using GPR in identifying void potential beneath spillways. The paper concludes by providing details of how the voiding issue was remediated, outlining the measures employed.

INTRODUCTION

In early 2014 Mott MacDonald Bentley Ltd (MMB) was contracted by Yorkshire Water Services (YWS) to carry out the detailed design and construction of spillway improvement works at Watersheddles Reservoir, an 886,000m³ impounding reservoir (IRE), constructed in 1877. The works were required in response to the Inspecting Engineer's recommendations as Matters in the Interests of Safety to confirm the overflow system is capable of safely conveying the Probable Maximum Flood (PMF) of 29.8m³/s along with an assessment of its structural integrity during the design flood.

During the most recent Section 10 Inspection (2011) the Inspecting Engineer also recommended that water loss observed through joints in the invert of the stone masonry tumblebay should be investigated and resolved as a Matter in the Interest of Safety. Furthermore, he recommended that water seen entering the spillway chute through the base of the spillway walls should be investigated to determine its source.

Figure 1. Views of Original Structure

A separate paper has been prepared for the works at Watersheddles IRE, documenting the solution to increase the capacity of the overflow system (Speirs, 2015). The solution employed involved the use of pre-cast concrete u-channels placed within the existing masonry channel. The u-channels were fixed to the underlying original structure, with the original channel left in place and used as the foundation for the new spillway. This solution did not remove the original spillway structure, so it was therefore still necessary to investigate the suitability of it as a foundation to the new channel and to identify and isolate any potential seepage paths.

FINDINGS OF STATUTORY INSPECTION

During the Section 10 Inspection the water which was observed flowing through joints in the invert of the stone masonry tumblebay raised concerns that water could have be tracking below the spillway, passing the cut-off and contributing to the flows observed entering the spillway further downstream. Grouting works had previously been undertaken in the tumblebay area in around 2005 with some success in reducing flow into the spillway, but further grouting was recommended.

The Inspecting Engineer required further investigation into the source of this seepage observed entering the spillway chute, with a view to establishing: -

- Was the flow linked to the water disappearing through the joints in the tumblebay?
- Was the flow linked to the reservoir?

- Was the flow associated with rainfall / groundwater from valley sides?

Pending the results of the investigation and modelling to confirm the hydraulic and structural capacity of the spillway, the Inspecting Engineer recommended that the reservoir level be managed at a lower level. This drawdown was later extended until the structural modifications could be completed.

INITIAL SITE INVESTIGATION

MMB undertook a visual survey to examine the structural condition of the spillway and to scope the investigation required to examine the reported water ingress.

Initially the project team planned to undertake a suite of intrusive ground investigation to determine the construction and condition of the spillway overflow with the investigation comprising: -

- Coring through the spillway and tumblebay invert
- Coring through the spillway and tumblebay walls
- Trial holes excavated behind the walls to prove construction and the presence of any back of wall drainage
- Trial holes to prove the embankment core level and interface with adjacent structures (i.e. spillway)

However, due to the remote and inaccessible nature of the spillway (poor access / 1 in 2.5 slope), it was considered impractical to undertake extensive intrusive investigations on the spillway under the investigation contract. It was therefore decided to defer intrusive investigations on the spillway chute until safe access had been created and a safe system of work could be put in place.

In the intervening period, physical modelling undertaken to confirm the hydraulic capacity of the existing spillway showed that the spillway was under-capacity and therefore unable to safely pass the design flood. Furthermore, visual inspection of the spillway raised concerns regarding its structural integrity, further supporting the need to replace the asset. In light of these findings, the requirement for a new spillway structure was confirmed.

Although a decision was taken to replace the existing spillway, the preferred solution was to install precast concrete u-sections within the existing stone masonry channel, retaining this structure to act as a foundation to the new spillway. Retaining the original spillway did, however, mean that any pre-existing seepage paths or voids beneath the spillway would continue to exist. It therefore became increasingly important to establish the source of the water ingress and undertake remedial works as necessary to isolate these flow paths.

Under this solution, a nominal allowance was made for localised grouting of any voids around the structure prior to installation of the new spillway. In advance of confirming this solution the existing ground conditions needed to be proved through on-site intrusive investigation.

GROUND INVESTIGATION

Upon establishment of a safe system of work the spillway and tumblebay were initially investigated by digging trial holes through the invert of the structures. These trial holes revealed that there were voids beneath the existing masonry structure but did not confirm the extent of the voids. It was at this stage that GPR was suggested as a possible means to map the voids and ascertain any linkages.

The embankment core was investigated by digging trial holes along the dam crest with measurements taken to confirm the top of core level with visual inspection of the core confirming that the watertight element was formed of good quality clay.

SEEPAGE PATH INVESTIGATION

Sensitivity analysis was undertaken to establish if there was any linkage between reservoir level and seepage observed. Dropping the reservoir level by 3m was seen to correspond with the seepage flows ceasing, suggesting a strong relationship between seepage observed and reservoir level. This was further confirmed when the seepage observed did not re-establish following subsequent periods of heavy rainfall. Although it was now confirmed that the seepage observed originated from the reservoir, what remained unclear was the seepage pathway. The possible routes were theorised as follows: -

- Seepage initiating upstream of the weir and passing through the joints in the masonry tumblebay invert between masonry and concrete backing.
- Seepage initiating upstream of the overflow, from the reservoir body and travelling around the outside of the overflow/spillway structure.
- Seepage initiating in the reservoir body travelling through a defect in the embankment core and surfacing through into the spillway.

GROUND PENETRATING RADAR

Methodology

This paper does not seek to provide a detailed technical review of Ground Penetrating Radar (GPR), but rather summarise how it can be applied to investigate void potential beneath existing spillways.

GPR survey utilises electromagnetic pulses to produce graphic depth sections. The technology is commonly used in the construction industry for the detection of buried services but also has applications in the mapping of shallow geological features (including soil or rock), examining the internal condition of concrete structures (identifying delamination and internal layering in concrete) and locating reinforcement in concrete structures.

The GPR system consists of an antenna unit, a signal control and processing unit with a built in monitor, and a graphic recorder. The system is powered from a 12 Volt DC battery source. During survey operations, only the antenna unit is in direct contact with the structure with the remaining equipment typically stored in a waterproof box (Figure 2).

Figure 2. Application of GPR Survey

Supplier information from Subsurface Geotechnical states that "*accurate depth profiles are obtained by moving the antenna along marked survey lines, producing a two-dimensional image. The horizontal scale is obtained by marking regular intervals of distance along the profile, using a marker switch. The vertical time axis is calibrated in nanoseconds per centimetre, and converted to depth by using standard equations. The radar record produced is therefore a graph of reflection time against distance along the survey line."*

GPR is most effective at detecting changes at transitions between two different and distinct materials where a clear contrast is present at the interface. The GPR survey method is typically less effective in identifying gradual or transitional boundaries. The reflected signal can be correlated with physical interfaces within the ground such as layering, air cavities or bedrock when interpreted against intrusive investigations. Figure 3 indicates the main components of a GPR survey showing how radar signals are transmitted and recorded.

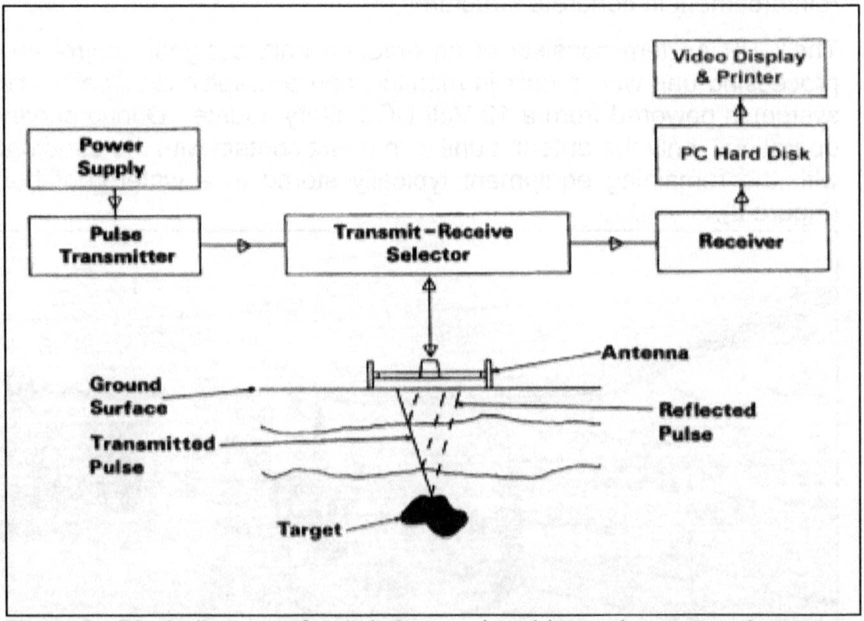

Figure 3. Block diagram of a digital ground probing radar system (courtesy of Subsurface Geotechnical).

<u>Limitations</u>

Although the GPR survey methodology can be a powerful tool in investigating void potential at existing spillways, the output produced is most effective in identifying changes in material with depth. GPR survey yields little information regarding the quality or density of each material encountered, so it is therefore necessary to supplement its results with intrusive investigations. Verifying the ground conditions by intrusive investigation allows the GPR data to be accurately interpreted. This is discussed in more detail later in this paper.

In subsequent use of GPR at other MMB sites it has been found that the GPR survey method is most successful in surveying spillways free of vegetation cover. Where the spillway has significant vegetation growth the performance of the antenna can be negatively impacted resulting in poor outputs being obtained. However, GPR can be used and will yield reasonable results where there is uniform grass cover, such as on a grassed spillway.

To collect the optimum quality GPR data the antenna should have intimate contact with the surface under investigation. Therefore stepped spillways create a potential problem, introducing an air void between the antenna and the surface as the GPR unit traverses the steps. To minimise the vertical displacement as the antenna is moved along each profile, the antenna can be housed in a long sled. However, there still are likely to be some data quality issues encountered.

For the purposes of determining detailed features within the spillway structure the GPR arrangement used was returning results to a depth of 2.4m. Greater depths can be surveyed, but the quality of the results, and ability to pick up detail is lost. It is also worth noting that the GPR may not pick up any detail below dense materials such as concrete and grout if it is more than about 1.5m thick.

The presence of water standing water can also present challenges to the GPR survey equipment. Ponded water (either on the spillway surface or in pockets beneath the masonry blocks) will reflect the GPR signal and return a signal similar to that obtained from a void. Undertaking GPR surveys during dry weather conditions is the most likely means of reducing the error associated with ponded water, reducing the number of false readings relating to void potential.

INTERPRETATION OF RESULTS

A typical output from the GPR survey is presented in Figure 4. Whilst the GPR supplier provided some initial interpretation of the results, this was done without further intrusive investigation so was not necessarily the correct interpretation.

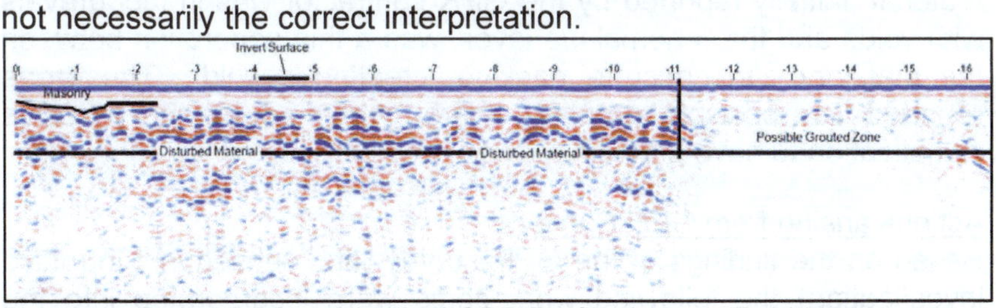

Figure 4. Example of Longitudinal GPR Profile through Tumblebay (courtesy of Subsurface Geotechnical).

The longitudinal profiles identified areas classified by the GPR supplier as "disturbed material", "areas of possible grouting" and "anomalous layers" requiring further investigation. An example of a GPR interpretation plan produced by the GPR contractor for Watersheddles' tumblebay is provided in Figure 5.

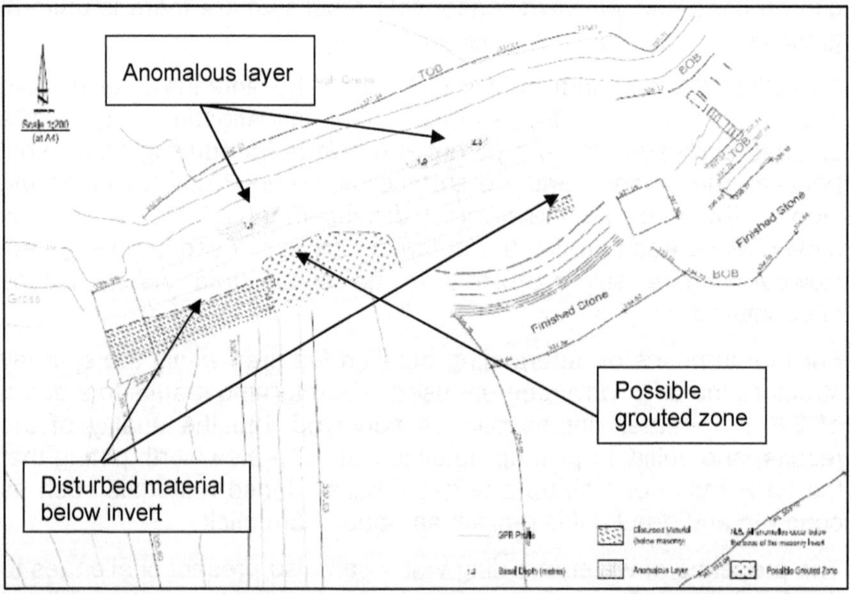

Figure 5 – Tumblebay GPR Interpretation Plan (courtesy of Subsurface Geotechnical).

Proving of Interpretation

It is necessary for a GPR survey to be supplemented with intrusive investigation to calibrate and validate the results obtained. At Watersheddles ground investigation findings were used alongside the GPR survey outputs to identify correlations between the various materials encountered beneath the spillway.

Pulling the two investigations together proved that the "disturbed material" initially reported by the GPR contractor was in fact gravels with voids and the "anomalous layer" was a thin separation between the masonry and concrete backing, creating a void. The areas reported as "possible grouting" were confirmed by the intrusive investigation to have previously been grouted.

Actions arising from GPR Survey

Based on the findings of the GPR survey (and subsequent intrusive investigation) the following work items were incorporated into the design of the new spillway: -

- Underdrainage to the new spillway section – designed to allow future monitoring of residual seepage flows
- Grouting beneath the existing spillway channel prior to installation of new spillway
- New cut-off below tumblebay at core location
- Grouting to fill shallow void identified between existing stone masonry and concrete backing within the tumblebay area.

SOLUTION IMPLEMENTATION

The design and implementation of the solution was undertaken under the supervision of a Qualified Civil Engineer, which in this case was the original Inspecting Engineer.

Spillway

Prior to the confirmation of voids beneath the spillway, a decision had been made to replace the existing spillway due to inadequate hydraulic capacity and concerns regarding its structural integrity. However, in developing the new spillway solution it was necessary to ensure any leakage paths were isolated and that the new spillway would be supported on a sound foundation.

Based on the results of the investigation the existing spillway was grouted with a cement based grout filling any voids beneath. Additionally, a new spillway underdrain was installed beneath the invert of the old structure to allow future monitoring of any residual seepage / groundwater from the adjacent hillside. This underdrainage was also used to verify that remediation works undertaken around the tumblebay had been successful.

Tumblebay

As the original masonry tumblebay upstream of the core was to be retained, and a potential seepage path had been verified through investigation at the interface between the masonry and concrete backing, again a cement based grout was specified to fill the voids. Low pressure grout injection was undertaken on a 1.5m x 1.5m grid pattern across the tumblebay invert, along with cleaning of masonry joints and repointing.

As the new spillway solution included installation of a new cut-off grouting was not initially carried out at the interface between the tumblebay and clay core. The new cut-off was designed to extend at least 500mm into the original clay core.

During the site works the original clay core was seen to have settled by up to 700mm resulting in the original cut-off becoming detached from the spillway structure as the clay had settled (Figure 6). The original cut-off was seen to be around 400mm deep x 400mm wide, formed in mass concrete. There had been a construction joint between it and the backing concrete of the masonry structure which is where it had become detached. The pulling down of the cut-off meant that there was a clear path for water to track beneath the tumblebay prior to the remedial works.

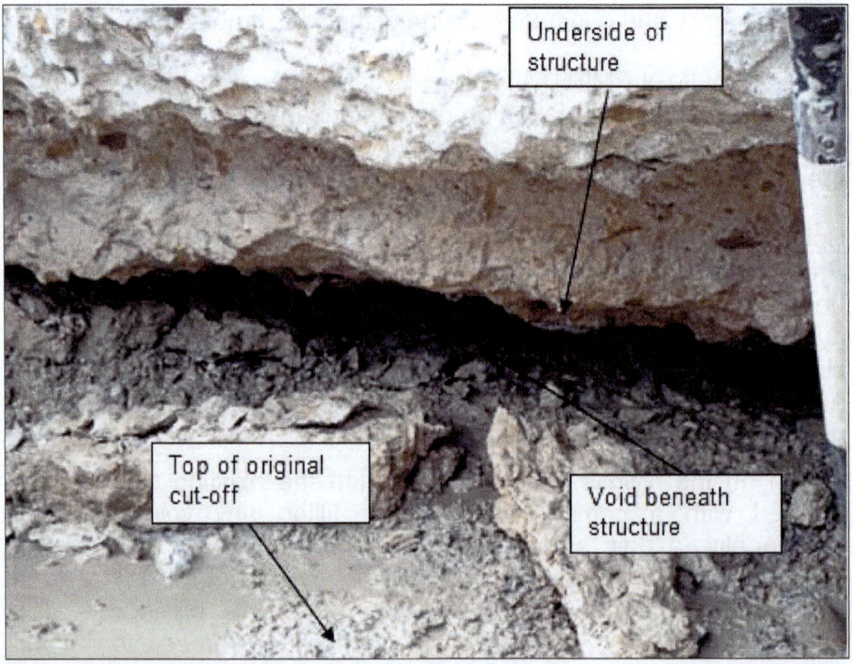

Figure 6 Excavation along the core

The voids upstream of the new cut-off were filled before the new cut-off would be finally excavated to depth across the full length. It was agreed with the QCE that the void would be filled with a bentonite grout instead of the cementitious grout as used in the remainder of the tumblebay. This was in part because of the volume of the void, but also due to the need to replace the lost clay within the embankment next to the structure. The grouting was also extended up the LHS wall although grout take was minimal. A pre-mixed material was used, Bentogrout, with water mixed prior to injection.

Trial excavations behind the tumblebay walls revealed an inverted cut-off with the puddle clay – whereby the puddle clay was placed within a large engineered notch formed in the back of the masonry wall. At the surface this showed good contact between the masonry and clay.

CONCLUSIONS

Observations on Ground Penetrating Radar

Through a number of recent investigations GPR survey has been found to offer a cost effective and useful means of identifying areas of increased "void potential" beneath an existing spillway structures. The results obtained can be used to inform the planning and scoping of an effective intrusive investigation to both verify the GPR results and obtain further information about the condition of the structure.

The initial GPR results alone were at Watersheddles were found to be inconclusive in determining the location of voids and required supplementary intrusive investigations to calibrate the survey outputs.

This paper demonstrates that undertaking GPR survey in advance of intrusive investigations can assist in providing an indication of suspect areas requiring further investigation. It is therefore considered as an effective tool in reducing the amount of intrusive investigation required across the whole survey area. The results should be used prior to on site intrusive investigations to target specific areas.

This paper acknowledges the limitations of the GPR survey methodology, highlighting issues pertaining to the presence of vegetation, stepped spillways, penetration of dense materials and the impact of water in confirming areas of void potential. However, even with these limitations, GPR survey is considered to provide enough information to allow dam engineers to focus their intrusive investigations and gain a high level understanding of void potential.

Costs

GPR survey offers a cost effective means of providing a broad overview of void potential and when used appropriately can help focus and reduce the cost of intrusive ground investigations. There are a number of GPR suppliers operating in the UK market providing healthy competition and helping to keep costs at a reasonable level. Based on a recent GPR survey commissioned for MMB, GPR survey and production of an interpretative report at 2015 prices cost approximately £3,000 for a spillway and tumblebay of around 900m^2 and was completed within the day.

Recommended Methodology of Overflow Investigations

Following the works at Watersheddles, the MMB team has gained valuable experience in the efficient execution of void potential investigation. The knowledge gained has since been put to use on subsequent spillway and tumblebay investigations for YWS and other Clients. Based on this experience, MMB project teams now implement the following set process for spillway investigation projects: -

- Procure GPR survey
- Clear spillway of debris and significant vegetation
- Undertake GPR survey
- Receive initial interpretative results (plans & long sections)
- Scope intrusive investigation (cores and trial pits) to verify / calibrate GPR interpretation
- If required, develop a repair strategy to target areas of potential void.

ACKNOWLEDGEMENTS
The authors acknowledge the guidance of the scheme's QCE, Mr Henry Hewlett, in specifying parameters for the works and the support provided by Yorkshire Water Services (YWS) 'the Client' in delivering the successful scheme and for allowing these experiences to be shared. Acknowledgement is also sought for Subsurface Geotechnical for sharing their results, figures and explanation on GPR.

Dams – Benefits and Disbenefits; Assets or Liabilities?
ISBN 978-0-7277-6133-0

doi: 10.1680/dbdal.61330.481

Control of moss on reservoir dam embankments

C A SPRING, STRI Ltd
T M YOUNG, STRI Ltd
R L MANN, STRI Ltd
M DIXON, United Utilities Water Plc
B TOLLITT, United Utilities Water Plc

SYNOPSIS Earth dam embankments need to maintain a year round high grass content sward to achieve a constant surface and ensure slope stability. Moss in the sward is considered to reduce potential stability, increase vulnerability to erosion and increase the risk of slips and trips. United Utilities and STRI have undertaken trials at Mitchell's House Reservoirs. The research objectives were to assess the degree of moss invasion, evaluate the optimum approach for controlling moss in sustainable ways, to identify optimum strategies of increasing grass dominance and to define practical safe and environmentally benign methods for delivery through basic grounds maintenance capabilities.

Moss control using several environmentally sensitive moss control products, physical removal by scarification and improving soil fertility to boost grass growth to outcompete moss, have all been successful strategies, with each treatment on its own significantly reducing moss content in treated plots. There was no advantage in combining the treatments. In terms of turf recovery, application of low levels of organic fertilisers and nitrogen were successful at speeding up the recovery of turf after moss removal, by scarification, resulting in almost full grass cover over all treated plots.

INTRODUCTION

Moss is a collective term for a diverse group of green plants in the phylum *Bryophyta* (Allott, 1954). Mosses are distributed widely across the United Kingdom and are adapted to grow in a wide variety of locations and under a diverse range of environmental conditions (Jefferson, 1947 and Allott, 1954). From a turf perspective, mosses are invasive turf weeds which are favoured by conditions where grass vigour is reduced, there are low levels of disturbance and where canopy gaps occur (Jefferson, 1947). They are indicative of underlying issues with growing conditions, which results in the moss having a competitive advantage over the grass within the sward.

On earth dam embankments the turf surface should have a high grass content. This is to promote surface stability and prevent erosion of the dam face. The presence of high proportions of moss in the grass sward has been highlighted as a potential issue. This is due to moss, being shallow rooting, having less of an effect on soil stabilisation compared to grass. Consequently, if moss were more likely to be removed from a dam embankment surface, thereby exposing the bare soil, there would be potential for an increased risk of soil erosion. Therefore the control of existing moss populations in turf on an earth dam embankment and the prevention of moss invasion is a key issue for the management of earth dam embankments.

The objectives of the research project were to:

- Investigate the extent of moss invasion in turf on a typical earth dam embankment.
- Characterise the soil conditions and soil mechanical strength under areas dominated by either moss or grass.
- Assess the efficacy of various moss control methods to remove moss from an established turf surface on an earth dam embankment.
- Evaluate the optimum method for reinstating ground cover after moss control operations.

MATERIALS AND METHODS

The research site was located Mitchell's House Impounding Reservoirs, situated near Baxenden in Lancashire (SD 790 276). There are two reservoirs supplied by a single stream and retained by a grassed earth bank, aligned roughly north-south.

The top water level of both reservoirs is 299mAOD. They were constructed for water supply, which remains their main function. The embankments are curved in plan, with the water on the east side

No 1 reservoir was built first in 1865 and is to the north. It has a maximum height of 19.5m with a length of 275m, the capacity is circa 493,000m^3 with a surface area of 7.12ha. No 2 was completed in

1892 and has a maximum height of 14.5m with a length of 250m, the capacity is circa 320,000m^3 with a surface area of 4.6ha.

The two reservoirs are connected only at top water level, at the inlet lodge, and there is a single overflow weir within the bank of dam No 1 at its north end. Each reservoir has its own draw-off for supply and scour with inlet valves located on the upstream slopes of each dam.

The dams are located on near horizontally bedded Crutchman Sandstone of the Lower Coal measures. Extensive coal workings exist beneath the reservoirs and there is a geological fault line trending NNE to SSW upstream.

These earth dam embankments were selected for the trial as they had a significant population of moss in the turf and were considered to be representative of embankments where moss control maybe required.

The research was run in three phases. The first phase was to carry out a site survey to assess the relationship between soil conditions, including rotational shear strength in turf dominated by either moss or grass.

The second phase was to set up a moss control trial to evaluate the optimum methodologies for removing moss from the embankments.

The third phase was a trial to investigate how best re-establish ground cover after moss had been removed.

<u>Phase 1: Initial site survey</u>

The site survey took place on 18 December 2014 under very wet ground conditions. The survey was specifically carried out under these conditions, as the intention was to assess the mechanical strength of the upper embankment soil, under water contents that would pose an increased risk of erosion or instability.

Three of the embankments were assessed during the visit. The orientations of banks were as follows:

- Bank 1 = West facing
- Bank 2 = South-West facing
- Bank 3 = West-North-West facing

On each bank 20 sample locations were selected at random (10 on turf dominated by moss and 10 dominated by grass). Half of the sampling points were located on the upper half of each embankment and the other half on the lower section.

At each sampling location two soil samples were taken using a 50mm diameter corer. One core was used to measure gravimetric water content, whilst the other was used for the determination of soil pH and available plant macro nutrients (phosphate, potassium, calcium and magnesium). Plant nutrients were extracted using an acetic acid extraction regime.

In addition to collecting soil samples, soil shear strength was measured either on moss dominant or grass dominant turf using a Geonor Shear Vane. The shear vane blades were inserted into the upper soil profile, to a depth of 50 mm. Increasing rotational force was applied to the shear vane until the soil failed. This measure gave an indication of combined mechanical strength of the soil and any stabilising effect of plant roots.

<u>Phase 2: Moss control trial</u>

An area on Bank 2 (South-West facing embankment) was selected for the trial, as there was significant moss invasion of the turf. The trial was set up with a three-way factorial design with three treatment groups (Table 1). This design allowed the overall effects of each treatment group to be assessed, as well as the effect of the various treatment combinations. Each treatment combination was replicated three times, with each plot measuring 2m x 2m.

Table 1. Treatments used in the moss control trial

Moss control product	Scarification	Fertiliser treatment
[1] Untreated control	[1] Untreated control	[1] Untreated control
[2] Granular moss killer	[2] Scarification	[2] Evolution fertiliser
[3] Liquid moss killer		[3] Mo Bacter fertiliser
		[4] Lime
		[5] Evolution + lime
		[6] Mo Bacter + lime

Granular moss killer was applied at 400kg ha^{-1}, whilst the liquid moss killer was applied at 150 l ha^{-1}. Evolution and Mo Bacter were both organic fertilisers, with application rates tailored to deliver a low level of nitrogen (35kg nitrogen ha^{-1}). Lime was applied as a ground limestone powder at a rate of 2000kg ha^{-1}. A soil carrier was used with the lime to prevent the powder from blowing away during application.

The treatments used in the trial were selected because they represented standard turf maintenance practices for controlling and preventing moss invasion. The moss control products were commercially available materials that were based on iron sulphate.

Scarification physically removed moss from the sward and is one of the main cultural practices for removing moss from turf. Scarification was carried out using a pedestrian light weight electrical scarifier fitted with wire tines.

Based on the initial site survey, the fertility treatments focussed on applying a low rate of a slow release balanced nutrient source to improve grass health and lime to raise soil pH. The aim of the fertility treatments was to improve turf health and vigour, but without causing significant growth flushes, which would necessitate more frequent mowing operations.

Moss control products were applied on 30 April 2015, whilst scarification and fertility treatments were carried out on 22 May 2015. This was in-line with standard turf maintenance practices, where moss control products would be applied to kill or weaken moss plants, which would then be physically scarified out of the turf. Fertility treatments were applied after scarification to prevent their physical removal by the scarification operation.

Visual assessments of turf composition, including percentage moss content, were carried out at the start of the trial and after 137 days. Additionally, percentage moss control was recorded 22 and 137 days after application of the moss control products.

Phase 3: Turf recovery trial

The turf recovery trial was set-up on the top of Bank 3 (West-North-West facing), as in this area the turf had a consistently high moss population in the sward. To allow the various turf recovery treatments to be evaluated, the whole trial area was scarified to remove the moss and create bare areas. The turf recovery treatments were then applied.

The treatment groups under investigation focussed on reseeding bare areas with an appropriate low input landscape seed mix and promoting grass growth through the application of various fertility treatments (Table 2). Each treatment combination was replicated three times, with individual plots measuring 2m x 2m. The trial was laid out as a two way factorial design.

Table 2. Treatments used in the moss control trial

Seeding treatments	Fertility treatments
[1] Untreated control	[1] Untreated control
[2] Landscape seed mix	[2] Evolution @ 35g m^{-2}
[3] Landscape seed mix + microclover	[3] Evolution @ 125g m^{-2}
	[4] Evolution @ 35g m^{-2} + lime
	[5] Evolution @ 125g m^{-2} + lime

The landscape seed mix was applied at a rate of 350kg ha^{-1}. For the treatment where microclover was incorporated into the mix, it was applied at a rate of 20kg ha^{-1}. The lower rate of Evolution organic fertiliser delivered 25kg nitrogen ha^{-1}, whilst the higher rate provided 50kg nitrogen ha^{-1}. To ensure an even application of seed and to prevent the powdered lime from blowing away, a topsoil carrier was used. The trial area was scarified and all treatments applied on 30 April 2015.

The composition of the turf was assessed before scarification and then 137 days after treatment. On 14 September, percentage ground cover was assessed visually as an indicator of turf recovery following the scarification and treatment operations.

<u>Statistical analysis</u>

The trials were setup to allow statistical analysis of data to identify if treatments had a consistent and significant effect. Data from the site survey were analysed using a combination of T-tests and analysis of variance (ANOVA). For the moss control and turf recovery trials, data were analysed using three-way ANOVA. Throughout the trial, a significance level of 95% was used and, where appropriate, least significant differences (LSD) were used to established differences between means.

RESULTS

<u>Phase 1: Initial site survey</u>

Gravimetric soil water content averaged 32% across the embankment. This was indicative of very wet soil conditions. When comparing soil characteristics between areas dominant in moss and grass, no statistically significant differences were measured (Table 3). Soil shear strength was not affected by vegetation type. However, when walking on the embankment it became clear that, whilst vegetation type did not affect soil strength, there were clear differences in the ease with which moss was dislodged from the surface. When a similar lateral force was applied to turf with a high moss content, it had a tendency to peel away from the soil surface, whilst grass dominant turf remained firmly attached. This has implications for the ease with which the soil surface of the dam would become exposed if water was to overtop the dam.

Table 3. Soil characteristics measured under grass and moss dominant turf

Soil characteristic	Vegetation type		
	Grass	Moss	
Shear strength (KPa)	5.9	5.9	NS
Soil pH	4.2	4.3	NS
Phosphate (mg l^{-1})	7.2	8.9	NS
Potassium (mg l^{-1})	91	87	NS
Magnesium (mg l^{-1})	56	59	NS
Calcium (mg l^{-1})	190	205	NS

NS= not statistically significant

There was also significant variation in soil characteristics across the three dam embankments (Table 4). This was likely the result of natural variation as a result of materials of differing origin being used to construct the embankment, as well as variations in environmental conditions around the reservoirs.

Table 4. Variation in soil characteristics measured on the three banks

Soil characteristic	Bank				LSD (5%)
	1	2	3		
Shear strength (KPa)	5.3	5.2	7.3	<0.001	0.70
Soil pH	4.1	4.8	3.9	<0.001	0.30
Phosphate (mg l^{-1})	7.3	8.0	8.9	NS	-
Potassium (mg l^{-1})	110	83	75	<0.001	12.5
Magnesium (mg l^{-1})	52	69	51	<0.001	8.9
Calcium (mg l^{-1})*	1.86 (102)	2.41 (432)	1.75 (59)	<0.001	0.219

* Data $\log_{10}$ transformed for analysis (values in parenthesis are untransformed means).

NS = not statistically significant

Data from the initial site survey indicated that soil pH was acidic and had low levels of certain nutrients, in particular phosphate. This may be part of the explanation of why moss became dominant on some areas of the embankments. The reduction of grass vigour due to nutrient and soil pH stress will have led to moss having a competitive advantage in exploiting any canopy gaps. Consequently, under normal turf management conditions, a key recommendation would be to improve grass vigour through controlled nutrient inputs and by increasing soil pH. This was the rationale for including organic fertilisers and lime in the management treatments in the moss control and turf recovery trials.

Phase 2: Moss control trial

At the start of the trial, the average grass, weed and moss content of the turf was assessed (Table 5). At the start of the trial, average moss content in the turf was 48%. However, 137 days after application of moss control products and 115 days after the fertility and scarification treatments were carried out, average moss content in the sward had dropped to 5%.

Table 5. Average turf composition at the start of the trial and 137 days after application of the moss control products (SD = standard deviation).

Vegetation type	30 April 2015		14 September 2015	
	Mean	SD	Mean	SD
Grass	25%	12.2	55%	16.0
Broadleafed weeds	27%	9.7	40%	15.0
Moss	48%	14.9	5%	4.6

Both scarification and application of moss control products resulted in significantly lower moss content in the turf (Figures 1 and 2). Overall, there was very little added benefit of carrying out both scarification and application of moss control products, as both were effective at significantly reducing moss content.

When no moss control products were applied or scarification carried out, the application of both fertiliser and lime, either on their own or in combination, resulted in significantly less moss in the turf (Figure 1).

Whilst there was no immediate benefit of applying fertiliser and/or lime if the turf had been scarified or moss control products used, there may be a longer term benefit during the winter months. If the turf was more vigorous and grass cover was thicker going into winter, then there would likely be fewer gaps for moss to exploit during a period of the year when moss has a competitive advantage, in terms of growth, over the less active grass plants. This would be most evident in Spring and therefore it is planned to assess the trial plots in Spring 2016.

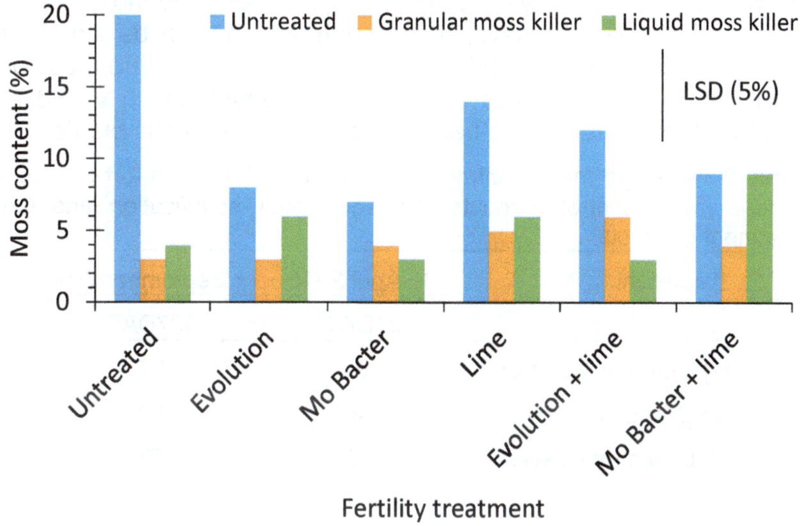

Figure 1. Moss content (%) on non-scarified turf 137 days after application of the moss control products.

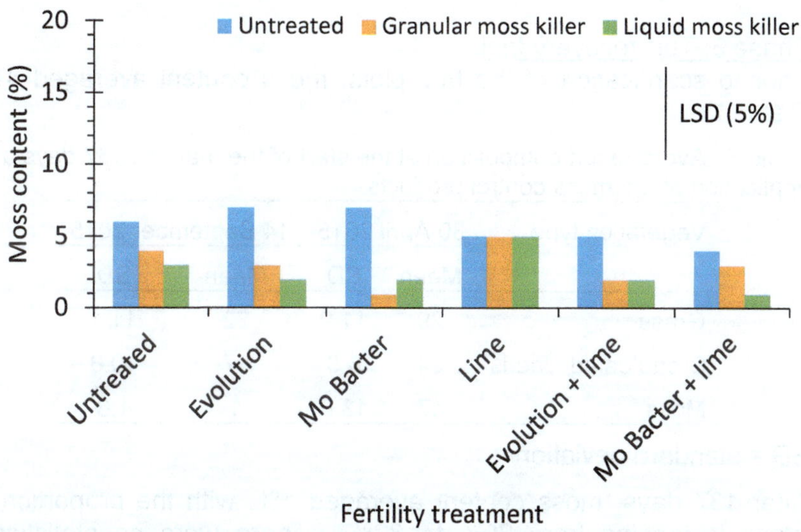

Figure 2. Moss content (%) on scarified turf 137 days after application of the moss control products.

When moss control was assessed, based on the condition and extent of the moss in the turf as compared to the untreated control plots, application of the moss killer products resulted in significant moss control (Table 6). The application of either granular or liquid moss killers resulted, 137 days after application, in average moss control of 73 to 83%. After 22 days, moss control tended to be significantly greater when the liquid product was used, as it turned the moss black more rapidly compared to the granular product, which would have had to dissolve first before being absorbed by the moss plants.

Table 6. Average moss control (%) on untreated turf and turf treated with various moss control products, averaged over scarification and fertility treatment products.

Treatment	30 April 2015	14 September 2015
	22DAT	137DAT
[1] Untreated control	0	27
[2] Granular moss killer	68	80
[3] Liquid moss killer	88	73
Significance (95% level)	<0.001	<0.001
LSD (5%)	6.8	9.7

xxDAT = days after treatment of moss control

Phase 3: Turf recovery trial

Prior to scarification of the trial plots, moss content averaged 47% (Table 7).

Table 7. Average turf composition at the start of the trial and 137 days after application of the moss control products.

Vegetation type	30 April 2015		14 September 2015	
	Mean	SD	Mean	SD
Grass	29	11.9	82	11.1
Broadleafed weeds	24	6.9	17	10.6
Moss	47	13.6	1	1.6

SD = standard deviation

After 137 days, moss content averaged 1%, with the proportion of grass increasing from 29% to 82%. There were no statistically significant differences among either the seeding or fertility treatments (data not shown). This was probably a reflection of the wet weather conditions experienced during the summer and autumn of 2015, which would likely have enhanced the growth of grass plants, which exploited the gaps produced by the physical removal of the moss.

In terms of the percentage ground cover after 137 days, plots treated with fertiliser, either on its own or with lime, had significantly greater ground cover (average = 97.5%), in comparison to the untreated control (average = 93.0%). This indicated that the application of additional nutrients helped to increase the recolonisation of bare areas with plants other than moss.

DISCUSSION

The invasion of moss into the turf of the earth dam embankments at Mitchell's House Reservoirs was likely the result of turf stress, either as a consequence of moisture stress due to elevation on the bank or nutrient or soil pH stresses.

The resistance of moss to lateral forces was lower than that observed for areas of turf dominated by grass. It was interesting that grass rooting, under wet soil conditions, did not significantly increase soil mechanical strength, when compared to shallow rooted moss plants.

The long-term management of moss in the turf of earth dam embankments should focus on preventative cultural practices. This is a fundamental part of an integrated pest management approach that aims to prevent a problem from occurring, rather than having to act curatively to remove the problem and reinstate the vegetation cover.

In situations where moss invasion has taken place the first stage of dealing with the issue would be to identify the underlying causes for why moss was able to establish in the turf in the first place.

Maintenance operations could then focus on resolving these issues to prevent moss spreading back into the turf after it has been removed.

Trial data showed how effective appropriate moss control was at significantly reducing moss populations in turf. Both the application of iron based moss control products and physical removal of moss proved successful at reducing moss content in the turf. Additionally, the application of organic fertiliser resulted in a significant reduction in moss by enabling the grass to be more vigorous and out-compete moss for the available space and resources.

The physical removal of moss can lead to bare areas of soil. This is undesirable as it could allow soil erosion of the dam embankment. However, if the right post scarification maintenance operations are carried out at the right time and under suitable conditions, the bare areas can re-vegetate.

As part of a moss management programme the optimal approach would be to develop a plan to prioritise treat problem areas of an embankment over a number years. The areas identified for treatment could be prioritised, so as to address areas with the highest moss populations first. The choice of removal technique and remedial management would have to be related to site specific requirements, but the treatments investigated as part of this research proved to be successful at controlling moss over a relatively short time period.

CONCLUSIONS

Moss in the turf of dam embankments was less firmly attached to the soil surface in comparison to grass. This could pose a problem for soil erosion should a dam overtop.

The use of dedicated moss control products, in either granular or liquid forms, was highly effective at controlling moss in dam embankment turf. Physical removal of moss through scarification was just as effective, but it did result in a thinning of ground cover, as vegetation was removed. However, with careful management and appropriate remedial maintenance, any bare or thin areas could be easily restored.

Identification of the root causes as to why moss invasion occurred is critical, as if they are not addressed the likelihood is that moss would re-invade the turf. Generally, this would involve improving the growing conditions for grass plants to prevent the creation of niches where moss can grow and spread.

ACKNOWLEDGEMENTS

The authors would like to acknowledge United Utilities Water Plc Innovation Fund for funding this research.

REFERENCES

Allott DJ (1954). *The moss plant: Its structure, habitats and life cycle*. J. Sports Turf Res. Inst. 8(30):pp. 387-395.

Jefferson P, (1947). *Moss in lawns and sports turf.* J. Sports Turf Res. Inst. 7(23):pp. 58-63.

Dams – Benefits and Disbenefits; Assets or Liabilities?
ISBN 978-0-7277-6133-0

doi: 10.1680/dbdal.61330.I1

Author Index